Programming in MATLAB®

Marc E. Herniter

Northern Arizona University

BROOKS/COLE

™

THOMSON LEARNING

Australia • Canada • Mexico • Singapore • Spain • United Kingdom • United States

Publisher: *Bill Stenquist*
Sponsoring Editor: *Heather Shelstad*
Marketing Team: *Samantha Cabaluna,*
 Christopher Kelly, and Ericka Thompson
Editorial Coordinator: *Shelley Gesicki*
Production Editor: *Mary Vezilich*
Manuscript Editor: *Luana Richards*

Permissions Editor: *Sue Ewing*
Interior Design: *Anne Draus*
Cover Design: *Denise Davidson*
Print Buyer: *Vena Dyer*
Typesetting: *Scratchgravel Publishing Services*
Printing and Binding: *Webcom Ltd.*

For more information about this or any other Brooks/Cole products, contact:

BROOKS/COLE
511 Forest Lodge Road
Pacific Grove, CA 93950 USA
www.brookscole.com
1-800-423-0563 (Thomson Learning Academic Resource Center)

Printed in Canada

10   9   8   7   6   5   4   3

Library of Congress Cataloging-in-Publication Data
Herniter, Marc E.
  Programming in MATLAB / Marc E. Herniter.
    p.   cm.
  Includes index.
  ISBN 0-534-36880-8 (pbk.)
    1. MATLAB. 2. Numerical analysis—Data processing. I. Title.

  QA297.H43 2000
  005.13'3—dc21   00-064228

For Corena

Preface

Motivation for This Text and Its Intended Audience

This text is intended for an introductory programming class that uses MATLAB as the programming language rather than traditional languages such as FORTRAN or C. With the programming enhancements introduced with version 5.0, MATLAB has similar programming capabilities to languages like FORTRAN, C, and BASIC, and includes both GUI tools and all of the toolboxes that make MATLAB a popular engineering, science, and math tool. With MATLAB, students can develop programs using the basic program constructs found in any programming language. However, students can also add the built-in MATLAB functions to their own programs to create nontrivial programs that would ordinarily take hundreds or thousands of lines to create.

At Northern Arizona University, we have struggled with how we should teach our introductory programming class and what language we should use in that course. Eventually, we settled on a traditional programming class using MATLAB as the language. Our choice solved the following problems:

- Our students need to understand the logic and thought process in order to write programs.
- Our students took a single programming course at the Freshman or Sophomore level and then never used it later in the curriculum.
- We cannot easily use a language like C, FORTRAN, or BASIC to solve complicated problems in higher level courses.
- With all the tools available for solving problems in engineering, the sciences, and mathematics, it seemed unwise to write a program in a traditional language.
- Because of the limited programming experience of students not in the Computer Science or Computer Engineering programs, problems we assigned in traditional programming languages had to be trivialized for our students to solve them in a reasonable amount of time.

We were thus looking for a solution that would give our students programming skills and then allow their skills to be used throughout the curriculum. With traditional programming languages, we could not simultaneously reinforce programming skills and solve nontrivial problems. Either the problems would be trivialized to make them easier to solve with a traditional language, or we would solve nontrivial problems and use another tool to solve the problem and never do any programming. We now teach a traditional programming course with MATLAB as the language, and then use MATLAB in as many high-level courses as possible. With this solution, students continually use the programming skills taught in the introductory course, and they use those skills to solve nontrivial problems.

v

Many excellent books show how to use MATLAB to solve engineering, science, and mathematics problems. This book is aimed at teaching an introductory programming class using MATLAB as the programming language.

Software and MATLAB Files Available at http://www.brookscole.com

The MATLAB files available at the publisher's website contain the example programs and functions used in this text. These programs and functions can be run with MATLAB and modified if desired. Also included are some differential equation solvers and numerical integration programs that do not come standard with the MATLAB student editions.

Comments and Suggestions

I would appreciate any comments or suggestions on this manual. Comments and suggestions from students are especially welcome. If you have examples you would like to suggest, please e-mail them to me or visit my website. I would appreciate any suggestions for additions to the coverage of the text, new problems, or example m-files. Please feel free to contact me using any of the methods listed here:

- **E-mail:** *Marc.Herniter@nau.edu.*
- www: www.cet.nau.edu/meh or http://www.cet.nau.edu/meh/Books/Matlab.htm
- **Phone:** (520) 523-4440
- **FAX:** (253) 369-9536
- **Mail:** Northern Arizona University, Electrical Engineering Department, Box 15600, Flagstaff, Arizona, 86011-1560.

Acknowledgments

I would like to thank my students at Northern Arizona University for their many suggestions for improving this text, including Christopher Adams, Michael Bautista, Caleb Breazeale, James Brietkrietz, David Briggs, Douglas Brown, Brian Callaway, Daniel Cervantez, Mingdon Chen, Aaron Cleaver, Scot Corapi, Mark Dilley, Joe Duenas, Brad Dugas, Michelle East, Kenneth Elliot, John Farris, Vanessa Flores, Michael Gallman, Jose Garcia, Craig Garrett, Sean Gilson, Vishal Golia, Scot Graham, Dwayne Gutierrez, Scott Hancock, Ahmed Hashim, Norman Heath, Randy Inch, Khurram Ismail, Rocio Jauregui, Abhishek Kar, Michael Kestler, Jason Klatt, Daniel Klein, Jay Larson, Roland Lee, Gene Legate, Daryl Leonard, Ryan Levine, Taylor Littlehat, Richard Minich, Tyler Mueller, Jeff Naylor., Luther Nez, Jon Parker, Nirakh Patel, Jason Phillips, Daniel Ruiz, Stafford Seaton, Jeremy Simpson, Gregory Sitrick, Jose Soto, Doug Southard, John Stoneberger, Christopher Talker, Bryan Taylor, Brett Thompsen, Jared Thompson, An Tran, Jonathan Tsoi, John Tugade, Victoria Upshaw, Robert Waltz, Patrick Worland, Jonah Yakel, Donnie Yazzie, Steven Young, and Calin Young. I would also like to thank David R. Scott of Northern Arizona University for contributing many of the problems contained in this text. Finally, I wish to thank my reviewers: Craig Borghesani, *Terasoft, Inc.*; Scott D. Gray, *Burgess Advanced Technologies;* and Kent L. Lawrence, *University of Texas at Arlington.*

—Marc E. Herniter

Before You Begin

MATLAB Versions

This book is intended to be used with MATLAB version 5.0 or higher. The examples shown in the screen captures were generated using MATLAB version 5.3 under Windows NT 2000 Professional. Most of the examples should be independent of the operating system and should work for all versions of Windows, Unix, and most other operating systems under which MATLAB runs. Most of the examples can also be used with MATLAB version 4.0. However, some of the newer programming constructs such as the **switch** statement, cell arrays, and structures were introduced with MATLAB version 5.0 and will not run under version 4.0. Also, certain functions such as the **input** function have changed with version 5.0. Approximately 90% of the examples will work with MATLAB 4.0. However, you should expect that some of the examples may generate errors if run under version 4.0.

Limitations of the Student Versions of MATLAB

The student versions have these array limitations:

- MATLAB 4.0, 8192 elements
- MATLAB 5.0, 16,384 elements
- MATLAB 5.3 (available from Prentice Hall), 16,384 elements

The student version of MATLAB 5.3 available from The MathWorks Incorporated, and the most recent version at the time of writing, has no limitations on array size. This version has all of the capabilities of the professional or full version of MATLAB 5.3.

Since the older student versions of MATLAB had array size limitations, some examples in this text use arrays whose size is limited to 16,384 elements. If your version does not have this limitation, you can increase the size of those arrays if you wish. If you are using MATLAB version 4.0, you will need to reduce the array size to 8192 elements.

The only examples that use arrays larger than 16,384 elements are the sound examples in Section 6.4. Long-playing sounds use arrays much larger than 16,384 elements, and these examples will not work with the versions of MATLAB that have array-size limitations. The examples will work if shorter sound files are used. A typical sound file that will fit in an array smaller than 16,384 elements is a sound approximately 1 second or shorter.

General Conventions

- This manual assumes that you have a two- or three-button mouse. The words *LEFT* and *RIGHT* refer to the left and right mouse buttons.
- This manual assumes that you have a color monitor. A color monitor is not necessary for running the software or following the manual. However, the manual refers to items highlighted in red when they are "selected." If you don't have a color monitor, you will still be able to see the selected items, but they may not be as obvious.
- Boldface text refers to menu selections. Examples are **File** and **Analysis.**
- Text in capital letters refers to keyboard selections. For example, press the ENTER key.
- `Text in this font will be text you will type at the MATLAB command prompt, or will be lines in a MATLAB program, or prefined MATLAB statements and functions.`
- `Text in this font refers to MATLAB output.`
- The predefined MATLAB statements and functions being used are noted in margin checklists throughout the text.
- *Text in this font are variables in a program.*
- The word "select" means "click the left mouse button on."

Keyboard Conventions

Throughout the manual, shortcut keyboard sequences are given for making menu selections. You will need to know the conventions that specify the sequences. Many control key sequences will be specified.

- For example, CTRL-R means hold down the "Ctrl" key and press the "R" key simultaneously. CTRL-A means hold down the "Ctrl" key and press the "A" key simultaneously. Not all keyboards are the same; some keyboards have a key labeled "Control" rather than "Ctrl."
- The keyboard sequence ALT-TAB in Microsoft Windows is used to toggle the active window. ALT-TAB means hold down the "Alt" key and press the "Tab" key simultaneously.

Windows Conventions Used in This Manual

This manual uses many terms associated with the Windows operating system. Some of the terms are shown here:

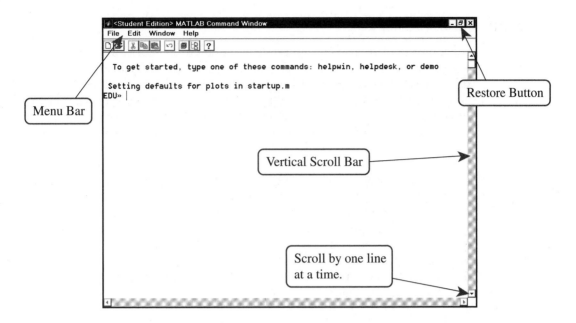

Note: The restore button shown above changes depending on the size of the window. The graphic ![restore] is the restore button and means restore the window to its last dimensions which were not full screen. The graphic ![maximize] is the maximize button and means expand a window to occupy the entire screen.

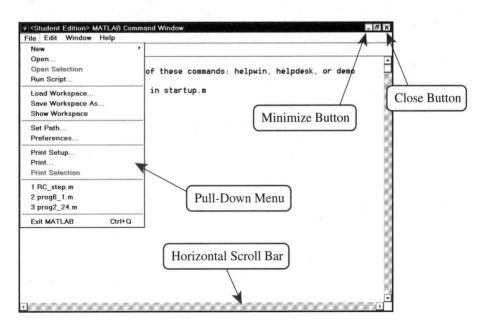

MATLAB **Program Interaction**

A large number of examples in this text show you how to use MATLAB with script files or to type commands at the command prompt. The MATLAB command prompt is **EDU»** for educational versions and **»** for the professional version. The command prompt indicates that MATLAB is waiting for input from the user. This is the typical interaction between MATLAB and the user: MATLAB displays the command prompt, the user types in a command, MATLAB executes the command, and then MATLAB displays the command prompt again and waits for the user to enter another command. In MATLAB, you can also place MATLAB commands in a text file and then run the file. MATLAB will execute the statements in the file sequentially with out displaying the command prompt. This type of text file is called a script file. Instead of typing a series of commands at the command prompt, you place the same commands in a file, and then have MATLAB execute the commands in a file. Most of the programs discussed in this text are script files which we refer to as programs. Any command that works at the MATLAB command prompt also works in a script file.

The following example shows the basic interaction between MATLAB and a user when the user enters commands at the command prompt.

```
EDU» A5=6
A5 =
     6
EDU» b17=27.987
b17 =
   27.9870
EDU» Dog_catcher=297.9
Dog_catcher =
   297.9000
```

The dialog shows the user entering a command and MATLAB executing the command. To follow along, you should enter the commands at the command prompt; the text is displayed in this manner to suggest that you enter the commands at the command prompt.

The next example also shows commands entered at the command prompt. These commands generate a plot and do not generate output on the command window as did the preceding example. In this case, you can write a script file that contains these commands and run the script file, or enter the commands at the prompt.

```
EDU» subplot(2,3,3)
EDU» x=logspace(-3,2,1000);
EDU» y=5.*10.^(3.*x);
EDU» loglog(x,y)
EDU» ylabel('y');
EDU» xlabel('x');
```

The example below shows commands without the command prompt. As shown, the **for** loop could not be entered at the command prompt.

```
x=logspace(1,2,10)';
y=log10(x)
for i = 1:9
  delta=y(i+1)-y(i);
  fprintf('y(%1d)-y(%1d)=%g\n', i+1, i, delta);
end
```

Any MATLAB command can be entered at the command prompt, even a **for** loop. To execute the code segment above, you can enter the entire command on a single line and separate the commands by a semicolon:

```
x=logspace(1,2,10)';y=log10(x);for i = 1:9   delta=y(i+1)-y(i);➡
fprintf('y(%1d)-y(%1d)=%g\n',   i+1,   i, delta);end
```

Note that this command does not display the same output as the previous code segment because we added a semicolon (;) after the line **y=log10(x)** —the semicolon prevented MATLAB from displaying the result of the command. Although we can type several MATLAB commands on a single line, and enter those lines at the command prompt, any example that contains several lines of code will be placed in a script file, and we will run the script file to test the code segment. Creating and running script files is covered in Section 1.13.

In the following session, commands are entered at the command prompt. We show it this way because it is possible to enter the commands one by one at the command prompt, and achieve the same results. The session is so long, however, that we recommend that these commands be placed in a script file, because the chances of making a typing mistake are good.

```
EDU»  subplot(2,1,1)
EDU»  FID=fopen('d:\EXAMPLES\ran_data.bin','r');
EDU»  [input, count] = fread(FID, [1 100], 'double');
EDU»  [output, count] = fread(FID, [1 100], 'double');
EDU»  status=fclose(FID);
EDU»  x=linspace(0,5,100);
EDU»  spline_out=spline(input, output, x);
EDU»  yp=spline(input, output,2.5);
EDU»  h1=plot(input, output, '+k', x, spline_out, 2.5, yp, 'ro');
EDU»  set(h1(3),'markersize',15);
EDU»  text_str=sprintf('At x = 2.5, y = %g',yp);
EDU»  h=text(2,200, text_str);
EDU»  set(h, 'FontSize', 15, 'FontName', 'Arial');
EDU»  title('Spline Fit');
EDU»  grid
EDU»  subplot(2,1,2)
EDU»  poly=polyfit(input,output,3);
EDU»  poly_out=polyval(poly,x);
EDU»  yp=polyval(poly,2.5);
EDU»  h2=plot(input, output, '+k', x, poly_out, 2.5, yp, 'ro');
```

```
EDU» set(h2(3),'markersize',15);
EDU» text_str=sprintf('At x = 2.5, y = %g',yp);
EDU» h=text(2,200, text_str);
EDU» set(h, 'FontSize', 15, 'FontName', 'Arial');
EDU» title('3rd Order Polynomial Fit');
EDU» grid
```

To summarize:

- **Text in this font** will always be MATLAB commands that you can enter at the command prompt or place in a script file.
- If the command prompt is shown, EDU» or », then you can enter the commands at the command prompt instead of placing them in a script file. You should obtain the same results with either method.
- If the command prompt is not shown, you must place the sequence of commands in a script file and then run the script file.

Matrices and Arrays

MATLAB was originally designed to easily handle complex mathematics using matrices. Indeed, the name MATLAB stands for "matrix laboratory." Consequently, a majority of The MathWorks documentation and other textbooks refer to variables as matrices and vectors. A vector is a matrix with one row or one column. A matrix with one row is a row vector, and a matrix with one column is a column vector. As MATLAB has added more programming capabilities and becomes more accepted as a programming language, variables that store multiple values can be thought of as either matrices or arrays. In traditional programming languages, we think of variables that store multiple values as arrays. Because in this text we use MATLAB as a programming language, we will also use the term "array." This is correct for the majority of the text since we are interested in data structures and an array is a data structure that stores many values. Only a brief section of the text uses matrices.

We will use the term "row array" to mean "one-dimensional array with one row," and the term "column array" to mean "one-dimensional array with one column." These terms obviously come from the terms "row vector" and "column vector." However, "row vector" and "column vector" have specific meanings in matrix algebra, and the author does not wish to use the term "vector" where it is not appropriate. Since "row vector" and "column vector" convey a specific structure when referring to arrays, the author has coined the terms "row array" and "column array" because they describe a specific shape of array.

Examples Used in the Text

We will revisit several streams of examples throughout the text. One stream is of a guessing game. The guessing game is introduced in Section 2.1 to illustrate the **IF** statement when used in the form **IF-END**. The game is used again in Sections 2.2, 2.5, and 2.6 to illustrate how the program can be written with **IF-ELSE-END** statements, and improved with **FOR** loops and **WHILE** loops. We will also take other examples and slowly improve on them by using different programming techniques; these include the Grade Calculation program, the Command Center shell, and the Factorial program.

Several examples illustrate the differences between using MATLAB as the programming language and conventional languages such as FORTRAN or C. One example is the sorting program developed in Section 4.4.2 and the Monte Carlo voltage divider gain program developed in Section 4.4.3. The examples first show the programming techniques that we would use to create the program in any programming language. Then we use a MATLAB-only approach that solves the problem with the matrix capabilities of MATLAB.

The goal of this text is to develop programming skills using MATLAB as the programming language. We will therefore develop some programs or functions that are already available in MATLAB. For most of these examples, we develop the program first as a programming exercise, then show how to solve the problem using built-in MATLAB functions. In later chapters, where we solve basic engineering problems, we will use MATLAB functions only and will not "recreate the wheel" by writing programs that already exist.

Contents

8 MATLAB Applications

List of Programs

List of Functions

Programming in MATLAB®

MATLAB Environment

OBJECTIVES

- ☐ Become familiar with the general use of the MATLAB interface.
- ☐ Perform simple calculations with MATLAB.
- ☐ Define variables and the limitations on naming variables.
- ☐ Observe the display formats available with MATLAB and select specific formats.
- ☐ List the predefined MATLAB variables.
- ☐ Perform complex arithmetic calculations.
- ☐ Introduce simple matrix operations.
- ☐ Use strings and perform simple string operations.
- ☐ Perform some basic input/output (I/O) operations using the `FPRINT` and `INPUT` statements.
- ☐ Create simple plots.
- ☐ Use the options for obtaining help in MATLAB.
- ☐ Create script files.
- ☐ Modify the MATLAB search path.

This chapter is a "quick start" guide to using MATLAB features. Our goal is to familiarize you with the MATLAB interface and show you how to use some of MATLAB's basic features. In this chapter we don't attempt to teach programming, as that is the goal of the remainder of the text. Instead, after completing this chapter, you should be able to use the MATLAB interface competently, perform some simple calculations, use the MATLAB editor to write script files, and find help when necessary.

1.1 MATLAB as a Calculator

MATLAB performs simple calculations as if it were a calculator. For example, type **5+3** and press the ENTER key:

```
EDU» 5+3
ans =
    8
```

TABLE 1.1 Mathematical operators in MATLAB

Operation	MATLAB Symbol	Example	Answer
Addition	+	5+3	8
Subtraction	−	5-3	2
Multiplication	*	6*2	12
Division	/	6/2	3
Power	^	6^2	36

You can do many calculations on a line:

```
EDU» 5+3-4*6+7/2
ans =
  -12.5000
```

The simple math operators in MATLAB are given in Table 1.1.

   Precedence is the order in which operations are executed by the computer. Higher-precedence operations are executed before lower-precedence operations. If two operators have the same precedence, then the expressions are executed from left to right. The power operation has the highest precedence, followed by multiplication and division, which have the same precedence. Addition and subtraction have the same precedence and are the lowest-precedence operators listed in Table 1.1. The order in which an expression is evaluated can be changed by using parentheses (). A few examples are given.

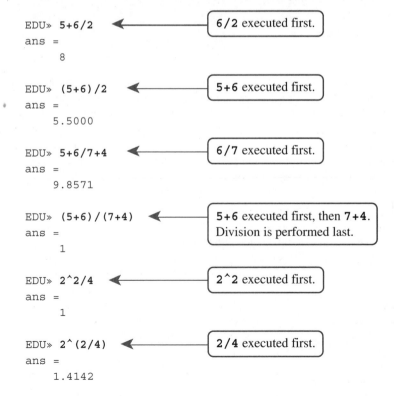

```
EDU» 5+6/2          ◄──────  6/2 executed first.
ans =
       8

EDU» (5+6)/2        ◄──────  5+6 executed first.
 ◦ ans =
     5.5000

EDU» 5+6/7+4        ◄──────  6/7 executed first.
ans =
     9.8571

EDU» (5+6)/(7+4)    ◄──────  5+6 executed first, then 7+4.
ans =                        Division is performed last.
     1

EDU» 2^2/4          ◄──────  2^2 executed first.
ans =
     1

EDU» 2^(2/4)        ◄──────  2/4 executed first.
ans =
     1.4142
```

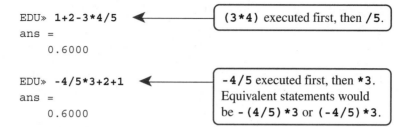

```
EDU» 1+2-3*4/5
ans =
     0.6000
```
(3*4) executed first, then /5.

```
EDU» -4/5*3+2+1
ans =
     0.6000
```
-4/5 executed first, then *3.
Equivalent statements would
be - (4/5) *3 or (-4/5) *3.

1.2 Defining Variables

We define variables by typing a variable name followed by the equals sign and then a value or a mathematical expression.

```
EDU» A=5
A =
     5
EDU» zoo=26/3.7
zoo =
     7.0270
```

Variable names can be up to 21 characters long and can contain letters, numbers, and the underscore character.

```
EDU» A5=6
A5 =
     6
EDU» b17=27.987
b17 =
    27.9870
EDU» Dog_catcher=297.9
Dog_catcher =
   297.9000
```

Variables longer than 21 characters are truncated to 21 characters.

```
EDU» This_is_a_very_long_variable_name=789
This_is_a_very_long_variable_na =
    789
```

Variables can be used to perform calculations:

```
EDU» a=5;
EDU» b=3;
EDU» c=8;
EDU» x=a+b/c
x =
    5.3750
```

To see the variables currently stored in memory, use the **WHO** command:

```
EDU» who

Your variables are:

A
A5
Dog_catcher
This_is_a_very_long_variable_na
a
ans
b
b17
c
x
zoo
```

A second form of the **WHO** command is **WHOS**. This command lists the variables stored in memory and the amount of memory used by each variable.

```
EDU» whos
Name                            Size            Bytes    Class

    A                           1x1                 8  double array
    A5                          1x1                 8  double array
    Dog_catcher                       1x1                 8  double array
    This_is_a_very_long_variable_na   1x1                         8  double array
    a                           1x1                 8  double array
    ans                         1x1                 8  double array
    b                           1x1                 8  double array
    b17                         1x1                 8  double array
    c                           1x1                 8  double array
    x                           1x1                 8  double array
    zoo                         1x1                 8  double array

Grand total is 11 elements using 88 bytes
```

This table shows that our variables all contain a single numerical value. The numbers are stored as double-precision floating-point numbers that occupy 8 bytes of memory. A variable with a single numerical value can be thought of as an array with a single element; thus, the table states that the variable is a 1×1 array. We will also use arrays with dimensions other than 1×1. For example, let's define a 3×4 array and then use the **WHOS** command:

```
EDU» m=[1 2 3 4; 66 76 88 44; 567 45 76 0]
m =
```

```
        1       2       3       4
       66      76      88      44
      567      45      76       0
```

```
EDU» whos
Name                    Size              Bytes      Class

    A                   1x1                   8   double array
    A5                  1x1                   8   double array
    Dog_catche                1x1                 8   double array
    This_is_a_very_long_variable_na   1x1             8   double array
    a                   1x1                   8   double array
    ans                 1x1                   8   double array
    b                   1x1                   8   double array
    b17                 1x1                   8   double array
    c                   1x1                   8   double array
    m                   3x4                  96   double array
    x                   1x1                   8   double array
    zoo                 1x1                   8   double array

Grand total is 23 elements using 184 bytes


EDU»
```

A third way to look at the variables currently stored in memory is to use the MATLAB command window menus. Select **File** from the command window menu bar:

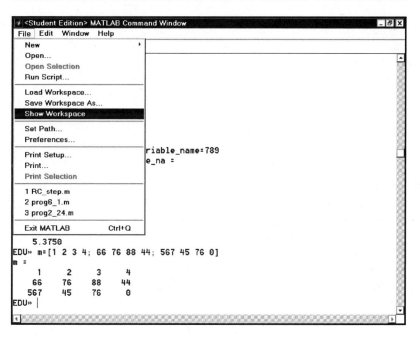

Selecting **Show Workspace** will open and display the variables stored in memory:

Name	Size	Bytes	Class
A	1x1	8	double array
A5	1x1	8	double array
Dog_catcher	1x1	8	double array
This_is_a_very_long_v...	1x1	8	double array
a	1x1	8	double array
b	1x1	8	double array
b17	1x1	8	double array
c	1x1	8	double array
m	3x4	96	double array
x	1x1	8	double array
zoo	1x1	8	double array

MATLAB Workspace

Grand total is 22 elements using 176 bytes

Delete   Close

This dialog box gives us the same information as the **WHOS** command. Click the **Close** button to close the window and return to the command window.

We can redefine the value of a variable by using it in an expression. Let's find the current value of *A* and then change it.

```
EDU» A

A =

     5

EDU» A=27

A =

    27

EDU» who

Your variables are:

A                          b
A5                         b17
Dog_catcher                c
This_is_a_long_vari        x
a                          zoo
ans
```

Now *A* has a different value, but it is still stored in memory.

Variable names in MATLAB are case-sensitive. For example, variables *a* and *A* are treated as different variables:

```
EDU» a=5;
EDU» A=7.9;
EDU» a
a =
      5
EDU» A
A =
      7.9000
```

To remove variables from memory, we can use the MATLAB **CLEAR** command. First let's look at the variables in memory:

```
EDU» who
Your variables are:

A                                b
A5                               b17
Dog_catcher                      c
This_is_a_long_vari              x
a                                zoo
ans
```

Suppose we wish to free up memory by removing variables *a*, *b*, and *c*.

```
EDU» clear a b c
```

Next, we use the **WHO** command to see what variables are stored in memory:

```
EDU» who
Your variables are:
A                                b17
A5                               x
Dog_catcher                      zoo
This_is_a_long_vari
ans
```

Variables *a*, *b*, and *c* are no longer listed. The memory they used is now available for other uses. To clear all variables from memory, type the MATLAB **CLEAR** command by itself:

```
EDU» clear
```

Now list the variables stored in memory:

```
EDU» who
Your variables are:
```

Now no variables are listed.

1.3 Functions

Matlab Predefined
sin function
abs function
fprintf function
log function

Except for basic commands such as addition, subtraction, multiplication and division, most MATLAB commands are functions. Examples are **sin**(*x*), **abs**(*x*), **fprintf**, **log**(*x*). Functions usually require an input argument (such as *x* above) and they return a value. An example would be the **sin**(*x*); *x* is the input value and the function returns a value that is the sine of *x*.

```
EDU» sin(2)
ans =
    0.9093
EDU» sin(1.5)
ans =
    0.9975
EDU» q=sin(1)
q =
    0.8415
»
```

In this example, the **SIN** function requires a single input and returns a single value. In general, a function can have several inputs and can return several values. Also, functions can have no inputs, and they do not have to return a value.

Built-in MATLAB functions are used throughout this text. We will discuss them when they first appear in the text. Chapter 3 covers how to create and use functions in detail.

1.4 Display Formats

Matlab Predefined
pi variable
format command

Values are displayed with MATLAB in several ways. Some of the format types are shown in Table 1.2.

If you cannot remember these formats when you are using MATLAB, you can type **help format** at the MATLAB prompt. A list of the available formats will be displayed:

```
EDU» help format

 FORMAT Set output format.
    All computations in MATLAB are done in double precision.
    FORMAT may be used to switch between different output
    display formats as follows:
       FORMAT          Default. Same as SHORT.
       FORMAT SHORT    Scaled fixed point format with 5 digits.
       FORMAT LONG     Scaled fixed point format with 15 digits.
       FORMAT SHORT E  Floating point format with 5 digits.
       FORMAT LONG E   Floating point format with 15 digits.
       FORMAT SHORT G  Best of fixed or floating point format with 5 digits.
       FORMAT LONG G   Best of fixed or floating point format with 15 digits.
       FORMAT HEX      Hexadecimal format.
```

TABLE 1.2 Displayed numeric value of π using different formats

MATLAB Command	Displayed	Comments
format short	3.1416	Default display
format long	3.14159265358979	16 digits
format short e	3.1416e+000	5 digits plus exponent
format long e	3.141592653589793e+000	16 digits plus exponent
format hex	400921fb54442d18	hexadecimal
format bank	3.14	2 decimal places (for currency)
format +	+	positive, negative, 0
format rat	355/113	ratio of whole numbers (fraction)

```
FORMAT +        The symbols +, - and blank are printed
                for positive, negative and zero elements.
                Imaginary parts are ignored.
FORMAT BANK     Fixed format for dollars and cents.
FORMAT RAT      Approximation by ratio of small integers.

Spacing:
FORMAT COMPACT  Suppress extra line-feeds.
FORMAT LOOSE    Puts the extra line-feeds back in.
EDU»
```

As an example, let's display the value of π using the different formats. Note that **pi** is a predefined variable in MATLAB numerically equal to π:

```
EDU» pi
ans =
    3.1416
```

To change the display format, type the **FORMAT** command followed by the format type you wish to use.

```
EDU» format long
EDU» pi
ans =
    3.14159265358979
EDU» format short
pi
ans =
    3.1416
EDU» format short e
EDU» pi
ans =
  3.1416e+000
EDU» format long e
EDU» pi
```

```
ans =
      3.141592653589793e+000
EDU» format hex
EDU» pi
ans =
      400921fb54442d18
EDU» format bank
EDU» pi
ans =
             3.14
EDU» format +
EDU» pi
ans =
+
EDU» format rat
EDU» pi
ans =
      355/113
```

Note that **format +** displays whether a number is positive or negative and **format rat** displays the number as a ratio of two whole numbers. Note that when you change the format, all following numbers are displayed using the format:

```
EDU» format long e
EDU» pi
ans =
      3.141592653589793e+000
EDU» 5/6
ans =
      8.333333333333334e-001
EDU» 6
ans =
      6
EDU» a=78
a =
      78
EDU» b=18.72
b =
      1.872000000000000e+001
```

1.5   Saving the Variables Stored in Memory

MATLAB Predefined
save command
load command

Suppose you've done a large amount of work and defined a number of variables. You need to quit using MATLAB for the moment, but you would like to use these variables in the future. This can be done in two ways. You can use the MATLAB menus and use the

File and **Save Workspace As** menu selections, or you can use the MATLAB **SAVE** command. Suppose a large number of variables are defined:

```
EDU» a=1;
EDU» b=2;
EDU» c=3;
EDU» d=4;
EDU» e=5;
EDU» f=6;
EDU» g=7;
EDU» who


Your variables are:

A         ans        c          e          g
a         b          d          f
```

You can save the variables to the default workspace file, or you can specify a name for the file. To save the variables to the default file, named matlab.mat, type the **SAVE** command by itself:

```
EDU» save
Saving to: matlab.mat
```

To see that we can restore the variables, let's clear the workspace:

```
EDU» clear
EDU» who
Your variables are:
```

No variables are in memory. Now let's use the **LOAD** command to load the default file, matlab.mat, and restore the variables:

```
EDU» load
Loading from: matlab.mat
EDU» who
Your variables are:
A         ans        c          e          g
a         b          d          f
```

If you wish to save the variables in a file with a different name, you can use the **LOAD** and **SAVE** commands followed by a file name:

```
EDU» save xyzzy
```

Let's clear the workspace and restore the variables using file xyzzy.

```
EDU» clear
EDU» who
Your variables are:
```

No variables are defined. To load the variables saved in file xyzzy, use the **LOAD** command with the file name:

```
EDU» load xyzzy
EDU» who
Your variables are:
a          b          d          f
ans        c          e          g
```

The **SAVE** command places the files in directory c:\matlab\bin and automatically attaches the extension ".mat" to the file name. Thus, the full pathname for file xyzzy is c:\matlab\bin\xyzzy.mat.

1.6 Predefined Variables

MATLAB has a number of predefined variables available for use (Table 1.3). You should not redefine these variables when using MATLAB.

We can display the value of these variables the same way we display values of variables that we define.

TABLE 1.3 Predefined variables in MATLAB

Variable	Definition	Value
ans	Holds the value of the most recent calculation if the calculation was not stored in a variable.	
pi	(π) Ratio of the circumference of a circle to its diameter.	3.141592653589793e+000
eps	Smallest number such that when added to 1 creates a floating-point number greater than 1 on the computer.	2.220446049250313e-016
inf	(∞) Infinity—for example, 1/0.	inf
NaN	Not a Number. Obtained from invalid operations such as 0/0, ∞/∞, and $\infty \cdot 0$.	NaN
i	The square root of -1.	i
j	The square root of -1.	j
realmin	The smallest usable positive real number.	2.225073858507202e-308
realmax	The largest usable positive real number.	1.797693134862316e+308

```
EDU» format long e
EDU» eps
eps =
    2.220446049250313e-016
EDU» inf
ans =
   Inf
```

Note that variable *Inf* does not display a numeric value, but it is treated as a number. If we perform a calculation that results in infinity, the value of the result will be *Inf*. As an example, let's divide 1 by zero:

```
EDU» a=1/0
Warning: Divide by zero
a =
   Inf
```

If we perform calculations with infinity, the result will be infinity:

```
EDU» 5+a
ans =
   Inf
EDU» a/29
ans =
   Inf
```

Note that we can have both positive and negative infinity.

```
EDU» -1*a
ans =
  -Inf
```

The variable *Inf* is provided so the computer can represent infinity internally. This allows the computer to use infinity in calculations and obtain infinite results instead of halting the program and generating error messages.

The variable *NaN* (Not a Number) resembles *Inf* in that its value is itself:

```
EDU» NaN
ans =
   NaN
```

NaN results from invalid operations such as $1/0$, $inf \times 0$, or $inf \div inf$. In mathematics, these operations are undefined. In MATLAB, an operation of this type is assigned the value *NaN*. This allows the computer to continue running the program even after a calculation generates a result of *NaN,* which is preferable to generating an error message and terminating the program. Any calculation with *NaN* results in *NaN*.

```
EDU» 5+NaN
ans =
    NaN
EDU» a=0*inf
a =
    NaN
EDU» y=a/25
y =
    NaN
```

1.7 Complex Numbers

MATLAB *Predefined*
sqrt function
real function
imag function
abs function
angle function

Pure imaginary and complex numbers are expressed in MATLAB by the letters i or j. Both i and j are defined as $\sqrt{-1}$. If we take the square root of –1, MATLAB displays a number containing the letter i. We will use the MATLAB **SQRT** function, which calculates the square root of a number. The **SQRT** function is equivalent to raising a number to the $\frac{1}{2}$ power.

```
EDU» format short
EDU» sqrt(-1)
ans =
        0 + 1.0000i
```

Note that the result is displayed as a complex number. The real part is zero and the imaginary part is $1 \times i$, or just i. As a second example, let's calculate the square root of –9:

```
EDU» (-9)^(1/2)
ans =
    0.0000 + 3.0000i
```

The square root of –9 is $3i$.

Complex numbers have a real and an imaginary part. To express a complex number, we express the number as the real part plus the imaginary part times i or j. Examples are $3 + 4i$ and $7 + 2j$. MATLAB automatically handles calculations with complex numbers. For example, we can calculate the ratio $(1 + 2i)/(3 + 5i)$ simply by typing in the expression as shown:

```
EDU» (1+2i)/(3+5i)
ans =
    0.3824 + 0.0294i
```

We can do any operation with complex numbers that we can do with real numbers. For example, let $a = 3 + 4i$ and $b = 1 + i$ and display the results for a^2, $a - b$, $a + b$, $a \times b$, a/b, and $\sqrt{a}$.

```
EDU» a=3+4i;
EDU» b=1+i;
EDU» a*a
ans =
  -7.0000 +24.0000i
EDU» a-b
ans =
   2.0000 +  3.0000i
EDU» a+b
ans =
   4.0000 +  5.0000i
EDU» a/b
ans =
   3.5000 +  0.5000i
EDU» a*b
ans =
  -1.0000 +  7.0000i
EDU» sqrt(a)
ans =
   2.0000 +  1.0000i
```

MATLAB has several functions for manipulating complex numbers. Let A be a complex number. We can express A in rectangular coordinates $A = a + bi$, or as a magnitude and a phase, $A = M \angle \Theta$ where $M = \sqrt{a^2 + b^2}$ and $\Theta = \tan^{-1}(b/a)$ if a is positive or $\Theta = \tan^{-1}(b/a) + 180$ if a is negative. Four of the MATLAB functions for manipulating complex numbers are discussed here. For our example, let $A = 3 + 4i$.

■ The **REAL** function returns the real part of a complex number.

```
EDU» A=3+4i;
EDU» real(A)
ans =
     3
```

■ The **IMAG** function returns the imaginary portion of the complex number:

```
EDU» imag(A)
ans =
     4
```

■ The **ABS** function returns the magnitude of the complex number:

```
EDU» abs(A)
ans =
     5
```

The **ANGLE** function returns the angle in radians:

```
EDU» angle(A)
ans =
    0.9273
```

The angle can be converted to degrees by multiplying the result of the **ANGLE** function by 360/2π:

```
EDU» angle(A)*360/(2*pi)
ans =
    53.1301
```

1.7.1   Rectangular and Polar Coordinates

Complex numbers can be represented in rectangular coordinates ($x = a + bi$) or in polar form as a magnitude and an angle, $x = M \angle \theta$ where M is the magnitude and θ is the angle in degrees. MATLAB will only do calculations of complex numbers in rectangular form, so we must convert numbers in polar form to rectangular form ($a + bj$). We can do this in two ways. The first is to remember the set of equations for converting between rectangular and polar notation. If $(a + bj) = M \angle \theta$, then

$$
\begin{aligned}
M &= \sqrt{a^2 + b^2} \\
\theta &= \begin{cases} \tan^{-1}(b/a) & \text{if } a \geq 0 \\ 180 + \tan^{-1}(b/a) & \text{if } a < 0 \end{cases} \\
a &= M\cos(\theta) \\
b &= M\sin(\theta)
\end{aligned}
$$

From this set of equations, we conclude that $a + bj = M\cos(\theta) + jM\sin(\theta)$.

A second way to convert from polar notation to rectangular notation is to remember that polar notation is short hand for the exponential equation shown here:

$$M \angle \theta \equiv Me^{i\theta}$$

where θ is in radians. In MATLAB, we use the equation $Me^{i\theta}$ directly, and MATLAB performs the calculation and returns the result in rectangular coordinates. As an example, let's find the rectangular coordinates of $5 \angle 30$. This next session uses the first method:

```
EDU» ang=30*2*pi/360;
EDU» mag=5;
EDU» mag*cos(ang)+j*mag*sin(ang)
ans =
    4.3301+ 2.5000i
```

Here, we do the same calculation using the exponential function:

```
EDU» ang=30*2*pi/360;
EDU» mag=5;
EDU» mag*exp(i*ang)
ans =
   4.3301+ 2.5000i
```

To convert from rectangular notation to polar notation, we simply use the **ABS** and **ANGLE** functions:

```
EDU» (360/(2*pi))*angle(4.3301+ 2.5000i)
ans =
   30.0002
EDU» abs(4.3301+ 2.5000i)
ans =
   5.0000
```

The angle function gives us the angle in radians, so we multiply the result by $\frac{360}{2\pi}$ to convert the result to degrees.

We can now easily perform calculations of numbers that are represented in either rectangular or polar notation. As an example, let's calculate the result of the equation:

$$x = \frac{(2.5 + i3.3)(7\angle 30)}{(4.4\angle 25) + (2.18 - i1.6)}$$

```
EDU» a=2.5+3.3i;
EDU» b=7*exp(i*30*(2*pi/360));
EDU» c=4.4*exp(i*25*(2*pi/360));
EDU» d=2.18-1.6i;
EDU» x=(a*b)/(c+d)
x =
   0.7794+ 4.6294i
EDU»
```

1.8   Matrices and Vectors

MATLAB Predefined
inv function
rand function
det function

MATLAB can store several numerical values in a single variable. For most of the text, we'll refer to these variables as arrays. MATLAB, however, treats these variables as either matrices or arrays. When treated as arrays, MATLAB addresses these variables like other conventional programming languages, by performing element-by-element operations. When these variables are treated as matrices, MATLAB performs matrix operations with the variables. Other texts cover the matrix capabilities of MATLAB in detail, so we will examine these capabilities only briefly. The majority of this text will focus on arrays and their uses from a conventional programming point of view, so you need not be familiar with matrix algebra.

In MATLAB, the default uses matrix algebra rather than array operations. Array operations are defined as element-by-element operations. For example, array multiplication would be the first element times the first element, the second element times the second element, the third element times the third, and so on. Matrix multiplication takes a row of one matrix and multiplies it with the column of another matrix and then adds the results. The matrix operators are *, /, ^, +, and – for matrix multiplication, division, power, addition, and subtraction. These operators perform matrix algebra if the operation is defined for the matrices being used with the operators. The array operators are .*, ./, .^, +, and –. Note that the multiplication, division, and the power operators are preceded by a dot (.) and that array addition and subtraction are the same as matrix addition and subtraction (element by element).

In this section, we briefly demonstrate a few of MATLAB's matrix capabilities. If you are not familiar with matrix algebra, you can skip this section with no loss of continuity in the remainder of the text. In this section we will use the terms *matrix* and *vector*. A vector is a matrix with one row (a row vector) or one column (a column vector).

Below is matrix A:

$$A = \begin{bmatrix} 1 & 2 & 3 \\ 4 & 5 & 6 \\ 7 & 8 & 9 \end{bmatrix}$$

In MATLAB, matrices and vectors are created using square brackets [ ]. Rows are separated by the semicolon character (;). To create the matrix above, type **A=[1 2 3;4 5 6;7 8 9]** and press the ENTER key. Note that entries in the same row are separated by spaces:

```
EDU» A=[1 2 3;4 5 6;7 8 9]
A =
       1       2       3
       4       5       6
       7       8       9
```

The elements in a row of the matrix can also be separated by commas (,) rather than spaces. For example, to create the matrix

$$B = \begin{bmatrix} 9 & 8 & 7 \\ 6 & 5 & 4 \\ 3 & 2 & 1 \end{bmatrix}$$

we type **B=[9,8,7;6,5,4;3,2,1]**. Note that rows are still separated by the semicolon (;) character:

```
EDU» B=[9,8,7;6,5,4;3,2,1]
B =
       9       8       7
       6       5       4
       3       2       1
```

MATLAB is rich in functions for manipulating matrices. We won't cover many of them here, but we will demonstrate a few.

1.8.1   Matrix Multiplication

Matrices are multiplied by using the * operator:

```
EDU» c=A*B
c =
      30      24      18
      84      69      54
     138     114      90
```

Note that with matrices, $A*B$ is not necessarily equal to $B*A$ because matrix multiplication is not commutative.

```
EDU» c2=B*A
c2 =
      90     114     138
      54      69      84
      18      24      30
```

The * operator can also be used to multiply a matrix by a scalar:

```
EDU» d=5*A
d =
       5      10      15
      20      25      30
      35      40      45
```

Note that since 5 is a scalar, every element in matrix A is multiplied by 5.

Row and column vectors can also be created with MATLAB. A row vector is a matrix with a single row and a column vector is a matrix with a single column. We will create the vectors

$$x = [1 \ 2 \ 3] \qquad \text{and} \qquad y = \begin{bmatrix} 3 \\ 2 \\ 1 \end{bmatrix}$$

To create x, type **x= [1,2,3]** and press the ENTER key:

```
EDU» x=[1,2,3]
x =
       1       2       3
```

To create y, type **y= [1;2;3]** and press the ENTER key. Note that a semicolon (;) separates the rows in a matrix. Thus y has three rows, each with one element.

```
EDU» y=[1;2;3]
```

```
y =
      1
      2
      3
```

We can now multiply these two quantities:

```
EDU» z=x*y
z =
       14
EDU» Z=y*x
Z =
      1      2      3
      2      4      6
      3      6      9
```

1.8.2   Matrix Addition and Subtraction

Matrices of the same size can be added and subtracted. Note that *A* and *B* are displayed before the operation so that we can see the elements of the arrays more easily.

```
EDU» A
A =
      1      2      3
      4      5      6
      7      8      9
EDU» B
B =
      9      8      7
      6      5      4
      3      2      1
EDU» q=A-B
q =
     -8     -6     -4
     -2      0      2
      4      6      8
EDU» b_sub_a=B-A
B_sub_a =
      8      6      4
      2      0     -2
     -4     -6     -8
```

Note that when we perform matrix addition or subtraction, the matrices are subtracted element by element. Addition and subtraction can be done between matrices and a scalar number as well. For example, we can add 5 to matrix *A*:

```
EDU» A+5
ans =
      6      7      8
      9     10     11
     12     13     14
```

We see that the scalar value of 5 has been added to each element of the array. This is quite different from adding matrices that have the same dimensions.

1.8.3 The Inverse of a Matrix

The inverse of a square matrix can be obtained by using the MATLAB **INV** function. To find the inverse of matrix A, type **inv(A)** and press the ENTER key:

```
EDU» inv(A)
Warning: Matrix is close to singular or badly scaled.
         Results may be inaccurate. RCOND = 2.937385e-018
ans =
   1.0e+016 *
     0.3152    -0.6304     0.3152
    -0.6304     1.2609    -0.6304
     0.3152    -0.6304     0.3152
```

MATLAB takes the inverse, but it also warns us that the result is probably not accurate. This is because our matrix A is close to a singular matrix. Let's create a new matrix and take its inverse:

```
EDU» A=rand(3)
A =
     0.2190     0.6793     0.5194
     0.0470     0.9347     0.8310
     0.6789     0.3835     0.0346
```

The **RAND**(m, n) function creates an $m \times n$ matrix and fills it with pseudo random numbers between 0 and 1. The **RAND** function generates a uniform distribution of numbers. That is, the generated numbers have an equal probability of being anywhere between 0 and 1. For a more detailed explanation of random numbers, see Sections 4.4.4 and 4.4.5.

In the preceding example, we only specified one dimension [we used the function as **RAND**(3) rather than **RAND**(2, 6)], so the **RAND** function creates a square matrix. We can now find the inverse of this matrix:

```
EDU» A_inv=inv(A)
A_inv =
   348.9189  -214.0972   -96.2296
  -685.3625   420.4187   191.9204
   751.1609  -459.5743  -210.4294
```

When we multiply a matrix by its inverse, we should obtain the identity matrix:

```
EDU» A*A_inv
ans =
      1.0000      0.0000      0.0000
      0.0000      1.0000      0.0000
      0.0000      0.0000      1.0000
```

1.8.4   The Determinant of a Matrix

The MATLAB **DET** function is used to find the determinant of a matrix. Let's find the determinant of a matrix we used earlier. First we display matrix B:

```
EDU» B
B =
      9      8      7
      6      5      4
      3      2      1
```

To calculate the determinant of the matrix, type **det(B)** and press the ENTER key:

```
EDU» det(B)
ans =
      0
```

In this case, the determinant is zero. Note in the previous section, when we calculated the inverse of this same matrix, the **inv** function stated that the matrix might be singular. Before taking the inverse of a matrix, you should check to see if the determinant is zero. If it is, do not take the inverse of the matrix. We will also calculate the determinant of matrix A:

```
EDU» A
A =
      0.2190      0.6793      0.5194
      0.0470      0.9347      0.8310
      0.6789      0.3835      0.0346
EDU» det(A)
ans =
  -8.2071e-004
```

1.8.5   Solving Systems of Equations

In many fields of engineering, you will end up with a system of n equations with n unknown variables. If the system of equations is linear, MATLAB solves it easily. Suppose we have the system of equations

$$
\textbf{Eq. 1.1} \qquad \begin{array}{rcrcrcr}
5x & + & 7y & + & 3z & = & 1 \\
-3x & + & 16y & - & 2.5z & = & -4 \\
15x & - & 12y & + & 3.9z & = & 16.8
\end{array}
$$

In matrix notation, we write this system of equations as $Ax = b$ where

$$
A = \begin{bmatrix} 5 & 7 & 3 \\ -3 & 16 & -2.5 \\ 15 & -12 & 3.9 \end{bmatrix} \qquad x = \begin{bmatrix} x \\ y \\ z \end{bmatrix} \qquad b = \begin{bmatrix} 1 \\ -4 \\ 16.8 \end{bmatrix}
$$

If you take a course in matrix algebra, you will develop the relation that the equation $Ax = b$ has a solution if the rank of A and the rank of the augmented matrix $[A \ b]$ are both equal to 3:

```
EDU» rank(A)
ans =

   3
EDU» rank([A b])
ans =

   3
EDU»
```

Since the rank of A and $[A \ b]$ are the same, we can solve the set of equations. We will show three ways.

If we multiply each side of the equation $Ax = b$ by the inverse of matrix A, we get

$$
\textbf{Eq. 1.2} \qquad\qquad A^{-1}Ax = A^{-1}b
$$

Any matrix times its inverse is equal to the identity matrix

$$
A^{-1}A = \begin{bmatrix} 1 & & \\ & 1 & \\ & & 1 \end{bmatrix} = I
$$

where I is referred to as the identity matrix. Substituting $A^{-1}A = I$ into Eq. 1.2 yields

$$
\textbf{Eq. 1.3} \qquad\qquad Ix = A^{-1}b
$$

From matrix algebra, the identity matrix times vector x is equal to x, so our solution is

$$
\textbf{Eq. 1.4} \qquad\qquad x = A^{-1}b
$$

If we write our system of equations as $Ax = b$, then the solution is $x = A^{-1}b$, where A^{-1} is the inverse of matrix A. We can use the MATLAB **INV** function to find the inverse of a matrix. We will now solve our system of equations in Eq. 1.1 using MATLAB. First, we define our A and b matrices:

```
EDU» A=[5 7 3; -3 16 -2.5; 15 -12 3.9]
```

```
A =
        5.0000       7.0000       3.0000
       -3.0000      16.0000      -2.5000
       15.0000     -12.0000       3.9000
EDU» b=[1; -4; 16.8]
b =
        1.0000
       -4.0000
       16.8000
EDU»
```

We now solve for x using $A^{-1}b$.

```
EDU» x=inv(A)*b
x =
        1.2921
       -0.2141
       -1.3206
EDU»
```

We defined $x = \begin{bmatrix} x \\ y \\ z \end{bmatrix}$ when we converted our system of equations to matrix form, so using the above result for x, we know that $x = 1.2921$, $y = -0.2141$, and $z = -1.3206$. We can use this method to solve any system of linear equations that we can write in matrix form. Note that if the determinant of matrix A is zero, MATLAB generates an error or warning message when calculating the inverse of A, and either you will not get a result for x or the result will be inaccurate.

The second method uses the MATLAB left-division operator or the backslash operator:

```
EDU» x=A\b
x =
        1.2921
       -0.2141
       -1.3206
EDU»
```

This method is the left division of matrix A into b. This method is preferred because it requires fewer operations and is more accurate than $x = A^{-1}b$.

1.9 Strings

MATLAB *Predefined*

length function

strcmp function

str2num function

upper function

lower function

Character strings in MATLAB are enclosed in single quotes. A character string is a series of alphabetic letters, numeric symbols, or other symbols. Examples of character strings are 'abcd', 'fish', 'x(12)*z{23}', and so on. A variable can be set to a string by enclosing the string in single quotes:

```
EDU» x='abcd'
x =
abcd
```

Note that variable *x* is set to the contents between the single quotes ('). Most characters can be placed in character strings.

```
EDU» y='abc123@#${}[ ]:;'
 y =
abc123@#${}[]:;
```

That is, to define a character string, we enclose a series of characters between two single quote characters—for example, **a='abcd'**. So how do we create a character string that already contains the single-quote character? As an example, let's specify the string *didn't*? In MATLAB, as in many other programming languages, a single quote is specified as a character in a string by using two single quotes next to each other. So now let's set the variable *word* to the string *didn't*. Type **word='didn''t'** and press the ENTER key:

```
EDU» word='didn''t'
word =
didn't
```
Press the ['] key *twice* here.

Two single quotes in a row (' ') give a different result than a double quote ("). They may look the same at a glance, but they are different. Let's repeat this example using a double quote instead of two single quotes. Type **word='didn"t'** and press the ENTER key:

```
EDU» word='didn"t'
word =
didn"t
```
The ["] character is typed here.

Now the string contains the double-quote character, and we don't get the result we want. As a final example, we will create a string that contains a single character, the single quote. Type **x=''''** and press the ENTER key:

```
EDU» x=''''
x =
'
```
Press the ['] key four times.

1.9.1 String Indexing

The first character of a string has an index of 1. To access any character, we specify the variable name with the index of the character enclosed in parentheses. We'll work with the string *a* that we defined previously:

```
EDU» a
a =
abcd
```

To look at the first character in the string, use index 1.

```
EDU» a(1)
ans =
a
```

Note that the first letter of the string contained in variable *a* is the letter a. We can look at each letter individually if we like:

```
EDU» a(2)
ans =
b
EDU» a(3)
ans =
c
EDU» a(4)
ans =
d
```

We can also specify a range of characters. The form *a*(first_index : second_index) gives us all the characters from the first index to the second index of string *a*. For example, to get the two middle letters of string *a*, we specify indices 2 to 3.

```
EDU» a(2:3)
ans =
bc
```

To get the last three characters, we specify indices 2 to 4.

```
EDU» a(2:4)
ans =
bcd
```

Note that individual characters of a string and substrings of a character string are themselves strings. That is, for variable *a*, *a*(1), *a*(2:3), and *a*(2:4) are also character strings, only shorter than the original string.

1.9.2 Concatenating Strings

Strings can be concatenated (that is, linked to make a larger string) by using the following notation:

new_string=[string1, string2, string3, …]

Let's create three strings *x*1, *x*2, and *x*3:

```
EDU» x1='abcd'
x1 =
abcd
EDU» x2='efgh'
x2 =
efgh
EDU» x3='ijklmnop'
x3 =
ijklmnop
```

We will now create a new variable that contains the concatenation of *x*1 and *x*2:

```
EDU» x=[x1, x2]
x =
abcdefgh
```

You can see that variable *x* is a new string that contains both strings *x*1 and *x*2. Here are some other examples:

```
EDU» y=[x1,x2,x3]
y =
abcdefghijklmnop
EDU» z=[x1,x2,x3,'qrstuvwxyz']
z =
abcdefghijklmnopqrstuvwxyz
EDU» mm=['xyzzy',x1,'123',x2]
mm =
xyzzyabcd123efgh
```

We can also concatenate substrings of larger strings:

```
EDU» a=[x1(1), x2(1), x3(1)]
a =
aei
EDU» b=[x1(1:2), x2(1:3),x3(1:5)]
b =
abefgijklm
```

1.9.3 String Functions

MATLAB contains many functions for manipulating strings, including the following:

- **LENGTH** Returns the number of characters in a string.
- **STRCMP** Compares two strings.
- **STR2NUM** Converts a string to a numerical value, for example, converts '123' to 123.

- ■ **STRREP** Replaces characters in a string with different characters.
- ■ **UPPER** Converts a string to uppercase.
- ■ **LOWER** Converts a string to lowercase.

Let's demonstrate these functions.

```
EDU» a='abcdefgh'
a =
abcdefgh
EDU» length(a)
ans =
     8
```

We see that character string *a* contains 8 characters. An easy way to find the last character of a string is

```
EDU» a(length(a))
ans =
h
```

Note that **a(length(a))** is the same as *a*(8) in this example:

```
EDU» a(8)
ans =
h
```

The **STRCMP** function is used to compare two strings. It returns the numeric value of 0 if the strings are different and the numeric value of 1 if the strings are the same. Note that a space is a valid character in a string. Suppose we have the following strings:

```
EDU» a='abcd'
a =
abcd
EDU» b='efgh'
b =
efgh
EDU» c='abcd'
c =
abcd
EDU» d='abcd '
d =
abcd
```

Here's how the comparisons work.

```
EDU» strcmp(a,b)
ans =
     0
```

The function returns 0 because the strings contain different characters.

```
EDU» strcmp(a,c)
ans =
     1
```

The function returns the value 1 because the strings are identical.

```
EDU» strcmp(a, 'abcd')
ans =
     1
```

The strings are identical.

```
EDU» strcmp(a,d)
ans =
     0
```

The function states that the strings contained in variables *a* and *d* are different. This is because *d* contains a space at the end of the string, *a* = 'abcd' and *d* = 'abcd '.

The **STR2NUM** function converts a string representation of a number to its corresponding numeric value. For example, if we add '123' to '321', we get a strange result:

```
EDU» '123' + '321'
ans =
   100    100    100
```

This result is actually the sum of the ASCII codes for the two numbers. The first 100 is the ASCII code for 1 plus the ASCII code for 3; the second 100 is the ACSII code for 2 plus the ASCII code for 2; and the third 100 is the ASCII code for 3 plus the ASCII code for 1. The **ABS** function gets the ASCII codes for a string:

```
EDU» abs('123')
ans =
    49     50     51
EDU»
```

Note that the ASCII code for '1' is 49, the ASCII code for '2' is 50, and the ASCII code for '3' is 51. We see that **'123' + '321'** is the same as **abs('123') + abs('321')**:

```
EDU» '123'+'321'
ans =
   100    100    100
EDU» abs('123')+abs('321')
ans =
   100    100    100
EDU»
```

If we want to interpret a string as a number, we use the **STR2NUM** function to convert the string to a number:

```
EDU» str2num('123') + str2num('321')
ans =
   444
```

Function **STR2NUM** converts the character string '123' to the number 123.

```
EDU» num2str('123')
ans =
123
```

Once we convert strings to their numerical representation, we can use the results in mathematical equations. **STR2NUM** is usually used when reading input from the keyboard. Information is read as character strings to prevent the input of invalid numbers. The program then converts the strings to numbers and uses the numbers in calculations. This allows the program to check the strings for errors and prevents the program from terminating because of bad input. As a final example, let's use **STR2NUM** with variables that contain strings.

```
EDU» a='111';
EDU» b='222';
EDU» a_val=str2num(a)
a_val =
   111
EDU» b_val=str2num(b)
b_val =
   222
```

Note that **a_val+b_val** adds the numerical values, but **a+b** manipulates the strings:

```
EDU» a_val+b_val
ans =
   333
EDU» a+b
ans =
    99    99    99
```

The last string function we will look at is **STRREP**. This function replaces characters in a string with different characters. Suppose we have a string in which we wish to replace all occurrences of the letter 'a' with an asterisk:

```
EDU» xx='String search and replacement';
EDU» yy=strrep(xx, 'a', '*')
yy =
String se*rch *nd repl*cement
```

We can also replace strings of characters with new strings:

```
EDU» line='supplement replacement excitement retirement';
EDU» yy=strrep(line, 'ent', '***')
yy =
supplem*** replacem*** excitem*** retirem***
```

Or, we can replace substrings with a different number of characters:

```
EDU» yy=strrep(line, 'ment', '*')
yy =
supple* replace* excite* retire*
EDU» yy=strrep(line, 'ment', '**********')
yy =
supple********** replace********** excite********** re-
tire**********
```

1.10 **Input and Output Statements**

MATLAB Predefined
input function
fprintf function
str2num function

Thus far we've used MATLAB as a calculator. Every time we make a calculation or assign a variable, MATLAB displays the result automatically. Everything we have entered has been a one-line calculation. We type in the equation and MATLAB gives us the result. Most programs, however, ask for input, perform a large number of calculations, and then at the end, display the final results. The intermediate results are not displayed. This raises the questions: How do we suppress the results of intermediate equations, how does a program request input from the user, and how does a program display results?

1.10.1 The Semicolon in MATLAB

To suppress the results of a calculation in MATLAB, we add a semicolon (;) to the end of a line. Here is an example:

```
EDU» a=5+3  ◄─────────  No semicolon.
a =
    8
```

Note that there is no semicolon in the line above. This instructs MATLAB to perform the calculation and display the result. Here is a similar calculation with the semicolon added:

```
EDU» a=5+6;
```

MATLAB performed the calculation but did not display the result. We can still use the variable later in the program and it will have the value assigned by the previous equation:

```
EDU» b=a+2
b =
      13
```

Thus, if we want to display the results of intermediate calculations, we write the equation without a semicolon. If we have no need to display the result of a particular equation, we add a semicolon to the end of the line.

The semicolon is also used to signal the end of a statement; it allows several commands to be entered on a single line. An example is

```
EDU» x=5; y=25; z=x+y/13;
EDU»
```

This line is equivalent to entering the three commands on individual lines:

```
EDU» x=5;
EDU» y=25;
EDU» z=x+y/13;
EDU»
```

The semicolon thus allows us to place several MATLAB commands on a single line. In both of these cases, the semicolon prevents MATLAB from displaying the result of the command.

Input and output from MATLAB are accomplished in several ways. Here we will look at the **INPUT** and **FPRINTF** functions.

1.10.2 MATLAB Programmed Output

Output to the MATLAB command window is accomplished using the **FPRINTF** function. This function produces formatted output. Here we will just touch upon some of the features required to use the function.

To generate a text line, all we do is use the **FPRINTF** command with the text string:

```
EDU» fprintf('This is a test.\n');
This is a test.
```

Note that the semicolon at the end of the line does not suppress the output of the function. Also note the **\n** at the end of the line. This special string means to start a new line. We could also have pieced together the line in several statements:

```
EDU» fprintf('This ');
EDU» fprintf('is ');
EDU» fprintf('a ');
EDU» fprintf('test.\n');
This is a test.
```

The strings are concatenated by the **FPRINTF** function—the **FPRINTF** function by itself does not start a new line. Also note that no output is generated until the **FPRINTF** function encounters the text **\n**. You must specifically tell **FPRINTF** to create a new line. Sometimes this feature causes problems. MATLAB will not display a line until it sees a **\n** in an **FPRINTF** function. Thus, if you issue an **FPRINTF** function near the beginning of a program and it does not contain a **\n**, the output will not be generated until a string that contains the text **\n** is used in the **FPRINTF** command somewhere else in your program. Suppose that the following code segment is contained in a program:

```
fprintf('This is a test!!!.');
a=5*6/2;
b=7+4/3;
c=a+b-27;
fprintf('End of calculations.\n');
This is a test!!!.End of calculations.
```

When the program runs and MATLAB executes this code segment, the output of both **FPRINTF** commands prints on the same line. This may or may not be what we intended. If we add a **\n** to the string of the first **FPRINTF** function, two lines are generated:

```
EDU» fprintf('This is a test!!!.\n');          ◄———— \n added to this line.
EDU» a=5*6/2;
EDU» b=7+4/ 3;
EDU» c=a+b-27;
EDU» fprintf('End of calculations.\n');
This is a test!!!.
End of calculations.
```

To display the value of a variable, we can include a **%g** format character in the text of the statement. The value of the variable will replace the **%g** when the line is printed:

```
EDU» fprintf('The value of pi is %g.\n', pi);
The value of pi is 3.14159.
```

We can place as many variables in an output line as we want, and place the variables anywhere in the output line:

```
EDU» a=3*pi;
EDU» b=pi/3;
EDU» fprintf('The value of a is %g, the value of b is %g, and that
is that.\n', a, b);
The value of a is 9.42478, the value of b is 1.0472, and that is that.
```

Here are some other examples:

```
EDU» who=25;
EDU» b=17;
EDU» c=27*pi;
EDU» d=b+c/who;
EDU» fprintf('%g      %g      %g      %g\n', who, b, c, d);
EDU» fprintf('\n\nWho = %g, b = %g\nThis is a test!\n',who, b);
25      17      84.823      20.3929


Who = 25, b = 17
This is a test!
```

The last application of the **FPRINTF** function we will look at is its use with strings. Instead of explicitly placing the text string in the **FPRINTF** function, such as

```
EDU» fprintf('This is a test!!!.\n');
This is a test!!!.
```

we can use a string variable to replace the character string:

```
EDU» xx='This is a test!!!.\n';
EDU» fprintf(xx);
This is a test!!!.
```

The two methods produce the same result. The second method can also be used to display the values of variables:

```
EDU» a=3*pi;
EDU» b=pi/3;
EDU» out='The value of a is %g, the value of b is %g, and that is that.\n';
EDU» fprintf(out, a, b);
The value of a is 9.42478, the value of b is 1.0472, and that is that.
EDU» who=25;
EDU» b=17;
EDU» c=27*pi;
EDU» d=b+c/who;
EDU» out1='%g      %g      %g      %g\n';
EDU» out2='\n\nWho = %g, b = %g\nThis is a test!\n';
EDU» fprintf(out1, who, b, c, d);
EDU» fprintf(out2, who, b);
25      17      84.823      20.3929


Who = 25, b = 17
This is a test!
```

Using a string variable in the **FPRINTF** function instead of typing the string directly in the **FPRINTF** function is convenient when you need to piece together a complicated output text string. Many MATLAB functions are available for manipulating character strings, so it may be easier to piece together a complicated string using string variables rather than the **FPRINTF** function.

1.10.3 MATLAB Programmed Input

Information from the user can be requested by using the MATLAB **INPUT** function. The form is

```
xx=input('Text string');
```

With this function, MATLAB displays the text string and then requests input from you in the form of a number. The value you enter is placed in the variable, *xx* in this case. Here is an example:

```
num=input('Specify a number.');
fprintf('\nThe number you typed was %g.\n',num);
Specify a number. 2

The number you typed was 2.
```

Note that the text string emitted by the **INPUT** function doesn't create a new line. That is, your response is typed on the same line as the text string emitted by the **INPUT** function. The **INPUT** function recognizes the new line character (**\n**) as does the **FPRINTF** function.

```
num=input('Specify a number.\n');
fprintf('\nThe number you typed was %g.\n',num);
Specify a number.
2
The number you typed was 2.
```

The **\n** can be used anywhere in the text string inside the input statement to generate a new line.

The form of the **INPUT** function shown previously requires the input of a numerical value. If we type in an invalid response, MATLAB generates an error message and then repeats the message:

```
EDU» xx=input('Specify a number.\n');
Specify a number.
test
??? Undefined function or variable 'test'.      ◄——— Error Message.
Specify a number.
Hi There
```

```
??? Hi There
       |
Unrecognized operand or partial expression.
```
◄──── Error Message.

```
Specify a number.
2
```

In this example, MATLAB expects a numerical value. Instead, you type in a text string and MATLAB generates error messages. A second form of the **INPUT** function allows you to input a character string:

```
xx_string=input('Text string','s');
```

Here, the **'s'** tells MATLAB to request input as a character string. The input is stored in variable *xx_string*. Since all characters compose valid text strings, this form will never generate an error message.

```
xx=input('Input something, anything.','s');
fprintf('The text  you typed was:\n');
fprintf(xx);
Input something, anything. Hi There.
The text you typed was:
Hi There.
```

We can read any input string without generating an error message. If we require a numerical value, we can test to see if the text string can be converted to a valid number. If so, we can use the **STR2NUM** function to convert the number to a numerical value:

```
xx_str=input('Specify a number','s');
%
% Some code will go here to check if the string can be converted
% into a valid number. If yes, execute the statement below.
% If not, generate an error message and ask again.
xx_val=str2num(xx_str);
fprintf('The value of the number is %g.\n', xx_val);
Specify a number: 2
The value of the number is 2.
```

The **%** character in MATLAB specifies a comment. Any text following the **%** character will be ignored by MATLAB. This allows you to place any text after the **%** character. Usually, this text explains what the program is doing or how the program works, and is referred to as a comment.

1.11 Plotting in MATLAB

MATLAB Predefined
plot *function*
linspace *function*
abs *function*

MATLAB has a plethora of functions dedicated to plotting. In Chapter 7, we cover a few of these functions in detail. In this section we demonstrate a quick way to generate a plot of some simple mathematical functions using the MATLAB **PLOT** function. In these examples we will use the array version of the basic mathematical operators (.*, ./, and .^). These

x **label** function
y **label** function
title function
legend function

operators differ from the matrix operators (*, /, and ^) in that the array operators perform element-by-element operations while the matrix operators perform matrix calculations. In the following examples, notice that we use only the array versions of the operators.

For a first example, suppose we want to plot the function

$$y = |x| \sin(x)$$

for $-100 \le x \le 100$. The first thing we do is generate the domain variable x. To do this, we use the **LINSPACE** function. The syntax of the **LINSPACE** function is $x =$ **LINSPACE**$(x1, x2, n)$. This function generates n equally spaced values from $x1$ to $x2$ and stores them in x as a one-dimensional array with one row. Here are some examples:

```
EDU» x=linspace(0,5,6)
x =
      0     1     2     3     4     5
EDU» x=linspace(0,5,11)
x =
  Columns 1 through 7
         0    0.5000    1.0000    1.5000    2.0000    2.5000    3.0000
  Columns 8 through 11
    3.5000    4.0000    4.5000    5.0000
EDU»
```

The syntax of the **PLOT** function is **plot**(x, y). This function plots the values stored in array y versus the values stored in array x. The two arrays must have the same number of values. We will now generate the x and y values and plot the function.

```
EDU» x=linspace(-100,100,5000);
EDU» y=abs(x).*sin(x);
EDU» plot(x,y)
```

The **ABS**(x) function returns the absolute value of x. The three lines above generate this plot:

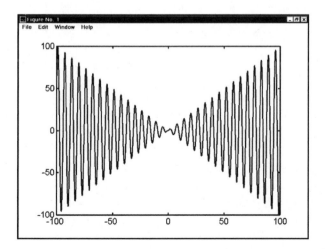

We can also generate the absolute value of x by taking the square root of x^2:

```
EDU» x=linspace(-100,100,5000);
EDU» y=sqrt(x.*x).*sin(x);
EDU» plot(x,y)
```

This code segment generates the same plot as above.

The **PLOT** function can also plot several traces on a single plot. As an example, let's plot the $\sin(x)$, $\cos(x)$, and $\sin^2(x)$ on the same plot for $0 \le x \le 2\pi$. In the following code segment, **pi** is a built-in MATLAB constant.

```
EDU» x=linspace(0,2*pi,100);
EDU» y1=sin(x);
EDU» y2=cos(x);
EDU» y3=sin(x).*sin(x);
EDU» plot(x,y1,x,y2,x,y3)
```

This code segment generates the plot below. In your plot, each trace should be a different color.

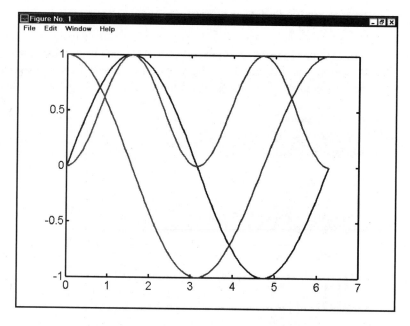

We use the functions **XLABEL**, **YLABEL**, **TITLE**, and **LEGEND** to label the axes and give the plot a title:

```
EDU» xlabel('x');
EDU» ylabel('y');
EDU» title('Various Trigonometric Functions');
EDU» legend('sin(x)', 'cos(x)', 'sin(x)*sin(x)');
```

These commands modify the plot as shown here:

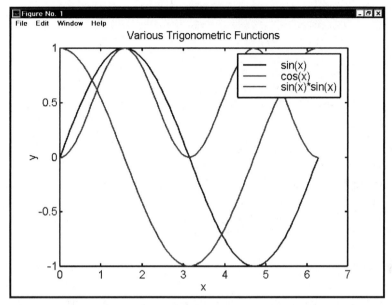

These examples show only a tiny fraction of the plotting capabilities of MATLAB. Other common plotting functions include **SEMILOGX**, which is used for semi-log plots, and **LOGLOG**, which is used for log-log plots. See Chapter 7 for more information on plotting.

1.12   MATLAB **Help Facilities**

MATLAB Predefined
help command
lookfor command
which command
roots function

1.12.1   Help (Text Version)

The MATLAB help facility can be used to find information on just about any function or operation that MATLAB provides. We start by typing **help** to get the general help screen:

```
EDU» help
HELP topics:

toolbox\local       -  Local function library.
matlab\datafun      -  Data analysis and Fourier transform functions.
matlab\elfun        -  Elementary math functions.
matlab\elmat        -  Elementary matrices and matrix manipulation.
matlab\funfun       -  Function functions - nonlinear numerical methods.
matlab\general      -  General purpose commands.
```

```
matlab\color        -  Color control and lighting model functions.
matlab\graphics     -  General purpose graphics functions.
matlab\iofun        -  Low-level file I/O functions.
matlab\lang         -  Language constructs and debugging.
matlab\matfun       -  Matrix functions - numerical linear algebra.
matlab\ops          -  Operators and special characters.
matlab\plotxy       -  Two dimensional graphics.
matlab\plotxyz      -  Three dimensional graphics.
matlab\polyfun      -  Polynomial and interpolation functions.
matlab\sounds       -  Sound processing functions.
matlab\sparfun      -  Sparse matrix functions.
matlab\specfun      -  Specialized math functions.
matlab\specmat      -  Specialized matrices.
matlab\strfun       -  Character string functions.
matlab\dde          -  DDE Toolbox.
matlab\demos        -  The MATLAB Expo and other demonstrations.
toolbox\symbolic    -  Symbolic Math Toolbox.
toolbox\sigsys      -  Signals and Systems Toolbox.
toolbox\wintools    -  GUI tools for MATLAB for MS Windows.

For more help on directory/topic, type "help topic".
```

To get more specific help, the last line of the general help file tells us to type **help** followed by the topic. Here the topic is the name of the directory. For example, to get help on the demos available with MATLAB, we type **help demos**:

```
EDU» help demos
 The MATLAB Expo and other demonstrations.

 MATLAB/Introduction.
    expo, demo- Start up The MATLAB Expo and display splash screen.
    expomap     - Open the MATLAB Expo Main Map (avoids Expo splash screen).

 MATLAB/Matrices.
    intro       - Introduction to MATLAB.
    inverter    - Demonstrate the inversion of a matrix.
    buckydem    - Connectivity graph of the Buckminster Fuller geodesic dome.
    sparsity    - Demonstrate effect of sparsity orderings.
    matmanip    - Introduction to matrix manipulation.
    delsqdemo   - Finite difference Laplacian on various domains.
    sepdemo     - Separators for a finite element mesh.

 MATLAB/Numerics.
    funfuns     - Demonstrate functions that operate on other functions.
    fitdemo     - Nonlinear curve fit with simplex algorithm.
    sunspots    - The answer is 11.08, what is the question?
    e2pi        - Which is greater, e^pi or pi^e?
```

```
    odedemo      - Ordinary differential equations.
    quaddemo     - Adaptive quadrature.
    zerodemo     - Zerofinding with fzero.
    fplotdemo    - Plot a function.
    eigmovie     - Symmetric eigenvalue movie.
    rrefmovie    - Computation of Reduced Row Echelon Form.
    fftdemo      - Use of the fast finite Fourier transform.
    census       - Try to predict the US population in the year 2000.
    spline2d     - Demonstrate GINPUT and SPLINE in two dimensions.

MATLAB/Visualization.
    graf2d       - Demonstrate XY plots in MATLAB.
    graf2d2      - Demonstrate XYZ plots in MATLAB.
    grafcplx     - Demonstrate complex function plots in MATLAB.
    lorenz       - Plot the orbit around the Lorenz chaotic attractor.
    xpklein      - Klein bottle demo.
    xfourier     - Graphics demo of Fourier series expansion.
    cplxdemo     - Maps of functions of a complex variable.
    peaks        - A sample function of two variables.
    membrane     - Generate MathWorks's logo.
    sqdemo       - Superquadrics using UIControls.
    colormenu    - Select color map.

MATLAB/Language.
    xplang       - Introduction to the MATLAB language.
    graf3d       - Demonstrate Handle Graphics for surface plots.
    hndlgraf     - Demonstrate Handle Graphics for line plots.
    hndlaxis     - Demonstrate Handle Graphics for axes.

Toolbox/Signal Processing.
    filtdem      - Signal Processing filter demo.
    filtdem2     - Demonstrate filter design techniques.
    sigdemo1     - Discrete-time Fourier transform of a signal.
    sigdemo2     - Continuous-time Fourier transform of a signal.
    phone        - Signal processing and the touch-tone phone.

Toolbox/Control System.
    ctrldems     - Set up Control System command line demos.

Toolbox/Symbolic Math.
    xpcalc       - Calculus operations.
    xpgiv        - Givens transformation.

Extras/Gallery.
    knot         - Tube surrounding a three-dimensional knot.
    quivdemo     - Demonstrate the quiver function.
    logo         - Display the MATLAB L-shaped membrane logo.
```

```
klein1      - Construct a Klein bottle.
cruller     - Construct cruller.
tori4       - Construct four linked tori.
spharm2     - Construct spherical surface harmonic.
```

```
Extras/Games.
  xpbombs     - Minesweeper game.
  life        - Conway's Game of Life.
  bblwrap     - Bubblewrap.
```

```
Extras/Miscellaneous.
  truss       - Animation of a bending bridge truss.
  travel      - Traveling salesman problem.
  makevase    - Generate and plot a surface of revolution.
  logospin    - Movie of The MathWorks' logo spinning.
  crulspin    - Spinning cruller movie.
  xpquad      - Superquadrics plotting demonstration.
  spinner     - Colorful lines spinning through space.
```

```
Extras/Contact Info.
  contact1    - How to reach The MathWorks, Inc.
  contact2    - How to reach The MathWorks, Inc. by email.
  contact3    - How to reach international agents for The MathWorks, Inc.
```

The topics listed are all demonstrations that you can run in MATLAB. You can get further information on a specific demo listed in this response by using the **HELP** command followed by the name of the function. Let's type **help xpbombs**:

```
EDU» help xpbombs

 XPBOMBS  Play the minesweeper game.
   There are 13 bombs hidden in the mine field. Try to flag them
   and uncover all of the other spaces without getting blown up.
   In each non-bomb square is printed the number of adjacent
   squares which contain bombs.

   Use the FLAG button to toggle in and out of flag mode. When in
   flag mode, clicking on any covered square places a flag on it.
   Clicking on any flag removes it.

   At any time during the game, the number of remaining unflagged
   bombs is shown in the upper left.

   NEW stops the game and creates a new minefield.

   CLOSE closes the game window.
```

As another example, we will obtain the general help again and list the topics available. Type **help** and press the ENTER key:

```
EDU» help

HELP topics:

toolbox\book              -  (No table of contents file)
egr222\fall 1998 homework -  (No table of contents file)
matlab\datafun            -  Data analysis and Fourier transforms.
matlab\datatypes          -  Data types and structures.
matlab\dde                -  Dynamic data exchange (DDE).
matlab\demos              -  Examples and demonstrations.
matlab\elfun              -  Elementary math functions.
matlab\elmat              -  Elementary matrices and matrix manipulation.
matlab\funfun             -  Function functions and ODE solvers.
matlab\general            -  General purpose commands.
matlab\graph2d            -  Two dimensional graphs.
matlab\graph3d            -  Three dimensional graphs.
matlab\graphics           -  Handle Graphics.
matlab\iofun              -  File input/output.
matlab\lang               -  Programming language constructs.
matlab\matfun             -  Matrix functions - numerical linear algebra.
matlab\ops                -  Operators and special characters.
matlab\polyfun            -  Interpolation and polynomials.
matlab\sparfun            -  Sparse matrices.
matlab\specfun            -  Specialized math functions.
matlab\specgraph          -  Specialized graphs.
matlab\strfun             -  Character strings.
matlab\timefun            -  Time and dates.
matlab\uitools            -  Graphical user interface tools.
toolbox\local             -  Preferences.
toolbox\control           -  Control System Toolbox.
control\obsolete          -  (No table of contents file)
toolbox\signal            -  Signal Processing Toolbox.
toolbox\symbolic          -  Symbolic Math Toolbox.

For more help on directory/topic, type "help topic".
```

Now let's look at the elementary functions toolbox to see what functions are available. From the list above, toolbox elfun contains the elementary math functions. Type **help elfun** and press the ENTER key:

```
EDU» help elfun

  Elementary math functions.
```

```
Trigonometric.
    sin         - Sine.
    sinh        - Hyperbolic sine.
    asin        - Inverse sine.
    asinh       - Inverse hyperbolic sine.
    cos         - Cosine.
    cosh        - Hyperbolic cosine.
    acos        - Inverse cosine.
    acosh       - Inverse hyperbolic cosine.
    tan         - Tangent.
    tanh        - Hyperbolic tangent.
    atan        - Inverse tangent.
    atan2       - Four quadrant inverse tangent.
    atanh       - Inverse hyperbolic tangent.
    sec         - Secant.
    sech        - Hyperbolic secant.
    asec        - Inverse secant.
    asech       - Inverse hyperbolic secant.
    csc         - Cosecant.
    csch        - Hyperbolic cosecant.
    acsc        - Inverse cosecant.
    acsch       - Inverse hyperbolic cosecant.
    cot         - Cotangent.
    coth        - Hyperbolic cotangent.
    acot        - Inverse cotangent.
    acoth       - Inverse hyperbolic cotangent.

Exponential.
    exp         - Exponential.
    log         - Natural logarithm.
    log10       - Common logarithm.
    sqrt        - Square root.

Complex.
    abs         - Absolute value.
    angle       - Phase angle.
    conj        - Complex conjugate.
    imag        - Complex imaginary part.
    real        - Complex real part.

Numeric.
    fix         - Round towards zero.
    floor       - Round towards minus infinity.
    ceil        - Round towards plus infinity.
    round       - Round towards nearest integer.
    rem         - Remainder after division.
    sign        - Signum function.
```

These are the functions used for standard mathematical calculations. Now that we know their names, we can find information about specific functions. For example, let's find out how to use the **ABS** function:

EDU» `help abs`

```
 ABS Absolute value and string to numeric conversion.
  ABS(X) is the absolute value of the elements of X. When
  X is complex, ABS(X) is the complex modulus (magnitude) of
  the elements of X.

  See also ANGLE, UNWRAP.


  ABS(S), where S is a MATLAB string variable, returns the
  numeric values of the ASCII characters in the string.
  It does not change the internal representation, only the
  way it prints.
  See also SETSTR.
```

1.12.2   Help (Window-Based Version)

Help information is also available in a window-based version. The information is the same as you would receive using the text commands show in Section 1.12.1. To access the help window, select **Help** and then **Help Window** from the MATLAB menus, or click on the ? icon:

The help window will appear:

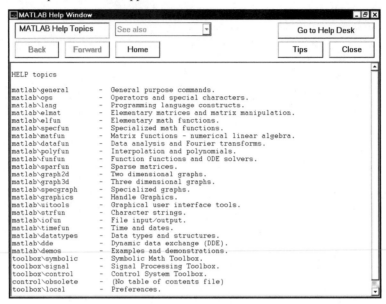

Notice that the topics listed in this window are the same as those generated using the **HELP** command on page 43.

To get more help on an item in the list, double click the *LEFT* mouse button on the text line that interests you. To get more information on the elementary math functions, double-click on the text matlab\elfun:

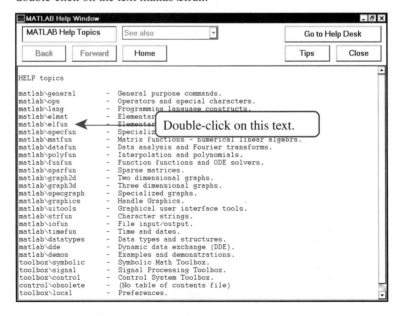

The help information for elementary math functions is displayed:

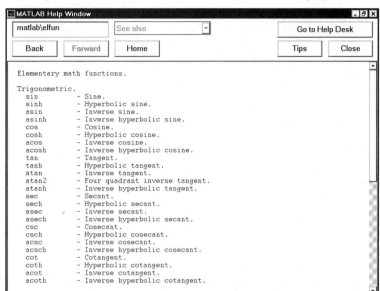

This screen lists the available functions in this area. For more information on a specific command, double-click on the name of the command. For example, double-click on the text "sin." You should get this window:

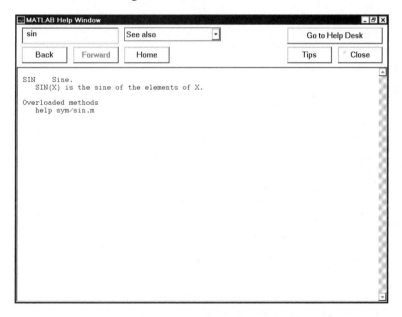

This is the help information for the **SIN** function.

We can also enter text directly in the top field, which will then display the information for the command entered. For example, enter the text **demos** as follows:

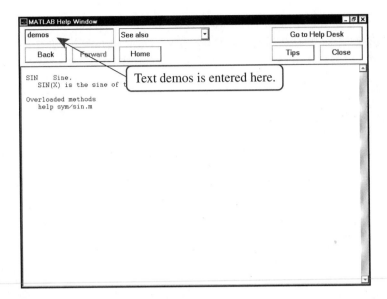

When you press the ENTER key, the help information, if any, for the entered text will be displayed:

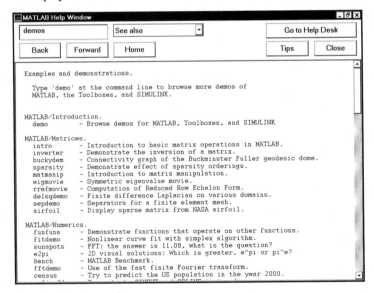

We can get more information on any of the topics listed in this window by using the methods discussed earlier.

1.12.3 The LOOKFOR Command

A second help function called **LOOKFOR** is also available with MATLAB. This function looks through the first line of the help description of each function and searches for the keyword you specify in the **LOOKFOR** command. For example, to find all functions that have the word "logarithm" in the first line of their help description, type **lookfor logarithm** and press the ENTER key:

```
EDU» lookfor logarithm
LOG        Natural logarithm.
LOG10      Common logarithm.
LOGSPACE Logarithmically spaced vector.
LOGM       Matrix logarithm.
BETALN     Logarithm of beta function.
```

We can now use the **HELP** command to find specific information on any of the functions listed above.

1.12.4 The **WHICH** Command

The MATLAB **WHICH** commands allows you to locate m-files on your computer. An m-file is a text file that contains a series of MATLAB statements designed to perform a specific function. In this text, you will learn how to write m-files to perform specific tasks. However, MATLAB provides thousands of m-files that perform functions commonly used by engineers. Since m-files contain readable text, you can look at the contents of the m-files to see how the function performs its task, or to modify the function to your specific application, if necessary.

Since MATLAB contains many m-files, and the m-files are located in different directories, the **WHICH** command helps you to locate them. For example, the MATLAB **ROOTS** function finds the roots of polynomials. If you want to look at the m-file for this function, you must first find the file and then open it with a text editor. To find the file, you can use the **WHICH** command:

```
EDU» which roots
d:\matlab\toolbox\matlab\polyfun\roots.m
EDU»
```

Now you see the full pathname to the file.

The **WHICH** command is also useful when creating new m-files. Whenever you create a new m-file, you need to give it a name. If you give it the same name as an m-file that already exists, problems may arise when you try to use the function. If two m-files have the same name but are located in different directories, MATLAB will use the function it finds first. This may give you the impression that MATLAB is executing the wrong function. Another problem also arises when you save multiple copies of an m-file with the same name but to different directories. Suppose you make changes to one of the m-files, but when you run MATLAB, no changes show. Again, MATLAB uses the copy of the m-file it finds first, so in this case it first found the copy that you did not change.

These problems can be avoided by using a unique name for all m-files. Since MATLAB comes with thousands of m-files, when you name your own m-file, you need to know if the name is already being used by another m-file. If the **WHICH** command finds an m-file with the same name, it will tell you the location of the file. If the **WHICH** command cannot find an m-file with the same name, it will also tell you.

Suppose we want to name our m-file roots. In the preceding example, we located of the m-file **ROOTS** and thus found that an m-file with this name already exists. So we shouldn't name our m-file roots.m. We want to give our m-file a similar name to roots, but the name must be unique. Let's see if there is an m-file named roots1:

```
EDU» which roots1
roots1 not found.
EDU»
```

We can now name our file roots1.m knowing that it is a unique name.

1.13 MATLAB Script Files

A script file is a text file that contains a sequence of MATLAB commands. Using a script file is equivalent to typing in a series of commands at the MATLAB command prompt, except that you don't need to type the commands each time you wish to use them. You can use any text editor to create script files. We'll use the MATLAB M-File Editor. If you use a word processor, such as Microsoft Word, make sure that you save the file as a text file rather than as a ".doc" file. If you save the document in the word processor's native file format, the file will contain formatting characters that could cause errors when used as script files for MATLAB.

Now let's create a script file using the MATLAB facilities. This facility will invoke the MATLAB M-File Editor for editing script files. From the command window, select **File**, **New**, and then **M-file**:

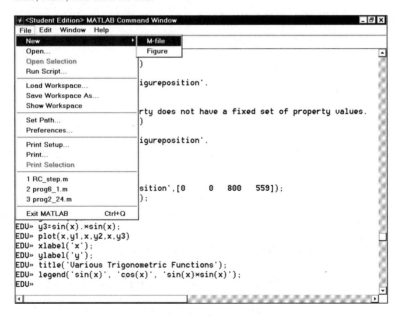

After selecting **M-file,** the M-File Editor/Debugger will open:

We enter a series of MATLAB commands:

Now let's save the file. All script files must be saved with the extension ".m"—this is why MATLAB refers to them as m-files. If you are using the MATLAB M-File Editor, the files will automatically be saved with a ".m" extension. If you are using another text editor, you must make sure that the file is saved with the ".m" extension, or MATLAB will not be able to find and run the script file.

Select **File** and then **Save** from the M-File Editor:

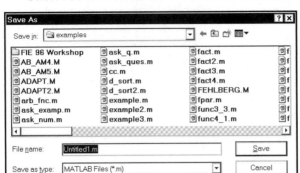

The directory shown in the screen capture above is D:\examples. The directory shown on your dialog box will be different. If you are using MATLAB on your own computer, you should store the files in directory D:\matlab\toolbox\local where D: is the drive label of the hard drive where MATLAB is installed:

D:\matlab\toolbox\local is the directory in which MATLAB assumes you will place your files. However, you can store the file in any directory on your system. If you choose a different directory, you will need to change the MATLAB path to run the script file. See Section 1.13.1 for instructions on changing the path.

The examples for this book are stored in directory d:\examples. Let's save our file in this directory. You can use the local directory, a directory specified by your system administrator, or any other directory. After selecting the directory, we specify the file name:

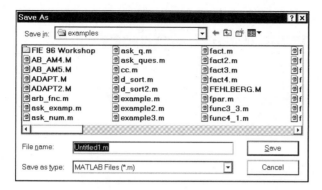

The present name of the file is untitled1.m. Name the file ex1.m and then click the **Save** button:

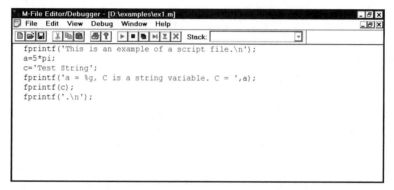

Select **File** and then **Exit Editor/Debugger** to exit the M-File Editor. Return to the MATLAB command window:

MATLAB automatically looks for m-files in the directory matlab\toolbox\local. If you saved the file in the local directory, you should be able to run the script file now. If you saved the file in a different directory, MATLAB may not be able to find your script file and will generate an error. Let's attempt to run the file now. To run a script file, all we need to do is type the name of the file. Type **ex1** and press the ENTER key:

The screen capture shows that MATLAB generated an error when I attempted to run the script file. This error was generated because I saved the file in a directory that was not on the MATLAB search path. MATLAB could not find the file, so it generated the error message. If you received the same error message, you will need to follow the next section, which shows how to change the MATLAB search path for your computer.

1.13.1 Changing the MATLAB Search Path

From the MATLAB menus, select **File** and then **Set Path**:

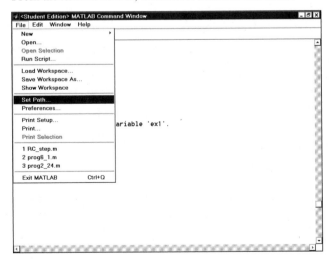

The following dialog box will appear:

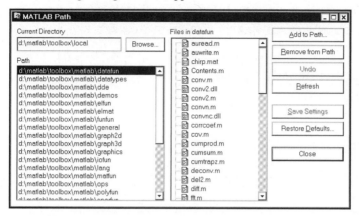

This dialog box displays all the directories currently in the MATLAB search path and also allows us to add or delete directories. Let's add a new directory to the path. Click the **Add to Path** button:

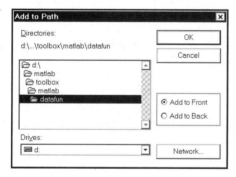

Use the dialog box to specify the directory you would like to add. Let's specify the example directory in which we saved the m-file of Section 1.13:

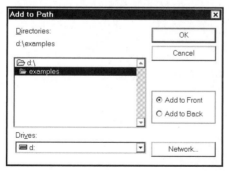

We can add the directory to the top or bottom of the list of directories. If we add the new directory to the top, MATLAB will search the new directory for m-files before it searches the others on the list. This is usually advantageous because we want MATLAB to find our m-file rather than another m-file of the same name in another directory. Add the directory to the top of the list by clicking the **OK** button:

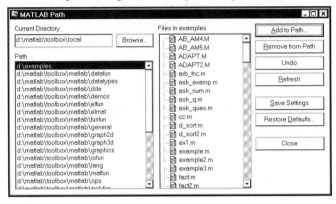

Click the **Close** button to close the dialog box:

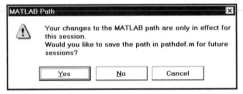

If you want your change in the path to work for all future uses, click the **YES** button. If you want your change to apply only to the current MATLAB session, click the **NO** button. If you click the **NO** button, the next time you run MATLAB, the path will be the list of directories before you modified it. Click the **YES** button. You will return to the MATLAB command window.

Now attempt to run the script file. Type the name of the script file and press the ENTER key:

```
EDU» a=5
a =
     5
EDU» b=19
b =
    19
EDU» c=66
c =
    66
EDU» ex1
??? Undefined function or variable 'ex1'.

EDU» ex1
This is an example of a script file.
a = 15.708, C is a string variable. C = Test String.
EDU»
```

The script file ran this time because MATLAB was able to find an m-file named ex1.m in the directory we just added to the MATLAB search path.

1.13.2 Saving and Running a Script File from a Floppy Disk

As a final example, let's save a script file on your floppy disk and then run the script file from the floppy disk. From the command window, select **File**, **New**, and then **M-file** to open the MATLAB Text Editor, and then add some MATLAB commands:

Save the file on the floppy disk.

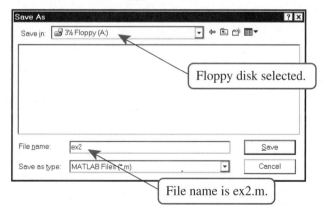

Now that we have named the file ex2.m and saved it on floppy drive A:, click the **Save** button to save the file. Exit the MATLAB Text Editor and return to the MATLAB command window:

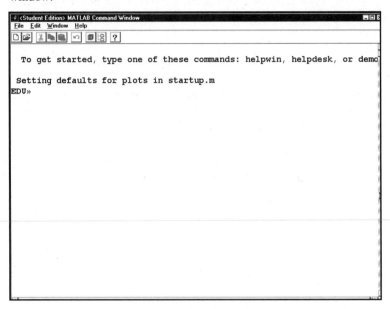

The floppy drive is not normally on the MATLAB search path. If we just typed the name of the file, MATLAB would not find it and would generate an error message. To run an m-file saved on a floppy disk, either we can add the floppy drive to the search path, or we can change the current working directory to the floppy disk. In Section 1.13.1 we showed how to modify the MATLAB search path. Here, we will change the working directory. To change the directory to a:\ type **cd a:** and press the ENTER key:

To run the file, type the name of the m-file without the ".m" extension; type **ex2** and press the ENTER key:

```
<Student Edition> MATLAB Command Window                    _ 8 X
 File   Edit   Window   Help
 
EDU» cd a:\
EDU» ex2
A MATLAB script file that will be saved on a floppy disk.
ans =
      8
ans =
    2.3039+ 1.5192i
ans =
    11.6204
End of script file.
EDU»
```

The script file runs successfully.

1.14 Problems

1-1

Plot the following functions using the **PLOT** function.

 a. $\sin(x)$ for $-10 \le x \le 10$
 b. $\sin(x)\cos(x)$ for $-10 \le x \le 10$
 c. $\cos(x)\sin(x)\exp(x/10)$ for $-1 \le x \le 1$

1-2

Write a script file in MATLAB that displays the results of the following calculations. You should have a single script file that displays all results. The script file should clearly identify each problem and pause after each answer. Answers should be displayed only after the user presses a key or clicks the mouse.

 a. $(5 + 6)/2$ in short format.
 b. The value of 5π to 16 digits.
 c. The value of e displayed as a fraction (ratio of two whole numbers).
 d. Use the MATLAB function **dec2hex** to find how the number 65535 would be displayed in base 16 (hexadecimal). Use the MATLAB help facility to learn how to use the **dec2hex** function.
 e. The result of $(3 + i)/(1 + i)$ in rectangular coordinates (numeric format short).
 f. The value of $\sqrt{27}$.
 g. The result of $(3 + i)*(2 + 3i)$ as a magnitude and a phase (numeric format short).

    h. The angle of $(1 + i)$ displayed in degrees.

    i. $\dfrac{3+4j}{\sqrt{-6.5}}$

    j. Find the real and imaginary parts, magnitude, and angle of $3 + 4i$.

1-3   Write a script file in MATLAB that displays the results of the following calculations. You should have a single script file that displays all results. The script file should clearly identify each problem and pause after each answer. Answers should be displayed only after the user presses a key or clicks the mouse.

    a. 23(6 + 2/3) in short general format.
    b. The value of π to 16 digits.
    c. Find the value of π to 64 digits.
    d. The value of π displayed as a fraction (ratio of two whole numbers).
    e. Use the MATLAB function **dec2base** to find how the number 16 would be displayed in base 2 (binary). Use the MATLAB help facility to learn how to use the **dec2base** function.
    f. The result of $(3 + i3)(5 \angle 35)$ in rectangular coordinates (numeric format short).
    g. The value of $\sqrt{-5j}$.
    h. The result of $(3 + i)$ times $5 \angle 27$ as a magnitude and a phase, numeric format short. Note that $5 \angle 27$ is a phasor (magnitude and angle in degrees).
    i. The angle of $(1 + i)$ displayed in degrees.
    j. $\dfrac{3+4j}{\sqrt{-65}}$
    k. Find the real and imaginary parts, magnitude, and angle of $5 \angle 35$.
    l. Find the real and imaginary parts of $5\,e^{-j0.567}$.

1-4   Write a script file in MATLAB that displays the results of the following calculations. You should have a single script file that displays all results. The script file should clearly identify each problem and pause after each answer. Answers should be displayed only after the user presses a key or clicks the mouse.

    a. $23*(6 + 2/3)^2$ in short general format.
    b. The value of e to 16 digits.
    c. Find the value of π to 64 digits.
    d. The value of e^{π} displayed as a fraction (ratio of two whole numbers).
    e. Use the MATLAB function **dec2base** to find how the number 31 would be displayed in base 2 (binary). Use the MATLAB help facility to learn how to use the **dec2base** function.
    f. The result of $(3 + i4)(5 \angle 35)$ in rectangular coordinates, numeric format short.
    f. The value of $\sqrt{-3j}$.
    h. The result of $(3 + 4i)(6 \angle 33)$ as a magnitude and a phase (numeric format short).
    i. The angle of $(2 + 7j)$ displayed in degrees.
    j. Find the value of $\dfrac{7+6i}{\sqrt{-3i}}$.
    k. Find the real and imaginary parts, magnitude, and angle of $2 \angle{-}23°$.
    l. Find the real and imaginary parts of $5e^{1}e^{-j\pi}$.

1-5   Write a script file that does the following:

a. Creates a random 5×5 matrix, displays the random matrix, calculates and displays the inverse of the random matrix, and displays the result of the matrix times its inverse. There should be no intermediate i/o displayed.

b. Asks the user to input three text strings with the **INPUT** function and prints out the concatenation of all three. There should be no intermediate i/o displayed. Use the **FPRINTF** and **INPUT** functions liberally. (Make the input and output look nice.)

c. Asks the user for a text string and changes the word "the" to "hey." There should be no intermediate i/o displayed. Use the **FPRINTF** and **INPUT** functions liberally. (Make the input and output look nice.)

d. Asks the user for a text string and then tells the user what the first character was, what the second character was, and what the third character was.

1-6   Create a script file that does the following:

a. Displays your name and ID number, homework number, class name and number, and the due date.

b. Pauses for 5 seconds after the above information is displayed. (Use the help command to find a function that will do this.)

c. Asks the user to input a number between 50 and 100. Call it x.

d. Takes the square root of the number and displays it in long format.

e. Pauses until a key is pressed.

f. Sets up an array of 100 values for time from 0 to 0.1 seconds. Do not display the values.

g. Calculates $y = \sin(2\pi x t)$ using the time array created in part f and the value of x from part c.

h. Plots y versus t using the **PLOT** function.

i. Labels the x and y axes using the MATLAB **xlabel** and **ylabel** functions. Use the help facility if you don't know how to use these functions.

1-7   Find the solution (A, B, C) to the system of equations:

$$10 = 400A - 175B$$
$$0 = -175A + 800B - 250C$$
$$5 = -300B + 600C$$

Control Flow

2

OBJECTIVES

- ☐ Study the different methods for control flow available in MATLAB.
- ☐ Introduce the three forms of the **IF** statement: **IF-END, IF-ELSE-END, IF-ELSEIF-ELSE-END**.
- ☐ See how the different forms of the **IF** statement change a program's structure.
- ☐ Use the **SWITCH-CASE** statement and compare it to the three forms of the **IF** statement.
- ☐ Use single and nested **FOR** loops.
- ☐ Use **WHILE** loops and compare them to **FOR** loops.

In this chapter we cover most of the control structures available in MATLAB. Control structures allow you to make decisions and to repeat operations a number of times. The **IF** and **SWITCH-CASE** statements allow you to specify conditions and take different actions depending on the result of a condition. The **FOR** loop allows you to loop through a set of commands a specified number of times, and the **WHILE** loop allows you to continue looping until a condition is met. After completing this chapter, you will be able to solve some fairly complicated problems and also be able to choose the control structure that best suits a specific problem.

Several streams of examples are revisited throughout the text, starting with this chapter. One stream is a guessing game. In the following section we use our guessing game to illustrate the **IF** statement in the form **IF-END**. We rewrite the game in Section 2.2 using **IF-ELSE-END** statements, then improve it with **FOR** loops and **WHILE** loops in Sections 2.5 and 2.6. Other streams that we will slowly improve using different programming techniques include the grade calculation program, the command center shell, and the factorial program.

2.1 IF-END

MATLAB Predefined
rand function
ceil function
isreal function
disp function
imag function

The **IF** statement can be used to test a condition and execute a block of instructions if the condition is true. There are three forms of the **IF** statement: **IF-END, IF-ELSE-END**, and **IF-ELSEIF-ELSE-END**. Here we will look at its simplest form:

TABLE 2-1 Relation operators in MATLAB

Relation Operator	Definition	Example
>	Greater than	A > B A > 5 abs(x) > 3
>=	Greater than or equal to	A >= B real(y) >= 7
<	Less than	A < B A > 5 abs(x) < 3
<=	Less than or equal to	A <= B real(y) <= 7
==	Equal to	A == 5 abs(x) == 27
~=	Not equal to	A ~= 5 abs(x) ~= 27

```
IF condition
     statement 1;
     statement 2;
     statement 3;
          ⋮
     statement n;
END
```

If the condition has a numeric value of 1, or is true, the statements between the **if** and **end** keywords will be executed. If the condition has a numeric value of 0 or is false, then the program will skip the statements between the **if** and **end** keywords and resume at the line after the **end** keyword. Conditions are typically of the form

value relop value

where the values may be constants, variables, or the results of functions or algebraic expressions. The term *relop* is short for relation operator, and can be any of the operators listed in Table 2-1. Let's use the **IF** statement in a few examples.

In the first example, Program 2-1, the computer generates a random number between 1 and 10 and then asks the user to guess what the number is. The computer then notifies the user of the result of the guess, and if incorrect, tells the user what the number was.

Program 2-1 Guessing program

```
%
% This is file guess0000.m
%
```

```
real_number=rand(1); % Generate a number between 0 and 1
real_number=10*real_number; % Scale the number between 0 and 10
number=ceil(real_number); % Round the number to the next highest integer.

% Ask the user to guess the number:
fprintf('I am thinking of a number between 1 and 10.\n');
guess = input('Guess a number: ');

% Test the number.

if guess == number
    fprintf('Congratulations - You guessed the correct number.\n');
end

if guess ~= number
    fprintf('Your guess was wrong.\n');
    fprintf('The number I was thinking of was %g.\n', number);
end
```

Program 2-1 contains two built-in MATLAB functions, **RAND** and **CEIL**, that we have not yet discussed. Function **RAND** generates random numbers between 0 and 1. In general, **RAND** generates an array of random numbers and is used in the form **rand(n,m)**. This form will generate an array of *n* rows and *m* columns and fill the array with random numbers between 0 and 1. For example,

```
EDU» rand(2,3)
ans =
     0.0668    0.6868    0.9304
     0.4175    0.5890    0.8462
EDU» rand(3,5)
ans =
     0.5269    0.4160    0.7622    0.7361    0.7564
     0.0920    0.7012    0.2625    0.3282    0.9910
     0.6539    0.9103    0.0475    0.6326    0.3653
```

When the **RAND** function is used with a single argument, it generates a square array:

```
EDU» rand(3)
ans =
     0.2470    0.7534    0.6316
     0.9826    0.6515    0.8847
     0.7227    0.0727    0.2727
EDU» rand(2)
ans =
     0.4364    0.4777
     0.7665    0.2378
```

RAND(1) generates a single number between 0 and 1. Note that each time we use the function, the number it produces is different.

```
EDU» rand(1)
ans =
    0.2749
EDU» rand(1)
ans =
    0.3593
EDU» rand(1)
ans =
    0.1665
```

Since the **RAND** function generates a random number between 0 and 1, multiplying the number by 10 generates a random number between 0 and 10:

```
EDU» 10*rand(1)
ans =
    4.8652
EDU» 10*rand(1)
ans =
    8.9766
```

Now that we have a real number between 0 and 10, we need to convert it to an integer between 1 and 10. The MATLAB function **CEIL** rounds a real number to the next largest integer:

```
EDU» ceil(5.6)
ans =
    6
EDU» ceil(5.1)
ans =
    6
EDU» ceil(5)
ans =
    5
EDU» ceil(-5.6)
ans =
    -5
EDU» ceil(10*rand(1))
ans =
    10
EDU» ceil(10*rand(1))
ans =
    1
EDU» ceil(10*rand(1))
ans =
    6
```

Now let's examine the remainder of Program 2-1:

```
if guess == number
     fprintf('Congratulations - You guessed the correct number.\n');
end

if guess ~= number
     fprintf('Your guess was wrong.\n');
     fprintf('The number I was thinking of was %g.\n', number);
end
```

The first **IF** statement tests whether the guess is equal to the number generated by the computer. If the numbers are equal, the lines between the first **if** and **end** are executed. That is, the line

```
fprintf('Congratulations - You guessed the correct number.\n');
```

is executed. Execution then resumes at the line following the **end** keyword. If the numbers are not equal, the computer skips the lines between the first **if** and **end** keywords and resumes execution at the line following the first **end** keyword.

The second **IF** statement works the same as the first except that it tests for unequal numbers. Note that we can have a single line between the **if** and **end** keywords, or we can have many lines between them.

In the preceding code segment, we see the **fprintf** statement

```
fprintf('The number I was thinking of was %g.\n', number);
```

Remember that when an **fprintf** statement contains a place holder **%g**, the place holder will be replaced by the value of a variable in the **fprintf** statement. In this case, the text **%g** is replaced by the value stored in variable *number*.

We will now run our program to test that it works correctly. Note that the program is saved as a MATLAB script file (guess0000.m) in directory c:\matlab\toolbox\local. Since this directory is in the MATLAB path, we simply type in the name of the script file to run the program:

```
EDU» guess0000
I am thinking of a number between 1 and 10.

Guess a number: 4
Your guess was wrong.
The number I was thinking of was 3.
EDU»
```

Let's run the program several times in the hope that we can guess the number:

```
EDU» guess0000
I am thinking of a number between 1 and 10.
```

```
Guess a number: 4
Your guess was wrong.
The number I was thinking of was 1.
EDU» guess0000
I am thinking of a number between 1 and 10.

Guess a number: 4
Your guess was wrong.
The number I was thinking of was 7.
EDU» guess0000
I am thinking of a number between 1 and 10.

Guess a number: 4
Your guess was wrong.
The number I was thinking of was 7.
EDU» guess0000
I am thinking of a number between 1 and 10.

Guess a number: 4
Your guess was wrong.
The number I was thinking of was 10.
EDU» guess0000
I am thinking of a number between 1 and 10.

Guess a number: 4
Congratulations - You guessed the correct number.
EDU»
```

The condition used in the **IF** statement can be enlarged by using the logical operators in conjunction with the relational operators to make more complicated tests. The logical operators are shown in Table 2-2.

TABLE 2-2 Logical operators available in MATLAB.

Operator	Description	Example
&	Logical AND. True if both conditions are true.	(a>b) & (x<5) (x~=5) & (y<=1)
|	Logical OR. True if either condition is true or both conditions are true.	(a>b) | (x<5) (x~=5) | (y<=1)
~	True if the result of a condition is false, and false if the result if the condition is true.	~(a=b) ~((a>b) | (x<5))
xor	Exclusive OR. True if either condition 1 is true or condition 2 is true. False if both conditions are true or both conditions are false.	xor((x>2), y==5)) xor((a>b). (y<=1))

In our second example, Program 2-2, we calculate a student's grade based on three exams. An exam average greater than or equal to 90 assigns an A; 89–80 assigns a B; 79–70 a C; 69–60 a D; and less than 60 an F.

Program 2-2 Grade calculation

```
%
% This is file grade00.m
%
% This program requests three exam scores from the user,
% calculates the average of the three exams, and assigns a grade.
%
%      Condition              Grade
%      ————                    —
%      Average >= 90            A
%      80 <= Avg <90            B
%      70 <= Avg <80            C
%      60 <= Avg <70            D
%      Average <  60            F

% Ask for the three exam scores.
exam1 = input(' Input Exam Score 1: ');
exam2 = input(' Input Exam Score 2: ');
exam3 = input(' Input Exam Score 3: ');

average=(exam1 + exam2 + exam3)/3;

% Determine the grade

fprintf(' Your average exam score is %g.\n', average);
fprintf(' This corresponds to a grade of ');

if average >= 90
     Grade = 'A';
end

if (average >= 80) & (average <90)
     Grade = 'B';
end

if (average >= 70) & (average <80)
     Grade = 'C';
end

if (average >= 60) & (average <70)
     Grade = 'D';
end
```

```
if average < 60
      Grade = 'F';
end

fprintf(Grade);
fprintf('.\n\n');
```

This function contains more complicated **IF** statements that use the logical operators. An example is

```
if (average >= 80) & (average <90)
   Grade = 'B';
end
```

This statement uses the logical AND operator (**&**). The statement **Grade = 'B'** is executed only if both conditions are true. That is, variable *Grade* is set to B only if the average is greater than or equal to 80 AND less than 90.

Program 2-2 uses the **FPRINTF** statement in a unique way. The line

```
fprintf(' This corresponds to a grade of ');
```

adds the text **This corresponds to a grade of** to the print buffer. Note that this **FPRINTF** statement does not contain a newline character (**\n**). The text is added to the print buffer, but since there is not a newline character, the text is not yet printed. The program then goes on to assign a value (**'A'**, **'B'**, **'C'**, **'D'**, or **'F'**) to variable *Grade*. Note that the values assigned to *Grade* are text strings that can be directly printed using the **FPRINTF** statement. As an example, let's suppose that after executing all of the **IF** statements, variable *Grade* has the value **'A'**. MATLAB next executes the statement **fprintf(Grade);**—this statement adds the text string contained in *Grade* to the print buffer. The print buffer now contains the text **This corresponds to a grade of A** since variable *Grade* contains the string **'A'**. The last statement executed is **fprintf('.\n\n');**—this statement adds a period and two newline characters to the print buffer. The buffer now contains the text **This corresponds to a grade of A.\n\n**. Note that **\n** is not the text **'\n'**, but a special character referred to as "newline." When the **FPRINTF** statement sees the text **'\n'**, it inserts a newline character into the print buffer.

Whenever the print buffer encounters the newline character (**\n**), the contents of the buffer are printed. Thus, when the line **fprintf('.\n\n');** occurs, the print buffer is output. In our example, the text **This corresponds to a grade of A.** and two carriage returns are printed on the MATLAB command window screen.

Let's test this program to see how it functions. Note that the tests shown are not exhaustive. You should do much more testing to show that the program works correctly for all the conditions it tests. A dialog of a program session is shown here:

```
EDU» grade00

  Input Exam Score 1: 60

  Input Exam Score 2: 70

  Input Exam Score 3: 80
Your average exam score is 70.

 This corresponds to a grade of C.
EDU» grade00

  Input Exam Score 1: 90

  Input Exam Score 2: 90

  Input Exam Score 3: 90
Your average exam score is 90.

 This corresponds to a grade of A.

EDU» grade00

  Input Exam Score 1: 80

  Input Exam Score 2: 80

  Input Exam Score 3: 80
Your average exam score is 80.

 This corresponds to a grade of B.

EDU» grade00

  Input Exam Score 1: 70

  Input Exam Score 2: 70

  Input Exam Score 3: 70
Your average exam score is 70.

 This corresponds to a grade of C.

EDU»
```

M<small>ATLAB</small> Built-In Function

> The M<small>ATLAB</small> **MEAN** function calculates the average of an array of numbers. Here are some examples:
> ```
> EDU» mean([10,20,30,40,50])
>
> ans =
>
>     30
>
> EDU» exam1=90;
> EDU» exam2=60;
> EDU» exam3=80;
> EDU» mean([exam1,exam2,exam3])
>
> ans =
>
>     76.6667
>
> EDU»
> ```

As a last example of this form of the **IF** statement, let's find the roots to the quadratic equation $ax^2 + bx + c = 0$. We must test for several conditions:

- If $a = 0$ and $b = 0$, then the equation reduces to $c = 0$. If this condition occurs, we call the equation degenerate.
- If $a = 0$ and $b \neq 0$, the equation reduces to $bx + c = 0$ with a single root at $-c/b$.
- If $a \neq 0$ and $c = 0$, the equation reduces to $ax^2 + bx = 0$. There are two roots, 0 and $-b/a$.
- If none of the preceding conditions apply, the equation is either $ax^2 + bx + c = 0$ with all coefficients nonzero, or $ax^2 + c = 0$ with neither $a = 0$ nor $c = 0$. In these cases, we use the quadratic equation to find the roots:

$$roots = \frac{-b \pm \sqrt{b^2 - 4ac}}{2a}$$

For this last case, we will determine whether the roots are complex or real. The term $b^2 - 4ac$ is called the discriminant of the equation. If the discriminant is positive, there will be two real roots. If the discriminant is negative, there will be two complex roots. Program 2-3 shows how we do this.

Program 2-3   Roots of a quadratic equation

```
%
% This is file quad_eq1.m
%
% This program calculates the roots of a quadratic equation.
%
```

```
array=input('Specify the coefficients of the quadratic as a row array ➡
"[a,b,c]":');
a=array(1);
b=array(2);
c=array(3);

if (a==0) & (b==0) % The quadratic is degenerate.
    fprintf('The equation is degenerate.\n');
end
if (a==0) & (b~=0) % There is a single root at -c/b.
    fprintf('There is a single root at %g.\n',-c/b);
end

if (a~=0) & (c==0) % There are two roots, 0 and -b/a
    root=-b\a;
    fprintf('There are two real roots at 0 and %g.\n',-b/a);
end

if (a~=0) & (c~=0) % Use the quadratic equation.
    discriminant=b*b - 4*a*c;
    if discriminant < 0  % The roots are complex.
        im=sqrt(-1*discriminant)/(2*a);
        re=-b/(2*a);
        fprintf('The roots are complex:\n');
        fprintf('        %g +i%g\n',re, im);
        fprintf('    and %g -i%g\n',re, im);
end
if discriminant >= 0  % The roots are real.
        re=-b/(2*a);
        im=sqrt(discriminant)/(2*a);
        root1=re+im;
        root2=re-im;
        fprintf('There are two real roots at %g and %g.\n', root1, root2);
    end
end
```

The MATLAB **INPUT** statement accepts numbers in many forms. Since MATLAB variables can be arrays, complex numbers, and scalar real numbers, the **INPUT** function is generalized to accept numbers in any of these forms. Some examples are shown here. In all cases, we use the **INPUT** function the same way.

```
EDU» x=input('Enter a number.\n');
Enter a number.
2  ◄─────────────────
EDU» x
x =
        2  ◄
```

User enters a real number and *x* is set to a real number.

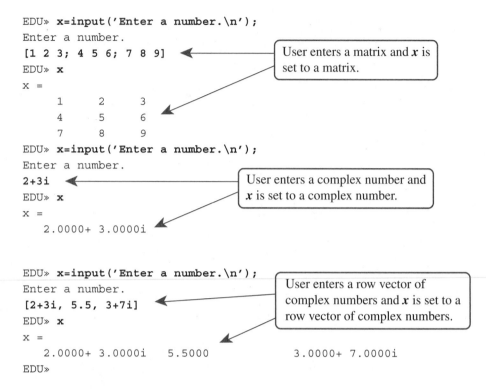

```
EDU» x=input('Enter a number.\n');
Enter a number.
[1 2 3; 4 5 6; 7 8 9]
EDU» x
x =
        1       2       3
        4       5       6
        7       8       9
EDU» x=input('Enter a number.\n');
Enter a number.
2+3i
EDU» x
x =
    2.0000+ 3.0000i
```

User enters a matrix and x is set to a matrix.

User enters a complex number and x is set to a complex number.

```
EDU» x=input('Enter a number.\n');
Enter a number.
[2+3i, 5.5, 3+7i]
EDU» x
x =
    2.0000+ 3.0000i    5.5000              3.0000+ 7.0000i
EDU»
```

User enters a row vector of complex numbers and x is set to a row vector of complex numbers.

Note that the form of numbers returned by the **INPUT** statement (complex, array, real number, and so on) is determined by what the user enters. The **INPUT** statement accepts numerical values in any of the forms supported by MATLAB. It is up to the programmer to determine if the number entered by the user is acceptable.

In Program 2-3, the statement **array=input('Specify the coefficients of the quadratic as a row array "[a,b,c]":');** instructs the user to enter a row array containing three numbers. This allows the programmer to use a single input statement to request three values at the same time.

MATLAB Built-In Function

> MATLAB has a number of built-in functions for determining the form of a number. The function **ISREAL**(x) returns a numerical value of 1 if x is a real number and a numerical value of zero if x is complex. The function **SIZE**(x) returns the dimensions of array x. If x is a scalar, the **SIZE** function would indicate that x is a 1×1 array.

Let's attempt an exhaustive test to verify Program 2-3:

```
EDU» quad_eq1

Specify the coefficients of the quadratic as a row array "[a,b,c]":[0 0 7]
The equation is degenerate.
EDU» quad_eq1
```

Specify the coefficients of the quadratic as a row array "[a,b,c]":[0 10 2]
There is a single root at -0.2.
EDU» **quad_eq1**

Specify the coefficients of the quadratic as a row array "[a,b,c]":[2 3 0]
There are two real roots at 0 and -1.5.
EDU» **quad_eq1**

Specify the coefficients of the quadratic as a row array "[a,b,c]":[1 5 6]
There are two real roots at -2 and -3.
EDU» **quad_eq1**

Specify the coefficients of the quadratic as a row array "[a,b,c]":[1 1 1]
The roots are complex:
                -0.5 +i0.866025
        and   -0.5 -i0.866025
EDU»

We can simplify Program 2-3 because MATLAB handles complex numbers automatically. Program 2-4 shows how this works.

Program 2-4 Roots of a quadratic equation

```
%
% This is file quad_eq2.m
%
% This program calculates the roots of a quadratic equation.
%
array=input('Specify the coefficients of the quadratic as a row
array "[a,b,c]":');
a=array(1);
b=array(2);
c=array(3);

if (a==0) & (b==0) % The quadratic is degenerate.
    fprintf('The equation is degenerate.\n');
end

if (a==0) & (b~=0) % There is a single root at -c/b.
    fprintf('There is a single root at %g.\n',-c/b);
end

if (a~=0) & (c==0) % There are two roots, 0 and -b/a
    root=-b\a;
    fprintf('There are two real roots at 0 and %g.\n',-b/a);
end
```

```
if (a~=0) & (c~=0) % Use the quadratic equation.
    root1=(-b+sqrt(b*b-4*a*c))/(2*a);
    root2=(-b-sqrt(b*b-4*a*c))/(2*a);
    if isreal(root1)
        fprintf('There are two real roots at %g and %g.\n',root1, root2);
    end
    if ~isreal(root1)
            fprintf('The roots are complex:\n');
            fprintf('                ');
            disp(root1);
            fprintf('        and    ');
            disp(root2);
    end
end
```

Because MATLAB automatically handles complex numbers, we can take the square root of a number and not worry if the number is negative. After we calculate the roots, all we need to do is determine whether the roots are real or complex. The statements

```
root1=(-b+sqrt(b*b-4*a*c))/(2*a);
root2=(-b-sqrt(b*b-4*a*c))/(2*a);
```

calculate the two roots. If **b*b-4*a*c** is negative, then root1 and root2 will have complex values. In programming languages like FORTRAN or C, taking the square root of a number generates an exception and the program halts.

The MATLAB **ISREAL** function determines whether a number is pure real or complex. The function returns a value of 0 if the number is complex and a value of 1 if the number is pure real. In **IF** statements, MATLAB treats a numerical value of 0 as false, and a numerical value of 1 as true. Here are a few examples:

```
EDU» a=3+5i;
EDU» isreal(a)
ans =
      0
EDU» b=5;
EDU» isreal(b)
ans =
      1
```

Since the MATLAB **IF** statement interprets a numerical value of 1 as true and a numerical value of 0 as false, the statements **if isreal(b)==1** and **if isreal(b)** are equivalent.

The MATLAB **DISP** function is used to display arrays easily. If you want to display an array nicely with the **FPRINTF** function, you must print each element of the array individually. If you use an **FPRINTF** function and blindly print an array, the results may be ugly:

```
EDU» a=rand(3);
EDU» fprintf('The array is %g\n',a);
The array is 0.218959
The array is 0.0470446
The array is 0.678865
The array is 0.679296
The array is 0.934693
The array is 0.383502
The array is 0.519416
The array is 0.830965
The array is 0.0345721
EDU»
```

Since the specified **FPRINTF** function has only one occurrence of **%g**, the **FPRINTF** function is used nine times to provide enough **%g** place holders to accommodate all the elements of the array. Here is a better way:

```
a=rand(3);
fprintf(' The array is:\n');
fprintf('       %g    %g    %g\n       %g    %g    %g\n       %g    %g    %g\n',a);
The array is:
        0.930436     0.846167     0.526929
        0.0919649    0.653919     0.415999
        0.701191     0.910321     0.762198
EDU»
```

The columns still don't line up however. Here's another attempt:

```
a=rand(3);
fprintf(' The array is:\n');
fprintf('\t\t%g\t\t%g\t\t%g\n\t\t%g\t\t%g\t\t%g\n\t\t%g\t\t%g\t\t%g\n',a);
 The array is:
            0.98255     0.72266   0.753356
            0.651519     0.0726859     0.631635
            0.884707     0.27271 0.436411
EDU»
```

The **\t** in the **FPRINTF** control string means go to the next tab stop. Tabs are useful in lining up columns; **\t\t** means move two tab stops. The output still doesn't look very good because of the different lengths of displayed numbers. To fix this problem we use the **%f** formatting character and specify the number of digits to be displayed. The formatting string **%8.7f** means allow for eight total digits with seven of the digits coming after the decimal point. This makes all numbers use the same number of digits.

```
a=rand(3);
fprintf('The array is:\n');
fprintf('%8.7f  %8.7f %8.7f\n%8.7f  %8.7f %8.7f\n%8.7f  %8.7f %8.7f\n',a);
The array is:
0.6449104   0.8179743   0.6602276
0.3419706   0.2897259   0.3411936
0.5340790   0.7271132   0.3092902
```

This method is a good way to display arrays, but we used the **DISP** function in Program 2-4. The **DISP** function easily displays arrays without having to create complicated control strings for the **FPRINTF** function.

```
EDU» a=rand(3);
EDU» disp(a);
    0.7665    0.2749    0.4865
    0.4777    0.3593    0.8977
    0.2378    0.1665    0.9092
```

Note that the columns automatically line up.

In Program 2-4, we used the **DISP** function to display a complex number because the **FPRINTF** function doesn't display complex numbers properly. But let's use the **FPRINTF** function and see what happens:

```
EDU» a=3+4i;
EDU» fprintf('The number is %g.\n',a);
The number is 3.
```

The **FPRINTF** function only displays the real portion of the complex number. We can display the whole number with a little more work:

```
EDU» a=3+4i;
EDU» fprintf('The number is %g +%gi.\n',real(a), imag(a));
The number is 3 +4i.
```

This method still doesn't work when the number has a negative imaginary part:

```
EDU» a=3-4i;
EDU» fprintf('The number is %g +%gi.\n',real(a), imag(a));
The number is 3 +-4i.
```

We can fix the problem by testing for a negative imaginary part:

```
a=3-4i;
if imag(a) >= 0
fprintf('The number is %g +%gi.\n',real(a), imag(a));
else
```

```
fprintf('The number is %g %gi.\n', real(a), imag(a));
end
The number is 3 -4i.
```

Next, let's test the same code for complex numbers with positive real parts:

```
a=3+4i;
if imag(a) >= 0
fprintf('The number is %g +%gi.\n',real(a), imag(a));
else
fprintf('The number is %g %gi.\n', real(a), imag(a));
end
The number is 3 +4i.
```

This last solution works well, but it requires a few lines of code.

Now contrast the preceding solution with the code for the **DISP** function in Program 2-4, which we used to display the same complex number:

```
EDU» a=3+4i;
EDU» disp(a)
   3.0000 + 4.0000i
```

Although the **DISP** function doesn't have much flexibility, it is easy to use:

```
a=3+4i;
fprintf('The number is ');
disp(a);
The number is    3.0000 + 4.0000i
```

Note that the **DISP** function automatically creates a new line after it displays the values:

```
a=3+4i;
fprintf('The number is ');
disp(a);
fprintf('This is a second line.\n');
The number is    3.0000 + 4.0000i
This is a second line.
```

Finally, here's how Program 2-4 works with complex numbers:

```
EDU» quad_eq2
Specify the coefficients of the quadratic as a row array ➡
"[a,b,c]":[1 1 1]
The roots are complex:
            -0.5000+ 0.8660i
      and   -0.5000- 0.8660i
EDU»
```

2.2   IF-ELSE-END

A second form of the **IF** statement includes the **else** clause. This allows the programmer to use a single **IF** statement to handle cases where the test in the **IF** statement is either true or false. The second form of the **IF** statement is

> **IF** condition
>> statement t_1;
>> statement t_2;
>> statement t_3;
>> $\vdots$
>> statement t_n;
> **ELSE**
>> statement f_1;
>> statement f_2;
>> statement f_3;
>> $\vdots$
>> statement f_m;
> **END**

If the condition is true, statements t_1 to t_n are executed. After statement t_n is completed, the program resumes operation at the statement following the **end** keyword. If the condition is false, statements f_1 to f_m are executed. When statement f_m is completed, the program resumes operation at the statement following the **end** keyword. The **IF-ELSE-END** form of the **IF** statement replaces two **IF-END** statements. The two **IF** statements below are logically equivalent to the **IF-ELSE-END** statement shown above.

> **IF** condition is true
>> statement t_1;
>> statement t_2;
>> statement t_3;
>> $\vdots$
>> statement t_n;
> **END**
> **IF** condition is false
>> statement f_1;
>> statement f_2;
>> statement f_3;
>> $\vdots$
>> statement f_m;
> **END**

You can use whichever form you find most convenient.

Now, we'll rewrite Programs 2-1, 2-2, and 2-3 using the **IF-ELSE-END** form of the **IF** statement. In Program 2-5, the computer generates a random number between 1 and 10 and then asks the user to guess what the number is. The computer then notifies the user of the result of the guess, and, if incorrect, tells the user what the number is.

Program 2-5 Rewrite of Program 2-1, the guessing program

```
%
% This is file guess000.m
%

real_number=rand(1); % Generate a number between 0 and 1
real_number=10*real_number; % Scale the number between 0 and 10
number=ceil(real_number); % Round the number to the next highest integer.

% Ask the user to guess the number:

fprintf('I am thinking of a number between 1 and 10.\n');
guess = input('Guess a number: ');

% Test the number.

if guess == number
      fprintf('Congratulations - You guessed the correct number.\n');
else
      fprintf('Your guess was wrong.\n');
      fprintf('The number I was thinking of was %g.\n', number);
end
```

Note that the two **IF-END** statements of Program 2-1 have been replaced by a single **IF-ELSE-END** statement in Program 2-5. Let's test the program for correctness:

```
EDU» guess000
I am thinking of a number between 1 and 10.
Guess a number: 7
Your guess was wrong.
The number I was thinking of was 3.
EDU» guess000
I am thinking of a number between 1 and 10.
Guess a number: 7
Your guess was wrong.
The number I was thinking of was 1.
EDU» guess000
I am thinking of a number between 1 and 10.
Guess a number: 7
Congratulations - You guessed the correct number.
EDU»
```

The program appears to work properly.

Next, we'll rewrite Program 2-2 using the **IF-ELSE-END** statement. Recall that the grade calculation program assigns an A for an exam average (three exams) of ≥ 90, a B for an average ≥ 80 and < 90, a C for an average ≥ 70 and < 80, a D for an average ≥ 60 and < 70, and an F for an average < 60.

Program 2-6 Rewrite of Program 2-2, grade calculation

```
%
% This is file grade0.m
%
% This program requests three exam scores from the user,
% calculates the average of the three exams, and assigns a grade.
%
%      Condition           Grade
%      ————                —
%      Average >= 90        A
%      80 <= Avg <90        B
%      70 <= Avg <80        C
%      60 <= Avg <70        D
%      Average <  60        F

% Ask for the three exam scores.
exam1 = input(' Input Exam Score 1: ');
exam2 = input(' Input Exam Score 2: ');
exam3 = input(' Input Exam Score 3: ');


average=(exam1 + exam2 + exam3)/3;

% Determine the grade

fprintf(' Your average exam score is %g\n', average);
fprintf(' This corresponds to a grade of ');

if average >= 90
    Grade='A';
else
    if (average >= 80)   %If we are here, we know grade < 90.
        Grade = 'B';
    else
        if (average >= 70)   %If we are here, we know grade < 80
            Grade = 'C';
        else
            if (average >= 60)   %If we are here, we know grade <70
                Grade = 'D';
            else
                if average <60   %If we are here, we know grade <60
                    Grade = 'F';
                end
            end
        end
    end
end

fprintf(Grade);
fprintf('.\n\n');
```

Note that the **else** portion of the **IF** statement has the luxury of knowing that the result of the first test was false. If you look at the first few lines of the **IF-ELSE-END** structure, you'll see how it simplifies the tests needed in the **else** portion of the statement.

```
if average >= 90
   Grade='A';
else
   if (average >= 80)    %If we are here, we know grade < 90.
         Grade = 'B';
   else
```

If the condition **average >= 90** is true, then grade is set to A. This is the only action taken by the entire **IF** structure because the remaining tests are under the **else** portion of the statement and are never executed.

If the condition **average >= 90** is false, then the program skips to the line after the first **else** keyword, **if (average >= 80)**. For the program to get to this line, we know that the condition **average >= 90** must be false. Thus, we know that the average is less than 90 and we don't need to test for it again. This property of the **IF-ELSE-END** structure allows us to simplify the tests in each of the **IF** statement conditions. Notice that the conditions used in the **IF** statements of Program 2-2 have more complicated tests than those of Program 2-6, and almost always use a logical AND to combine two conditions (**if (average >= 80) & (average <90)**, for example).

Before we move on, look at the indentation used in Program 2-6. Although the indentation is not required, it makes the program more readable and helps us determine which **else** belongs to which **IF** statement and which **end** belongs to which **IF** statement. It's usually a good idea to use indentation on complicated programs to make them easier to decipher.

Now let's run Program 2-6 to test its operation. Note that the tests shown here check specific portions of the program, but are not exhaustive.

```
EDU» grade0
 Input Exam Score 1: 95
 Input Exam Score 2: 95
 Input Exam Score 3: 95
 Your average exam score is 95
 This corresponds to a grade of A.

EDU» grade0
 Input Exam Score 1: 85
 Input Exam Score 2: 85
 Input Exam Score 3: 85
 Your average exam score is 85
 This corresponds to a grade of B.
```

```
EDU» grade0
  Input Exam Score 1: 75
  Input Exam Score 2: 75
  Input Exam Score 3: 75
  Your average exam score is 75
  This corresponds to a grade of C.

EDU» grade0
  Input Exam Score 1: 65
  Input Exam Score 2: 65
  Input Exam Score 3: 65
  Your average exam score is 65
  This corresponds to a grade of D.

EDU» grade0
  Input Exam Score 1: 55
  Input Exam Score 2: 55
  Input Exam Score 3: 55
  Your average exam score is 55
  This corresponds to a grade of F.

EDU»
```

2.3   ELSEIF

The third form of the **IF** statement uses the **elseif** keyword. The **elseif** portion of the **IF** statement lets us check a large number of different conditions. As an example, we will write a program that asks the user for a command number between 1 and 5. If the number specified is valid, the computer performs the task. If the user types in an invalid number (that is, one that is not an integer between 1 and 5 inclusive), an error message is generated. First let's write this program using **IF-END** statements.

Program 2-7   Command center using **IF-END**

```
%
% This is file command00.m
%
% This program asks for a number between 1 and 5 inclusive.
% The program then performs the tasks assigned to the number.
% The tasks are not specified for this program.

command=input('Specify your command [1-5]: ');
```

```
if command==1
     fprintf('You chose option 1.\n');
     % Command for option 1 entered here.
end
if command==2
     fprintf('You chose option 2.\n');
     % Command for option 2 entered here.
end
if command==3
     fprintf('You chose option 3.\n');
     % Command for option 3 entered here.
end
if command==4
     fprintf('You chose option 4.\n');
     % Command for option 4 entered here.
end
if command==5
     fprintf('You chose option 5.\n');
     % Command for option 5 entered here.
end
if (command<1)|(command >5)
     fprintf('Error - You must enter an integer from 1 to 5.\n');
end
```

Note that if the user enters a number such as 3.1, this program does not detect the error nor does it interpret the number as a valid response. The error checking portion of the program

```
if (command<1)|(command >5)
   fprintf('Error - You must enter an integer from 1 to 5.\n');
end
```

only emits an error message if the input is less than 1 or greater than 5. It does not check if the number is a noninteger between 1 and 5. The program does nothing for an input like 3.1. This logic error is corrected in Programs 2-8 and 2-9.

Now let's write the same program using the **IF-ELSE-END** statement.

Program 2-8 Command center using IF-ELSE-END

```
%
% This is file command0.m
%
% This program asks for a number between 1 and 5 inclusive.
% The program then performs the tasks assigned to the number.
% The tasks are not specified for this program.

command=input('Specify your command [1-5]: ');
```

```
if command==1
    fprintf('You chose option 1.\n');
    % Command for option 1 entered here.
else
    if command==2
        fprintf('You chose option 2.\n');
        % Command for option 2 entered here.
    else
        if command==3
            fprintf('You chose option 3.\n');
            % Command for option 3 entered here.
        else
            if command==4
                fprintf('You chose option 4.\n');
                % Command for option 4 entered here.
            else
                if command==5
                    fprintf('You chose option 5.\n');
                    % Command for option 5 entered here.
                else
                    fprintf('Error - You must enter an integer from 1 to 5.\n');
                end
            end
        end
    end
end
```

Program 2-8 fixes the logic error in Program 2-7 where it could not detect an incorrect response such as 3.1. However, the structure of the **IF-ELSE-END** statement leads to a lot of indentation that makes the code hard to read. The **elseif** variation of the **IF** statement shown in Program 2-9 fixes this problem.

Program 2-9   Command center using **ELSEIF**

```
%
% This is file command1.m
%
% This program asks for a number between 1 and 5 inclusive.
% The program then performs the tasks assigned to the number.
% The tasks are not specified for this program.

command=input('Specify your command [1-5]: ');

if command==1
    fprintf('You chose option 1.\n');
    % Command for option 1 entered here.
elseif command==2
    fprintf('You chose option 2.\n');
```

```
      % Command for option 2 entered here.
elseif command==3
      fprintf('You chose option 3.\n');
      % Command for option 3 entered here.
elseif command==4
      fprintf('You chose option 4.\n');
      % Command for option 4 entered here.
elseif command==5
      fprintf('You chose option 5.\n');
      % Command for option 5 entered here.
else
      fprintf('Error - You must enter an integer from 1 to 5.\n');
end
```

Program 2-9 contains one **IF** statement. All the **elseif** keywords and the **else** keyword are part of the **IF** statement. This **IF** statement works as follows. If the condition **command==1** is true, then the commands

```
fprintf('You chose option 1.\n');
% Command for option 1 entered here.
```

are executed. The program then jumps to the statement following the **end** keyword, if any. If the condition **command==1** is false, then control continues with the first **elseif** portion of the statement. If the first **elseif** condition is true, then the commands for that **elseif** are executed. When those commands are finished, the remainder of the **IF** statement is ignored and execution continues with the statement after the **end** keyword, if any. If the first **elseif** condition is false, then the second **elseif** condition is checked, and so on. The operation of the **IF-ELSEIF-ELSE-END** structure is that it checks each condition sequentially from top to bottom. Program control goes to the first portion of the statement with a true condition. The statements for that condition are executed, and then the program continues at the statement after the **end** keyword. Note that the **elseif** structure does not "fall through" when a true condition is found. That is, if the condition for the **IF** keyword is true, the program never tests the **elseif** or **else** conditions. If the condition for the first **elseif** keyword is true, the program never tests the following **elseif** and **else** conditions. The **elseif** conditions are tested sequentially, and they are only tested if the previous tests were false. Once a true condition is found, the statements for the portion of the **elseif** are executed and then the program continues at the statement following the **end** keyword.

As a final example, let's rewrite Program 2-2 using the **elseif** structure.

Program 2-10 Rewrite of Program 2-2 using ELSEIF

```
%
% This is file grade1.m
%
% This program requests three exam scores from the user,
```

```
% calculates the average of the three exams, and assigns a grade.
%
%       Condition               Grade
%       ————                    ——
%       Average >= 90           A
%       80 <= Avg <90           B
%       70 <= Avg <80           C
%       60 <= Avg <70           D
%       Average <  60           F

% Ask for the three exam scores.
exam1 = input(' Input Exam Score 1: ');
exam2 = input(' Input Exam Score 2: ');
exam3 = input(' Input Exam Score 3: ');

average=(exam1 + exam2 + exam3)/3;

% Determine the grade

fprintf(' Your average exam score is %g\n', average);
fprintf(' This corresponds to a grade of ');

if average >= 90
     Grade='A';
elseif (average >= 80)
     Grade = 'B';
elseif (average >= 70)
     Grade = 'C';
elseif (average >= 60)
     Grade = 'D';
else
     Grade = 'F';
end

fprintf(Grade);
fprintf('.\n\n');
```

2.4 SWITCH-CASE

The **SWITCH-CASE** statement is convenient for testing whether an expression is equal to a number of different values. It cannot be used to test a condition such as if $a \geq 5$. However, it is very useful in applications such as if $a = 1$, do something; if $a = 2$, do something else; if $a = 3$, do another thing; and so on.

The syntax of the **SWITCH-CASE** statement is

> **SWITCH** expression
> > **CASE** value$_1$
> > > statement 1a;
> > > statement 1b;

statement 1c;

⋮

statement 1*n*;

CASE value$_2$

statement 2a;

statement 2b;

statement 2c;

⋮

statement 2*n*;

⋮

CASE value$_i$

statement *I*a;

statement *I*b;

statement *I*c;

⋮

statement *I*n;

OTHERWISE

statement *O*a;

statement *O*b;

statement *O*c;

⋮

statement *O*n;

END

The value can be a scalar or a text string. One feature of the **SWITCH-CASE** statement is that each case can have multiple values:

CASE {value$_1$, value$_2$, value$_3$, ...}

statement a;

statement b;

statement c;

⋮

statement *n*;

In this statement, if the expression is equal to any of the values listed, statements a to *n* will be executed.

A good example of the **SWITCH-CASE** statement is the command center of Program 2-9, which we modify in Program 2-11 to use the **SWITCH-CASE** statement.

Program 2-11 Command center using **SWITCH-CASE**

```
%
% This is file command2.m
%
% This program asks for a number between 1 and 5 inclusive.
% The program then performs the tasks assigned to the number.
```

```
% The tasks are not specified for this program.

command=input('Specify your command [1-5]: ');

switch command
case 1
        fprintf('You chose option 1.\n');
        % Command for option 1 entered here.
case 2
        fprintf('You chose option 2.\n');
        % Command for option 2 entered here.
case 3
        fprintf('You chose option 3.\n');
        % Command for option 3 entered here.
case 4
        fprintf('You chose option 4.\n');
        % Command for option 4 entered here.
case 5
        fprintf('You chose option 5.\n');
        % Command for option 5 entered here.
otherwise
        fprintf('Error - You must enter an integer from 1 to 5.\n');
end
```

The **SWITCH-CASE** statement allows each **case** to test multiple expressions. As an example, we will modify the command center so that some responses execute the same command.

Program 2-12 Command center using SWITCH-CASE

```
%
% This is file command3.m
%
% This program asks for a number between 1 and 10 inclusive.
% The program then performs the tasks assigned to the number.
% The tasks are not specified for this program.

command=input('Specify your command [1-10]: ');

switch command
case 1
        fprintf('You chose option 1.\n');
        % Command for option 1 entered here.
case {2, 3}
        fprintf('You chose either option 2 or option 3.\n');
        % Command for options 2/3 entered here.
case {4,5,6}
        fprintf('You chose option 4, 5, or 6.\n');
```

```
            % Command for options 4/5/6 entered here.
case 7
        fprintf('You chose option 7.\n');
        % Command for option 7 entered here.
case {8,9,10}
        fprintf('You chose option 8, or 9, or 10.\n');
        % Command for options 8/9/10 entered here.
otherwise
        fprintf('Error - You must enter an integer from 1 to 10.\n');
end
```

A good application of the **SWITCH-CASE** statement comes from the MATLAB Users Guide.[1] Program 2-13 requests a number and its units. It then converts the units to centimeters and changes the numerical value accordingly.

Program 2-13 Units conversion using **SWITCH-CASE**

```
%
% This is file units1.m
%
% This program asks for a number and then asks for the units of the number
% as a text string. If the units are known, the units of the number are
% converted to Centimeters.


Number=input('Specify the value of the number: ');
Units = input('Specify the units of the number: ', 's');
error=0;

switch Units
    case{'inch', 'in', 'Inch', 'INCH', 'In', 'IN'}
        New_number=Number*2.54;
    case{'feet' , 'Feet', 'FEET', 'foot', 'Foot', 'FOOT', 'ft', 'FT', 'Ft'}
        New_number=Number*12*2.54;
    case {'Meter', 'meter', 'METER', 'm', 'M'}
        New_number=Number*100;
    case {'millimeter', 'Millimeter', 'MILLIMETER', 'mm',}
        New_number=Number/10;
    case {'centimeter', 'cm'}
        New_number=Number;
    otherwise
        error=1;
end
```

[1]*The Student Edition of MATLAB, Version 5 Users Guide,* The MATLAB Curriculum Series, Prentice Hall, Upper Saddle River, N.J., 1997, p. 120.

```
if error
    fprintf('A unit of ');
    fprintf(Units);
    fprintf(' is unknown to this program.\n');
else
    fprintf('%g ', Number);
    fprintf(Units);
    fprintf(' is equal to %g cm.\n', New_number);
end
```

Program 2-13 uses two forms of the **INPUT** statement. The line **Number=input('Specify the value of the number: ');** requests a numerical value from the user and stores the value in variable *Number*. If the user enters something other than a numerical value, MATLAB generates an error message and then repeats the entire statement. Here is an example of a poor response to this form of the input statement:

```
EDU» Number=input('Specify the value of the number: ');
Specify the value of the number: jdjd
??? Undefined function or variable 'jdjd'.

Specify the value of the number: jk
??? Undefined function or variable 'jk'.

Specify the value of the number: 5
EDU»
```

The user enters a text string in response to the question.

Program 2-13 also uses the form of the **INPUT** statement that accepts input as a text string. In the line **Units = input('Specify the units of the number: ', 's');** the **'s'** instructs the **INPUT** statement to read the input as a text string. The user can then type in anything and the **INPUT** statement won't generate an error message. The text string is stored in variable *Units*. The **SWITCH-CASE** statement compares the text string stored in *Units* to known text strings such as **'mm'**, **'INCH'**, and **'Meter.'** If the text string stored in *Units* matches one of the listed text strings, a numerical conversion is performed. If the text string stored in *Units* doesn't match any of the strings, the program jumps to the **otherwise** portion of the **SWITCH-CASE** statement and sets variable *error* to 1, indicating that the unit was not recognized. Variable *error* is a flag and is used by the program to indicate whether the unit entered by the user was recognized or not.

Remember that the statements **if error** and **if error == 1** are equivalent in MATLAB. As noted earlier, a numerical value of 1 is treated as true, and a numerical value of 0 is treated as false. In our program, variable *error* will only have values of 0 or 1.

An example session of the program is shown here:

```
EDU» units1
Specify the value of the number: 1
Specify the units of the number: in
```

```
1 in is equal to 2.54 cm.
EDU» units1
Specify the value of the number: 1
Specify the units of the number: INCH
1 INCH is equal to 2.54 cm.
EDU» units1
Specify the value of the number: 1
Specify the units of the number: INch
A unit of  INch is unknown to this program.
EDU»
```

EXERCISE 2-1   In Program 2-13, we noted that the program doesn't work for all possible combinations of upper- and lowercase permutations of units. Modify the program to make it independent of upper- and lowercase letters. *Hint:* MATLAB has built-in functions for converting between upper- and lowercase letters.

2.5   FOR **Loops**

2.5.1   Single FOR Loops

MATLAB *Predefined*
zeros function
length function
floor function
int2str function
vpa function
prod function
linspace function
max command
fprintf function
strcmp function
str2num function
isempty command

FOR loops are used to execute a set of statements a known number of times. The syntax of the **FOR** statement is

> **FOR** variable = expression
>> statement$_1$
>> statement$_2$
>> statement$_3$
>> $\vdots$
>> statement$_n$
> **END**

An example would be

> **FOR** $i = 1:10$
>> statement$_1$
>> statement$_2$
>> statement$_3$
>> $\vdots$
>> statement$_n$
> **END**

In this example, i is set to 1 and then statements 1 to n are executed. Next i is set to 2 and statements 1 to n are executed, then i is set to 3 and statements 1 to n are executed. This operation is repeated until i equals 10. When i is set to 10, statements 1 to n are executed, and the program continues at the line following the **end** keyword.

To quickly demonstrate, let's create a simple program:

```
fprintf('Testing the FOR loop structure.\n\n');
for i = 1:4
   fprintf('Testing the FOR loop. i = %g\n',i);
end
fprintf('\nEnd of loop.\n');
Testing the FOR loop structure.

Testing the FOR loop. i = 1
Testing the FOR loop. i = 2
Testing the FOR loop. i = 3
Testing the FOR loop. i = 4

End of loop.
```

The loop executes four times, and each time through the loop *i* has a different value. We can also make for loops countdown:

```
fprintf('Testing the FOR loop structure.\n\n');
for i = 4:-1:1  ◄─────────────────────────────   Note the –1
   fprintf('Testing the FOR loop. i = %g\n',i);    step here.
end
fprintf('\nEnd of loop.\n');
Testing the FOR loop structure.

Testing the FOR loop. i = 4
Testing the FOR loop. i = 3
Testing the FOR loop. i = 2
Testing the FOR loop. i = 1

End of loop.
```

Note the syntax of the previous **FOR** statement is a little different. In the preceding example, we have the line **for i = 4:-1:1**. This line means count *i* from 4 down to 1 using a step of –1. If we leave out the –1 step, the loop doesn't work as we intended:

```
fprintf('Testing the FOR loop structure.\n\n');
for i = 4:1  ◄──────────────────────────────   There is no
   fprintf('Testing the FOR loop. i = %g\n',i);   –1 step here.
end
fprintf('\nEnd of loop.\n');
Testing the FOR loop structure.


End of loop.
```

If we omit the –1 step, MATLAB automatically assumes a step of +1. Since the ending condition is smaller than the starting condition, the **FOR** statement assumes that it has completed the loop and completely ignores the statements inside the **FOR** structure. In order to have a **FOR** statement count down, you must specify a negative step.

The default step size is 1. However, the step size doesn't need to be an integer. In the following example, the step is **0.25**.

```
fprintf('Testing the FOR loop structure.\n\n');
for i = 1:0.25:2
    fprintf('Testing the FOR loop. i = %g\n',i);
end
fprintf('\nEnd of loop.\n');
Testing the FOR loop structure.

Testing the FOR loop. i = 1
Testing the FOR loop. i = 1.25
Testing the FOR loop. i = 1.5
Testing the FOR loop. i = 1.75
Testing the FOR loop. i = 2


End of loop.
```

Initially, this loop sets *i* to 1 and then counts up to 2 in 0.25 increments. We can also count down in noninteger-size steps.

```
fprintf('Testing the FOR loop structure.\n\n');
for i = 10:-2.25:1
    fprintf('Testing the FOR loop. i = %g\n',i);
end
fprintf('\nEnd of loop.\n');
Testing the FOR loop structure.

Testing the FOR loop. i = 10
Testing the FOR loop. i = 7.75
Testing the FOR loop. i = 5.5
Testing the FOR loop. i = 3.25
Testing the FOR loop. i = 1


End of loop.
```

To show how a **FOR** loop works, let's modify the guessing program of Program 2-1 so that it gives the user three guesses. We'll use the **FOR** statement to count the guesses. To give the user a better chance, the program will tell the user whether the number is higher or lower than the guess.

Program 2-14 Guessing game with three guesses

```
%
% This is file guess00.m
```

```
%
real_number=rand(1); % Generate a number between 0 and 1
real_number=10*real_number; % Scale the number between 0 and 10
number=ceil(real_number); % Round the number to the next highest integer.

% Ask the user to guess the number:

fprintf('I am thinking of a number between 1 and 10.\n');
fprintf('You have three guesses.\n');
correct=0;

for i = 1:3
    guess = input('Guess a number: ');

    if guess == number
        fprintf('Congratulations - You guessed the correct number.\n');
        correct=1;
    else
        if guess > number
            fprintf('Your guess was too high.\n');
        else
            fprintf('Your guess was too low.\n');
        end
    end
end

if ~correct
    fprintf('Sorry - The number I was thinking of was %g.\n\n', number);
end
```

In Program 2-14, the variable *correct* keeps track of whether the user guessed the correct number. The variable is initially set to 0. The only way *correct* is set to 1 is if the user guesses the correct number. The last **IF** statement in the program tests the value of *correct*. Remember that with **IF** statements, a numerical value of 0 is interpreted as false and a numerical value of 1 is treated as true. Also remember that the tilde character (~) means logical NOT in MATLAB. Thus, the statements **if correct** is equivalent to **if correct == 1**, and the statement **if ~correct** is equivalent to **if correct ==0**. The following statements are equivalent; that is,

```
if ~correct
    fprintf('Sorry - The number I was thinking of was %g.\n\n', number);
end
```

is equivalent to

```
if correct  == 0
    fprintf('Sorry - The number I was thinking of was %g.\n\n', number);
end
```

Now let's look at the program execution:

```
EDU» guess00
I am thinking of a number between 1 and 10.
You have three guesses.

Guess a number: 5
Your guess was too low.

Guess a number: 8
Your guess was too high.

Guess a number: 6
Congratulations - You guessed the correct number.
EDU» guess00
I am thinking of a number between 1 and 10.
You have three guesses.

Guess a number: 5
Your guess was too low.

Guess a number: 8
Your guess was too low.

Guess a number: 9
Congratulations - You guessed the correct number.

EDU» guess00
I am thinking of a number between 1 and 10.
You have three guesses.

Guess a number: 1
Your guess was too low.

Guess a number: 9
Your guess was too high.

Guess a number: 5
Your guess was too low.
Sorry - The number I was thinking of was 6.

EDU»
```

Program 2-14 does, however, have a problem. If the user guesses the number on the first or second guess, the program continues to ask for the remaining guesses.

```
EDU» guess00
I am thinking of a number between 1 and 10.
You have three guesses.

Guess a number: 2
Your guess was too high.

Guess a number: 1
Congratulations - You guessed the correct number.

Guess a number: 0
Your guess was too low.
```

Note that we guessed the correct number on the second guess. The program acknowledged our accomplishment but continued to ask for another guess. There are a few ways to fix this problem. We will show two ways. The first way uses the MATLAB **BREAK** statement when we guess the correct number. When MATLAB encounters a **BREAK** statement within a loop, MATLAB jumps to the **end** keyword for that loop and continues execution at the statement following the **end** keyword. Essentially the **BREAK** statement jumps immediately out of the **FOR** loop.

Program 2-15 Fix 1 of Program 2-14

```
%
% This is file fix1.m
%

real_number=rand(1); % Generate a number between 0 and 1
real_number=10*real_number; % Scale the number between 0 and 10
number=ceil(real_number); % Round the number to the next highest integer.

% Ask the user to guess the number:

fprintf('I am thinking of a number between 1 and 10.\n');
fprintf('You have three guesses.\n');
correct=0;

for i = 1:3
    guess = input('Guess a number: ');

    if guess == number
        fprintf('Congratulations - You guessed the correct number.\n');
        correct=1;
        break;          ◄──────────────────────  This line is added.
    else
        if guess > number
            fprintf('Your guess was too high.\n');
```

```
            else
                fprintf('Your guess was too low.\n');
            end
        end
end

if ~ correct
    fprintf('Sorry - The number I was thinking of was %g.\n\n', number);
end
```

Let's test the program:

```
EDU» fix1
I am thinking of a number between 1 and 10.
You have three guesses.

Guess a number: 5
Congratulations - You guessed the correct number.
EDU»
```

Instead of using the **BREAK** statement to jump out of the loop, we can use an **IF** statement to bypass most of the program if we guess the number.

Program 2-16 Fix 2 of Program 2-14

```
%
% This is file fix2.m
%

real_number=rand(1); % Generate a number between 0 and 1
real_number=10*real_number; % Scale the number between 0 and 10
number=ceil(real_number); % Round the number to the next highest integer.

% Ask the user to guess the number:

fprintf('I am thinking of a number between 1 and 10.\n');
fprintf('You have three guesses.\n');
correct=0;

for i = 1:3
    if ~correct          ←——————————————————  This line is added.
        guess = input('Guess a number: ');

        if guess == number
            fprintf('Congratulations - You guessed the correct number.\n');
            correct=1;
```

```
        else
            if guess > number
                fprintf('Your guess was too high.\n');
            else
                fprintf('Your guess was too low.\n');
            end
        end
    end                                           ┌──────────────────┐
end                               ◄───────────────┤ This line is added.│
                                                  └──────────────────┘

if~correct
    fprintf('Sorry - The number I was thinking of was %g.\n\n', number);
end
```

Note that when we make a correct guess in Program 2-16, the **FOR** loop still loops three times. Because of the added **IF** statement, the statements inside the **FOR** loop are ignored for the remaining loops, but the loop still executes three times. The operation of the program is as follows:

```
EDU» fix2
I am thinking of a number between 1 and 10.
You have three guesses.

Guess a number: 5
Your guess was too low.

Guess a number: 8
Congratulations - You guessed the correct number.
EDU»
```

As another example, let's modify the grade calculation program of Program 2-10 so that it uses a **FOR** loop to request the exam scores. We'll modify the program to ask the user to enter the number of exams to average, and then use a **FOR** loop to request the exam scores. Instead of storing the grade information in three variables *exam1*, *exam2*, and *exam3*, as was done in Program 2-10, we'll store the information in a row array. First, let's create a row array of arbitrary length. The MATLAB **ZEROS** function creates an array with the specified number of rows and columns and fills the array with zeros:

```
EDU» zeros(3,2)
ans =
        0      0
        0      0
        0      0
```

This command creates an array with three rows and two columns. The values of all elements in the array are set to zero. Here's another example:

```
EDU» zeros(5,8)
ans =
     0     0     0     0     0     0     0     0
     0     0     0     0     0     0     0     0
     0     0     0     0     0     0     0     0
     0     0     0     0     0     0     0     0
     0     0     0     0     0     0     0     0
```

This array has five rows and eight columns. Below is a one-dimensional array with one row and five elements:

```
EDU» zeros(1,5)
ans =
     0     0     0     0     0
```

We can create row arrays of arbitrary length:

```
EDU» zeros(1,10)
ans =
     0     0     0     0     0     0     0     0     0     0
```

The elements of a one-dimensional array can be addressed by their position in the array. Since a row array has only one row, we don't have to specify the row number and column number to address an element. We only have to specify the column number of the element. As an example, let's set the third element of a row array to 5.

```
EDU» x=zeros(1,10)
x =
     0     0     0     0     0     0     0     0     0     0
EDU» x(3)=5
x =
     0     0     5     0     0     0     0     0     0     0
```

We can address any element in the array:

```
EDU» x(1)=2;
EDU» x(7)=8;
EDU» x(10)=1;
EDU» x
x =
     2     0     5     0     0     0     8     0     0     1
```

Next, we need to find the sum of the elements in the array. We will show two ways find the sum. In the first one, we address each element individually, and add the elements:

```
EDU» sum=x(1)+x(2)+x(3)+x(4)+x(5)+x(6)+x(7)+x(8)+x(9)+x(10)
sum =
    16
```

This way requires you to know the number of elements in the array. It is only useful if you know the number of elements in the array, the number of elements in array is never changed, and there are not too many elements in the array.

The second way to add the elements uses a **FOR** loop. Since we know the number of elements in the array, we can use the **FOR** statement to step through the elements:

```
for i = 1:10
   fprintf(' Element %g of array x is %g.\n', i, x(i));
end
 Element 1 of array x is 2.
 Element 2 of array x is 0.
 Element 3 of array x is 5.
 Element 4 of array x is 0.
 Element 5 of array x is 0.
 Element 6 of array x is 0.
 Element 7 of array x is 8.
 Element 8 of array x is 0.
 Element 9 of array x is 0.
 Element 10 of array x is 1.
```

These lines of code increment i from 1 to 10 and then display the value of i and the ith element of x, $x(i)$. Now let's use the loop to sum the elements:

```
sum = 0;
for i = 1:10
   sum=sum+x(i);
end
fprintf('The sum is %g.\n', sum);
The sum is 16.
```

This example sums the values of an array for a fixed and known number of elements.

In the next improvement we rewrite the program to sum the elements for an arbitrary and unknown number of elements. First, we use the **LENGTH** function to determine the number of elements in the array:

```
EDU» n=length(x)
n =
    10
```

Once we know the number of elements in the array, we can use a **FOR** loop to step through the array:

```
sum=0;
n=length(x);
for i = 1:n
   sum=sum+x(i);
end
fprintf('The number of elements in the array was %g.\n', n);
fprintf('The sum of the elements was %g.\n', sum);
The number of elements in the array was 10.
The sum of the elements was 16.
```

To show that the program works for an arbitrary number of elements, let's create a row array with 123 elements and fill it with random numbers:

```
EDU» x=rand(1,123)
x =
  Columns 1 through 7
    0.2190    0.0470    0.6789    0.6793    0.9347    0.3835    0.5194
  Columns 8 through 14
    0.8310    0.0346    0.0535    0.5297    0.6711    0.0077    0.3834
  Columns 15 through 21
    0.0668    0.4175    0.6868    0.5890    0.9304    0.8462    0.5269
  Columns 22 through 28
    0.0920    0.6539    0.4160    0.7012    0.9103    0.7622    0.2625
  Columns 29 through 35
    0.0475    0.7361    0.3282    0.6326    0.7564    0.9910    0.3653
  Columns 36 through 42
    0.2470    0.9826    0.7227    0.7534    0.6515    0.0727    0.6316
  Columns 43 through 49
    0.8847    0.2727    0.4364    0.7665    0.4777    0.2378    0.2749
  Columns 50 through 56
    0.3593    0.1665    0.4865    0.8977    0.9092    0.0606    0.9047
  Columns 57 through 63
    0.5045    0.5163    0.3190    0.9866    0.4940    0.2661    0.0907
  Columns 64 through 70
    0.9478    0.0737    0.5007    0.3841    0.2771    0.9138    0.5297
  Columns 71 through 77
    0.4644    0.9410    0.0501    0.7615    0.7702    0.8278    0.1254
  Columns 78 through 84
    0.0159    0.6885    0.8682    0.6295    0.7362    0.7254    0.9995
  Columns 85 through 91
    0.8886    0.2332    0.3063    0.3510    0.5133    0.5911    0.8460
  Columns 92 through 98
    0.4121    0.8415    0.2693    0.4154    0.5373    0.4679    0.2872
  Columns 99 through 105
    0.1783    0.1537    0.5717    0.8024    0.0331    0.5344    0.4985
```

```
Columns 106 through 112
   0.9554    0.7483    0.5546    0.8907    0.6248    0.8420    0.1598
Columns 113 through 119
   0.2128    0.7147    0.1304    0.0910    0.2746    0.0030    0.4143
Columns 120 through 123
   0.0269    0.7098    0.9379    0.2399
```

An array of this size can't be displayed on a single line, so MATLAB breaks the display into seven columns per line. We can now sum the elements:

```
sum=0;
n=length(x);
for i = 1:n
   sum=sum+x(i);
end
fprintf('The number of elements in the array was %g.\n', n);
fprintf('The sum of the elements was %g.\n', sum);
The number of elements in the array was 123.
The sum of the elements was 62.5593.
```

Now let's modify Program 2-10. We'll change the program so that it asks the user for the number of exams, creates an array to store the exam scores, and then uses a **FOR** statement to find the sum.

Program 2-17 Grade calculation

```
%
% This is file grade2.m
%
% This program requests a variable number of exam scores from the user,
% calculates the average of the exams, and assigns a grade.
%
%      Condition            Grade
%      ─────────            ─────
%      Average >= 90          A
%      80 <= Avg <90          B
%      70 <= Avg <80          C
%      60 <= Avg <70          D
%      Average <  60          F


% Ask for the number of exams:

num_of_exams=input('Specify the number of exams to average:');

% Create a row array to store the exams.
scores=zeros(1,num_of_exams);

% Request the exam scores from the user.
```

```
for i = 1:num_of_exams
     fprintf('Specify the score for Exam %g:',i);
     scores(i)=input(' ');
end

% calculate the sum of the scores

sum=0;
for i = 1:num_of_exams
     sum=sum+scores(i);
end

% calculate the average

average=sum/num_of_exams;

% Determine the grade

fprintf('The number of exams was %g\n', num_of_exams);
fprintf('Your average exam score is %g\n', average);
fprintf('This corresponds to a grade of ');

if average >= 90
     Grade='A';
elseif (average >= 80)
     Grade = 'B';
elseif (average >= 70)
     Grade = 'C';
elseif (average >= 60)
     Grade = 'D';
else
     Grade = 'F';
end

fprintf(Grade);
fprintf('.\n\n');
```

Let's execute the program.

```
EDU» grade2
Specify the number of exams to average:2
Specify the score for Exam 1:
100
Specify the score for Exam 2:
80
The number of exams was 2
Your average exam score is 90
This corresponds to a grade of A.
```

```
EDU» grade2
Specify the number of exams to average:7
Specify the score for Exam 1:
100
Specify the score for Exam 2:
96
Specify the score for Exam 3:
75
Specify the score for Exam 4:
58
Specify the score for Exam 5:
87
Specify the score for Exam 6:
76
Specify the score for Exam 7:
100
The number of exams was 7
Your average exam score is 84.5714
This corresponds to a grade of B.
EDU»
```

EXERCISE 2-2   Modify Program 2-17 to use built-in MATLAB functions to calculate the exam average. Use the MATLAB **LOOKFOR** command to find a function that calculates the average of a list of numbers.

Our last example of a single **FOR** loop asks the user for an integer that will be stored in variable n, and then calculates the factorial of that number, $n!$. Recall that the factorial of a number n is

$$n! = \begin{cases} n(n-1) \cdot (n-2) \cdot (\cdots) \cdot 2 \cdot 1 & \text{for} \quad n \geq 1 \\ 1 & \text{for} \quad n = 0 \end{cases}$$

Here are some examples:

$0! = 1$
$1! = 1$
$2! = 2 \cdot 1 = 2$
$5! = 5 \cdot 4 \cdot 3 \cdot 2 \cdot 1 = 120$

To calculate a factorial, we must check whether n is a positive integer and for the special case of $n = 0$. Let's use the MATLAB **FLOOR** function to do this. This function rounds a number to the next lowest integer:

```
EDU» floor(5.9)
ans =
      5
```

```
EDU» floor(5)
ans =
     5
EDU» floor(0)
ans =
     0
EDU» floor(-5.1)
ans =
    -6
EDU» floor(-5)
ans =
    -5
```

You can see from these examples that the **FLOOR** function has no effect on integers. Thus, if floor(n) = n, then n must be an integer:

```
n=5.6;
if floor(n) ~= n
   fprintf('n is not an integer.\n');
else
   fprintf('n is an integer.\n');
end
n is not an integer.
n=12;
if floor(n) ~= n
   fprintf('n is not an integer.\n');
else
   fprintf('n is an integer.\n');
end
n is an integer.
```

As you can see, it's easy to check whether a number is an integer. Before we write the entire program, let's calculate a factorial using a **FOR** loop. Below, we calculate 5!. Note that the **FOR** loop uses the expression 5:–1:2. This means count from 5 down to 2 in steps of –1.

```
fact=1;
for i = 5:-1:2
   fact=fact*i;
end
fprintf('5 factorial is %g.\n', fact);
5 factorial is 120.
```

This program can be made general for any positive integer n.

```
n=16;
fact=1;
```

```
for i = n:-1:2
  fact=fact*i;
end
fprintf('%g factorial is %16.0f.\n', n, fact);
16 factorial is    20922789888000.
```

Now let's write the factorial program. Note that the program must check for these things: that the number is an integer, that the number is positive, and for the special case of 0! = 1.

Program 2-18 Factorial program

```
%
%   This is file fact_prog.m
%
% This program calculates the factorial of a number.
%

n=input('Specify a non-negative integer: ');

error_flag=0;
if floor(n) ~= n
      fprintf('The number you entered is not an integer.\n')
      error_flag=1;
end

if n < 0
      fprintf('The number you entered is not positive.\n')
      error_flag=1;
end


if error_flag==0  % The number entered was a good number
      if n == 0
            fact=1;
      else
            fact=1;
            for i = n:-1:2
                  fact=fact*i;
            end
      end
      fprintf('%g factorial equals %20.0f.\n', n, fact);
end
```

In Program 2-18, variable *error_flag* indicates an error. Initially, *error_flag* is set to zero, indicating that there is not an error. The error conditions are then checked and if an error is found, *error_flag* is set to 1. The factorial of *n* is only calculated if, after checking all of the error conditions, variable *error_flag* is still equal to zero.

Note that the **FPRINTF** statement, `fprintf('%g factorial equals %20.0f.\n', n, fact);`, uses the `%20.0f` formatting string, which means display a number with 20 digits and use 0 of the digits for the decimal portion of the number. Since the result of a factorial is an integer, this will display an integer with up to 20 digits. As we shall see, the problem with this format is that if the number contains less than 20 digits, the unused places print as blanks.

Let's test the program and see how it displays results. We'll test all of the error conditions and then check how the result of the factorial calculation is printed.

```
EDU» fact_prog

Specify a non-negative integer: 5.6
The number you entered is not an integer.
EDU» fact_prog

Specify a non-negative integer: -4
The number you entered is not positive.
EDU» fact_prog

Specify a non-negative integer: -5.9
The number you entered is not an integer.
The number you entered is not positive.
EDU» fact_prog

Specify a non-negative integer: 0
0 factorial equals                     1.
EDU» fact_prog

Specify a non-negative integer: 5
5 factorial equals                   120.
EDU» fact_prog

Specify a non-negative integer: 16
16 factorial equals       20922789888000.
EDU» fact_prog

Specify a non-negative integer: 32
32 factorial equals 263130836933693600000000000000000000.
EDU»
```

As you can see, the program properly checks for all of the error conditions and for the special case of 0!. The program also displays the result as an integer. Note that the result of 32! has a lot of zeros in it,

$$263130836933693600000000000000000000$$

This is because 32! is a very large number and must use a floating-point number to represent it. The floating-point version of 32! is $2.631308369336936 \times 10^{35}$. MATLAB only performs calculations with double-precision arithmetic, which gives only about 16 digits of accuracy. Note that both

$$26313083693336936000000000000000000000 \quad \text{and} \quad 2.631308369336936 \times 10^{35}$$

have 16 digits of accuracy. When MATLAB converts $2.631308369336936 \times 10^{35}$ to an integer, it fills in the lower digits with zeros. Thus, you should realize that 32! is not exactly equal to

$$26313083693336936000000000000000000000$$

This result is as precise as we can get with double-precision arithmetic.

We would like the function to calculate factorials for both large and small numbers. However, as you can see, the format we used to display the result padded small results with a lot of blanks:

```
0 factorial equals                    1.
```

We wish to remove these spaces.

This happened because the format we used allowed for 20 spaces, **%20.0f**. If we change this to a smaller format, **%3.0f**, small numbers will be OK and larger numbers that don't fit into the format will be displayed with the appropriate number of digits:

```
EDU» fact_prog

Specify a non-negative integer: 0
0 factorial equals    1.
EDU» fact_prog

Specify a non-negative integer: 5
5 factorial equals 120.
EDU» fact_prog

Specify a non-negative integer: 16
16 factorial equals 20922789888000.
EDU» fact_prog

Specify a non-negative integer: 32
32 factorial equals 26313083693336936000000000000000000000.
EDU»
```

Abusing a format string in this manner is not considered good programming technique. Let's look at a different method.

MATLAB provides a function for converting an integer to a text string. The string has no leading blank characters.

```
EDU» int2str(5)
ans =
5
```

```
EDU» int2str(0005)
ans =
5
EDU» int2str(-5)
ans =
-5
EDU» int2str(2.631308369336936e+035)
ans =
263130836933693600000000000000000000
```

Let's modify Program 2-18 to use the **INT2STR** function.

Program 2-19 Revised factorial program

```
%
%  This is file fact_x.m
%
% This program calculates the factorial of a number.
%

n=input('Specify a non-negative integer: ');

error_flag=0;
if floor(n) ~= n
     fprintf('The number you entered is not an integer.\n')
     error_flag=1;
end

if n < 0
     fprintf('The number you entered is not positive.\n')
     error_flag=1;
end


if error_flag==0  % The number entered was a good number
    if n == 0
          fact=1;
    else
          fact=1;
          for i = n:-1:2
                fact=fact*i;
          end
    end
    str=int2str(fact);    ◄──────    This line is added and the
    fprintf('%g factorial equals ',n);          following lines modified.
    fprintf(str);
    fprintf('.\n');
end
```

The line `str=int2str(fact);` converts the number stored in **fact** into a text string and saves the result in **str**. The three **FPRINTF** lines produce one line of output text. Since only the last **FPRINTF** statement has a new line character (**\n**), the text produced by all three **FPRINTF** statements appears in a single line of output text.

This program produces nicer output:

```
EDU» fact_x

Specify a non-negative integer: 0
0 factorial equals 1.
EDU» fact_x

Specify a non-negative integer: 5
5 factorial equals 120.
EDU» fact_x

Specify a non-negative integer: 10
10 factorial equals 3628800.
EDU» fact_x

Specify a non-negative integer: 32
32 factorial equals 263130836933693600000000000000000000.
EDU»
```

EXERCISE 2-3   An easy way to calculate the factorial of a number is to use the MATLAB **PROD** function. Show that prod(1:*n*) calculates *n*! for *n* ≥ 1. Explain why this method works.

EXERCISE 2-4   Modify Program 2-19 so that it uses the prod(1:*n*) method to calculate the factorial rather than a **FOR** loop.

MATLAB Built-In Function

The MATLAB **VPA** function performs variable-precision arithmetic and is not limited to double-precision calculations. For example, we can display the value of π to an arbitrary number of digits:

```
EDU» vpa(pi,60)
ans =
3.14159265358979323846264338327950288419716939937510582097494
EDU»
```

EXERCISE 2-5   Use the **PROD** function and the **VPA** function to calculate the exact value of 32!. The answer is 263130836933693517766352317727113216.
(Note that the **VPA** function may not be available in all MATLAB installations.)

As another example of a **FOR** loop and an **IF** statement, let's plot the response of the following switched *RC* circuit:

FIGURE 2-1   *RC* circuit with switch

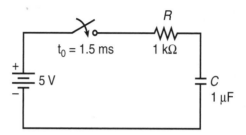

We will assume that the capacitor is initially uncharged, so its initial voltage starts at 0 V. The switch will remain open from $t = 0$ ms to $t = 1.5$ ms. At $t = 1.5$ ms, the switch will close and the capacitor will charge through the resistor. The capacitor voltage as a function of time is:

$$V_c(t) = \begin{cases} V_{ci} & \text{for } 0 \le t < 1.5 \text{ ms} \\ V_{cf} + (V_{ci} - V_{cf})e^{-(t-t_0)/RC} & \text{for } t \ge 1.5 \text{ ms} \end{cases}$$

where V_{ci} is the initial capacitor voltage, zero in our example; V_{cf} (final capacitor voltage) is the voltage the capacitor will reach if it charges for an infinite amount of time. In our example, V_{cf} is equal to 5 V, and t_0 is the time the switch closes, 1.5 ms.
Let's evaluate the equation for $V_c(t)$ for $0 \le t \le 8$ ms using MATLAB. The code segment is shown in Program 2-20.

Program 2-20 Simulation of a switched *RC* circuit

```
% This is m-file rc_step.m
% Example calculation of the response of an RC circuit with a switch.
% We would like to evaluate the equation
% Vc = Vci      for t < to
% Vc = Vci + (Vci-Vcf)*exp[-(t-to)/RC] for tf >t >= to
% where to is the time the switch closes and tf is the length of
% we would like to evaluate the equation. Independent of the
% time interval 0 to tf, we would like to generate 1000 points of data.

Vcf = 5; R = 1000; C=1e-6; to=0.0015; tf = 0.008; Vci = 0;
```

```
t=linspace(0,tf,1000);
Vc = zeros(1,1000);
for i = 1:1000
      if t(i) < to
            Vc(i) = Vci;
      else
            Vc(i) = Vcf + (Vci -Vcf)*exp(-(t(i)-to)/(R*C));
      end
end
plot(t*1000,Vc);
title('RC Step Response.');
ylabel('Capacitor Voltage (volts)');
xlabel('Time (milliseconds)')
grid
```

The syntax of the **LINSPACE** function is $x = $ **LINSPACE**$(x1, x2, n)$. This function generates n equally spaced values from $x1$ to $x2$ and stores them in x as a one-dimensional array with one row. Here are some examples:

```
EDU» x=linspace(0,5,6)
x =
      0      1      2      3      4      5
EDU» x=linspace(0,5,11)
x =
  Columns 1 through 7
          0     0.5000     1.0000     1.5000     2.0000     2.5000     3.0000
  Columns 8 through 11
     3.5000     4.0000     4.5000     5.0000
EDU»
```

The line **t=linspace(0,tf,1000)** creates an array of 1000 values for t. These values range from 0 to t_f. The line **Vc = zeros(1,1000)** creates an array of 1000 values for V_c and sets all the values to zero. The **FOR** loop then steps through each element in the t array and calculates a corresponding value of V_c. Since the t array has 1000 elements, we are calculating 1000 values for V_c. We then plot the values in V_c versus the values in t.

Note that the values of t range from 0 to 0.008 seconds. $1000 \times t$ ranges from 0 to 8, which can be interpreted as 0 to 8 ms. We are plotting V_c versus $1000 \times t$. The plot generated from the code segment is as follows:

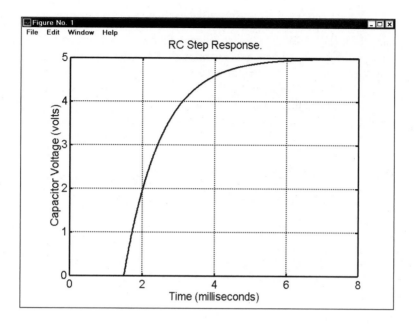

2.5.2 Nested **FOR** Loops

FOR loops can be nested. In the following code segment, every time i is increased by 1, j will be counted from 1 to 10.

```
fprintf('Demonstrating nested FOR loops.\n\n');
for i = 1:5
   fprintf(' i = %g    -    j = ', i);
   for j = 1:10
         fprintf('%g   ', j);
   end
   fprintf('\n');
end
fprintf('\nEnd of demonstration.\n');
Demonstrating nested FOR loops.
 i = 1    -    j = 1   2   3   4   5   6   7   8   9   10
 i = 2    -    j = 1   2   3   4   5   6   7   8   9   10
 i = 3    -    j = 1   2   3   4   5   6   7   8   9   10
 i = 4    -    j = 1   2   3   4   5   6   7   8   9   10
 i = 5    -    j = 1   2   3   4   5   6   7   8   9   10
End of demonstration.
```

The nested loops work as follows. First, i is set to 1. The program displays the value of i and the text j =. While i is 1, the j loop executes and counts j from 1 to 10. Each time through the j loop, the value of j is printed. When the j loop is complete, the new line character is emitted and then the **end** keyword for the i loop is encountered. The program returns to the beginning of the i loop and i is set to 2. The program again displays the value of i and the text j =, and the j loop executes as before. This looping continues until i reaches 5. Note that for each value of i, the j loop counts from 1 to 10.

The statements inside the j loop are executed $i \times j$ times, or 50 times in this example. If we want to count how many times the statements inside the j loop are executed, we add a counter to the program.

```
fprintf('Demonstrating nested FOR loops.\n\n');
count=0;                                              Initialize counter to 0.
for i = 1:5
    fprintf(' i = %g    -    j = ', i);
    for j = 1:10
            fprintf('%g   ', j);
            count=count+1;                            Increment counter by 1.
    end
    fprintf('\n');
end
fprintf('\nEnd of demonstration.\n');
fprintf('The value of count is %g.\n',count);
Demonstrating nested FOR loops.
    i = 1    -    j = 1    2    3    4    5    6    7    8    9    10
    i = 2    -    j = 1    2    3    4    5    6    7    8    9    10
    i = 3    -    j = 1    2    3    4    5    6    7    8    9    10
    i = 4    -    j = 1    2    3    4    5    6    7    8    9    10
    i = 5    -    j = 1    2    3    4    5    6    7    8    9    10

End of demonstration.
The value of count is 50.
```

As an example, in Program 2-21 we create a multiplication table using nested **FOR** loops.

Program 2-21 Multiplication table using nested **FOR** loops

```
%
%   This is file multab.m
%
% This program creates a multiplication table for 1 to 10.
%

fprintf('   1    2    3    4    5    6    7    8    9    10\n');
fprintf('  ---  ---  ---  ---  ---  ---  ---  ---  ---  --- \n');
```

```
for i = 1:10
    fprintf(' %2.0f   ',i);
    for j = 1:10
        fprintf('%3.0f   ',i*j);
    end
    fprintf('\n');
end;
```

The **FPRINTF** statement in this program uses the **%f** format instead of the **%g** format. The **%g** is the general format and is the easiest to use. We use the **%f** to display numbers with a fixed number of digits. To show how the **%f** formatting string works, let's display the value of π using several different values of the **%f** formatting string. The form **%3.0f** means allocate a total of 3 digits for the number and use zero digits after the decimal.

```
EDU» fprintf('Testing — %3.0f\n', pi);

Testing —   3
```

The format **%3.1f** means allocate a total of 3 digits and use 1 digit after the decimal.

```
EDU» fprintf('Testing — %3.1f\n', pi);
Testing — 3.1
```

The format **%3.2f** means allocate a total of 3 digits and use 2 digits after the decimal.

```
EDU» fprintf('Testing — %3.2f\n', pi);
Testing — 3.14
```

The format **%3.3f** means allocate a total of 3 digits and use 3 digits after the decimal.

```
EDU» fprintf('Testing — %3.3f\n', pi);
Testing — 3.142
```

In this last example, the format **%3.3f** does not allocate digits for a number greater than or equal to 1. Format **%3.3f** means 3 total digits with all of them following the decimal point. Note also that the number π cannot fit into this format and is displayed with 4 digits.

We can specify a large number of digits and a large number of decimal places:

```
EDU» fprintf('Testing — %10.8f\n', pi);
Testing — 3.14159265
```

This format string means allocate 10 total digits and use 8 of them for the decimal portion of the number. Note that unused spaces are replaced with blanks.

Now let's test Program 2-21:

```
EDU» multab
         1    2    3    4    5    6    7    8    9   10
        ──   ──   ──   ──   ──   ──   ──   ──   ──   ──
    1    1    2    3    4    5    6    7    8    9   10
    2    2    4    6    8   10   12   14   16   18   20
    3    3    6    9   12   15   18   21   24   27   30
    4    4    8   12   16   20   24   28   32   36   40
    5    5   10   15   20   25   30   35   40   45   50
    6    6   12   18   24   30   36   42   48   54   60
    7    7   14   21   28   35   42   49   56   63   70
    8    8   16   24   32   40   48   56   64   72   80
    9    9   18   27   36   45   54   63   72   81   90
   10   10   20   30   40   50   60   70   80   90  100
EDU»
```

As another example of a nested **FOR** loop, we'll calculate the sum

$$\sum_{i=1}^{n} i! = 1! + 2! + 3! + \cdots + (n-1)! + n! = 1 + (2 \cdot 1) + (3 \cdot 2 \cdot 1) + (4 \cdot 3 \cdot 2 \cdot 1) + \cdots + (n-1)! + n!$$

This sum is essentially the repeated use of the factorial program we created in Program 2-19. We'll modify Program 2-19 to perform the sum. Note that we add the restriction that n must be greater than or equal to 1. If the user enters a value less than 1, an error message is generated.

Program 2-22 Sum of factorials using a nested **FOR** loop

```
%
%   This is file sum_fact1.m
%
% This program calculates the sum from i = 1 to n of n!
%

n=input('Specify a non-negative integer: ');

error_flag=0;
if floor(n) ~= n
     fprintf('The number you entered is not an integer.\n')
     error_flag=1;
end

if n < 1
     fprintf('The number you entered is not greater than 1.\n')
     error_flag=1;
end


if error_flag==0  % The number entered was a good number
```

```
    sum=0;
    for j=1:n
            fact=1;
            for i = j:-1:2
                    fact=fact*i;
            end
            sum=sum+fact;
    end
    str=int2str(sum);
    fprintf('The sum equals ');
    fprintf(str);
    fprintf('.\n');
end
```

Outer loop performs the sum from 1 to *n*.

Inner loop finds *j*!

The outer loop

```
    sum=0;
    for j=1:n
        ....
            sum=sum+fact;
    end
```

implements the sum from 1 to *n*. The inner loop

```
        fact=1;
        for i = j:-1:2
            fact=fact*i;
        end
```

finds *j*! for each value of *j*. Each time *j* is incremented by the outer loop, the inner loop finds *j*! After the inner loop finds j! the value is added to the sum by the outer loop, **sum=sum+fact**.

The operation of the program is shown below.

```
EDU» sum_fact1

Specify a non-negative integer: 0
The number you entered is not greater than 1.
EDU» sum_fact1

Specify a non-negative integer: 1
The sum equals 1.
EDU» sum_fact1

Specify a non-negative integer: 2
The sum equals 3.
EDU» sum_fact1
```

```
Specify a non-negative integer: 3
The sum equals 9.
EDU» sum_fact1

Specify a non-negative integer: 5
The sum equals 153.
EDU» sum_fact1

Specify a non-negative integer: 10
The sum equals 4037913.
EDU»
```

A good application of a nested **FOR** loop is the current-voltage characteristic (*I-V* characteristic) of a transistor such as a MOSFET (metal oxide semiconductor field effect transistor). The equations that govern a MOSFET are

$$
\begin{aligned}
I_D &= K(V_{GS} - V_T)^2(1 + \lambda V_{DS}) && \text{for} \quad V_{DS} \geq V_{GS} - V_T \quad \text{and} \quad V_{GS} > V_T \\
&= K\left[2(V_{GS} - V_T)V_{DS} - V_{DS}^2\right](1 + \lambda V_{DS}) && \text{for} \quad V_{DS} < V_{GS} - V_T \quad \text{and} \quad V_{GS} > V_T \\
&= 0 && \text{for} \quad V_{GS} < V_T
\end{aligned}
$$

where K, V_T, and λ are constants. These equations describe how the current I_D varies with the voltages V_{GS} and V_{DS}. Typically, V_{GS} is held constant and I_D is plotted against V_{DS}. As an example, we'll use the constants $K = 25 \ \mu A/V^2$, $V_T = 2 \ V$, $\lambda = 0.05$, and let $V_{GS} = 5 \ V$. We'll use a **FOR** loop to generate values of I_D versus V_{DS}. Let's plot I_D for values of V_{DS} from 0 to 20 V.

Program 2-23 *I-V* characteristic of a MOSFET for a single value of V_{GS}

```
% This is m-file IV_char1.m
% This script calculates and plots the IV characteristic
% of MOSFET for a single value of Vgs.

K = 0.00025; Vt=2; Lambda = 0.005; Vgs = 5;
Vds = linspace(0,20,500);
Id = zeros(1,500);
for i = 1:500
    if Vgs < Vt
        id(i) = 0;
    elseif Vds(i) >= Vgs-Vt
        Id(i) = K*(Vgs-Vt)*(Vgs-Vt)*(1+Lambda*Vds(i));
    else
        Id(i) = K*(2*(Vgs-Vt)*Vds(i) - Vds(i)*Vds(i))* (1+Lambda*Vds(i));
    end
end
plot(Vds,1000*Id);
xlabel('V_{DS} (Volts)');
ylabel('I_D (mA)');
title('MOSFET I-V Characteristic');
```

Here is what the plot looks like.

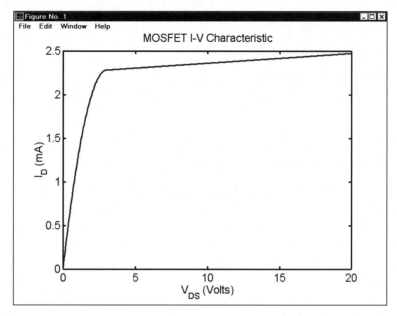

When we use a MOSFET, we typically use it for many different values of V_{GS}. One thing we want to know is how the *I-V* characteristic changes for different values of V_{GS}. We would like to generate several different *I-V* characteristics for different values of V_{GS}, and plot all of the *I-V* characteristics on the same plot. For example, we'll use $V_{GS} = 0$ V and generate an *I-V* characteristic. Then we'll let V_{GS} equal 1 V and generate another *I-V* characteristic, then $V_{GS} = 2$ V, and so on. We'll repeat this procedure for values of V_{GS} from 2 to 8 V and then plot all of the traces on the same plot. This program requires a second **FOR** loop. Essentially, we have to enclose the previous code segment for generating the single *I-V* characteristic inside a **FOR** loop that sweeps V_{GS} from 2 V to 8 V in 1-V steps. The new code segment is shown in Program 2-24.

Program 2-24 Program to calculate the complete *I-V* characteristic of a MOSFET

```
% This is m-file IV_char2.m
% This script calculates and plots the IV characteristic
% of MOSFET for several values of Vgs.
K = 0.00025; Vt=2; Lambda = 0.005; Vgs = 5;
Vds = linspace(0,20,500)';
Id = zeros(500,7);
for Vgs = 2:8
   for i = 1:500
      if Vgs < Vt
         Id(i,Vgs-1) = 0;
      elseif Vds(i) >= Vgs-Vt
         Id(i,Vgs-1) = K*(Vgs-Vt)*(Vgs-Vt)*(1+Lambda*Vds(i));
      else
```

```
            Id(i,Vgs-1) = K*(2*(Vgs-Vt)*Vds(i) - Vds(i)*Vds(i))* (1+Lambda*Vds(i));
        end
    end
end
plot(Vds,1000*Id);
xlabel('V_{DS} (Volts)');
ylabel('I_D (mA)');
title('MOSFET I-V Characteristic');
```

This is the resulting screen.

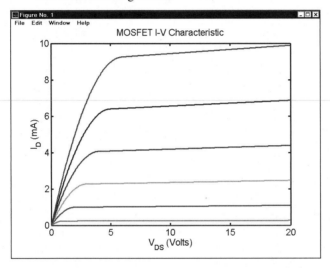

EXERCISE 2-6   In Program 2-20, we plotted the response of a switched *RC* circuit. Use nested **FOR** loops to plot the response for values of *C* from 1 µF to 10 µF. An example plot is

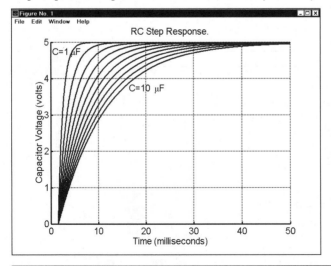

2.5.3 Special Cases of the FOR Loop

A few special cases of the **FOR** loop are worth a closer examination. The **FOR** loop can sweep through any fixed set of elements. Suppose we want a variable to take on the values 1, 5, 99, –23, and 4 using a **FOR** loop. The following code segments implement this task:

```
for i = [1, 5, 99, -23, 4]
   fprintf('The value of i is %g.\n', i);
end
The value of i is 1.
The value of i is 5.
The value of i is 99.
The value of i is -23.
The value of i is 4.
```

Note that variable *i* takes on all of the specified values and sweeps through them in the order in which they are listed. We can also sweep through values specified by a string:

```
for i = 'abcbca123xya'
   fprintf('The value of i is %c.\n', i);
end
The value of i is a.
The value of i is b.
The value of i is c.
The value of i is b.
The value of i is c.
The value of i is a.
The value of i is 1.
The value of i is 2.
The value of i is 3.
The value of i is x.
The value of i is y.
The value of i is a.
```

Or, we can sweep a variable using the letters of the alphabet:

```
for i = 'a':'e'
   fprintf('The value of i is %c.\n', i);
end
The value of i is a.
The value of i is b.
The value of i is c.
The value of i is d.
The value of i is e.
```

These three features of **FOR** loops are unique to MATLAB.

To show how we code a **FOR** loop using this technique, let's calculate the maximum power transfer to a load resistor from a voltage source that has a fixed series resistance. Here is the circuit:

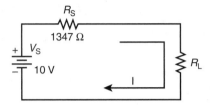

where V_S is fixed at 10 V and R_S is fixed at 1347 Ω. The question is, what value of R_L yields maximum power to R_L? This is a series circuit, so the current (I) through all elements is the same and is equal to

$$I = \frac{V_S}{R_S + R_L}$$

The power absorbed by a resistor is:

$$P_R = I^2 R$$

so

$$P_{R_L} = \left(\frac{V_S}{R_S + R_L} \right)^2 R_L$$

We can use these two equations to calculate the power absorbed by R_L for several different values of R_L. We'll add the constraint that the only available resistors are standard 5% resistors in the series 1, 1.2, 1.5, 1.8, 2.2, 2.7, 3.3, 3.9, 4.7, 5.6, 6.8, and 8.2. Note that R_S is the internal resistance of the voltage source and can be any value. Because R_L is limited to resistors that you can purchase, it can only take on a few limited values. The analytic solution is that maximum power occurs when R_L is equal to R_S. Since the choices of R_L are limited to available resistors, R_L will be close to R_S. The hard part will be to find out which standard value gives maximum power. An example solution for $V_S = 10$ V and $R_S = 1347$ Ω is shown in Program 2-25.

Program 2-25 Program to calculate the maximum power transfer to a load resistor from a voltage source with series resistance

```
% This is m-file max_pow1.m
% This file finds the maximum power transfer to a load resistor
% given a fixed source voltage and impedance.
% Rs is the source impedance and Vs is the source voltage.
% The allowed load resistors are standard 5% resistors in the series
% 1, 1.2, 1.5, 1.8, 2.2, 2.7, 3.3, 3.9, 4.7, 5.6, 6.8, and 8.2
Vs = 10; Rs = 1347;
Load_Power = [ ];
```

```
std = [1, 1.2, 1.5, 1.8, 2.2, 2.7, 3.3, 3.9, 4.7, 5.6, 6.8, 8.2];
Resistors = [std, 10*std, 100*std, 1000*std, 10000*std, 100000*std];

for RL = Resistors
    Current = Vs/(Rs+RL);
    Load_Power = [Load_Power, Current*Current*RL];
end

[max_power,i]=max(Load_Power);
RL_max = Resistors(i);

str=sprintf('Max Power is %g and occurs \nat a load resistance of ➡
%g.',max_power,RL_max);
semilogx(Resistors, Load_Power);
ylabel('Power (Watts)');
xlabel('Load Resistance (Ohms)');
text(2,0.016,str);
```

The line $std = [1, 1.2, 1.5, 1.8, 2.2, 2.7, 3.3, 3.9, 4.7, 5.6, 6.8, 8.2]$ creates array *std* that contains the standard resistor values from $1\,\Omega$ to $8.2\,\Omega$. The line $Resistors = [std, 10*std, 100*std, 1000*std, 10000*std, 100000*std]$ uses concatenation to create array *Resistors* that contains standard resistor values from $1\,\Omega$ to $820{,}000\,\Omega$:

```
EDU» Resistors
Resistors =
  Columns 1 through 6
         1          1.2        1.5        1.8        2.2        2.7
  Columns 7 through 12
       3.3          3.9        4.7        5.6        6.8        8.2
  Columns 13 through 18
        10           12         15         18         22         27
  Columns 19 through 24
        33           39         47         56         68         82
  Columns 25 through 30
       100          120        150        180        220        270
  Columns 31 through 36
       330          390        470        560        680        820
  Columns 37 through 42
      1000         1200       1500       1800       2200       2700
  Columns 43 through 48
      3300         3900       4700       5600       6800       8200
  Columns 49 through 54
     10000        12000      15000      18000      22000      27000
  Columns 55 through 60
     33000        39000      47000      56000      68000      82000
```

```
Columns 61 through 66
     1e+005      1.2e+005      1.5e+005      1.8e+005      2.2e+005      2.7e+005
Columns 67 through 72
   3.3e+005      3.9e+005      4.7e+005      5.6e+005      6.8e+005      8.2e+005
EDU»
```

The **FOR** loop

```
for RL = Resistors
     Current = Vs/(Rs+RL);
     Load_Power = [Load_Power, Current*Current*RL];
end
```

sweeps variable R_L through the standard resistor values contained in array *Resistors*. Thus, R_L can only take on values of available resistors. The loop calculates the current and power for each value of R_L and then adds the power to the array *Load_Power*. For each value of R_L, there will be a value of power stored in array *Load_Power*.

The MATLAB **MAX** function is used to find the maximum value stored in an array:

```
EDU» max([1   2   4   578   6532   1298   378921   65728   7   277])
ans =
       378921
EDU»
```

The result shows us that the maximum value contained in the array is 378921. When the **MAX** command is used in the form **[m, i] = max(array)**, the function returns the maximum value contained in the array and *i* indicates which element is the maximum:

```
EDU» [m,i]=max([1 2 4 578 6532 1298 378921 65728 7 277])
m =
       378921
i =
       7
EDU»
```

The result tells us that the maximum value is 378921 and that the seventh element in the array is the maximum.

In Program 2-25, the line **[max_power,i]=max(Load_Power)** finds the maximum element in array *Load_Power*, and *i* tells us which element in the array has the maximum power. Since a value of *Load_Power* is calculated for each value of R_L, if the *i*th value of *Load_Power* is the maximum value, then the *i*th value of R_L is the value of resistance that yields the maximum value. Thus, the **line RL_max = Resistors(i)** picks out the *i*th element of *Resistors* as the resistor value that yielded maximum power.

The remainder of the program plots the load power against R_L using a semilog plot and plots the power and value of R_L that yields maximum power on the graph. The plot generated by Program 2-25 is shown here:

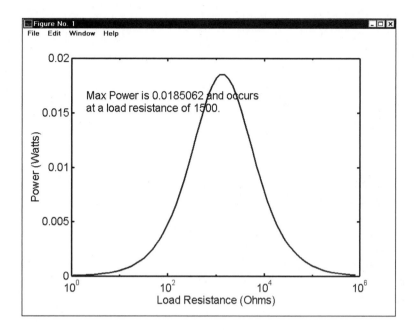

2.6 **WHILE Loops**

WHILE loops repeat a set of statements an unknown number of times. They differ from **FOR** loops in that **FOR** loops repeat a specified number of times. A **WHILE** loop will loop until a condition is met. The syntax is

WHILE condition
    statement 1
    statement 2
    ⋮
    statement n

END

While the condition is true or equal to 1, statements 1 through n are executed. The statement works as follows. Suppose a program is running and it encounters the **WHILE** statement for the first time. The condition is tested. If the condition is false or 0, MATLAB jumps to the statement after the **end** keyword and continues with the rest of the program, if any. In this case, statements 1 to n are never executed. If the condition is true, statements 1 through n are executed. When statement n is finished, the program jumps back to the **while** keyword and tests the condition again. If the condition is false, MATLAB jumps to the statement following the **end** keyword and continues with the rest of the program, if any. If the condition is true, statements 1 through n are executed again. When statement n is finished, the program jumps back to the **while** keyword and tests the condition again. The loop repeats indefinitely until the condition becomes false.

    If the condition never becomes false, an infinite loop results. Here are a few examples of infinite loops. Note that the conditions given will never become false:

```
while 1                                          Condition is constant and true.
end


i = -5;
while i <= 0                          Condition is checking the wrong thing.
    % Do some stuff
    % i=i-1;                                      i is made more negative.
    % Do some more stuff
end


i = 0
while i >= 0
    % Do some stuff
    % Do some more stuff                                  i never changes.
end
```

Note that with a **WHILE** statement, the statements between the **while** and **end** keywords must modify the variables in the condition so that the condition becomes false some time in the future. If this does not happen, the program produces an infinite loop.

Here are a few short examples.

```
fprintf('Testing the WHILE loop structure.\n\n');
i=1;
while i <= 4
    fprintf('Testing the WHILE loop. i = %g\n',i);
    i=i+1;
end
fprintf('\nEnd of WHILE loop. i=%g.\n', i);


Testing the WHILE loop structure.

Testing the WHILE loop. i = 1
Testing the WHILE loop. i = 2
Testing the WHILE loop. i = 3
Testing the WHILE loop. i = 4

End of WHILE loop. i=5.
```

Note that when the loop is complete, the value of i is 5. Suppose we want the loop to exit with i equal to the last value it displayed with the **FPRINTF** function inside the loop. That is, we would like the same output as above, but i will have the value of 4 when we are done with the **WHILE** loop. This can be done by changing the place we increment i.

```
fprintf('Testing the WHILE loop structure.\n\n');
i= 0;                                          These lines are changed.
while i <= 4
    i=i+1;
```

```
    fprintf('Testing the WHILE loop. i = %g\n',i);
end
fprintf('\nEnd of WHILE loop. i=%g.\n', i);
Testing the WHILE loop structure.

Testing the WHILE loop. i = 1
Testing the WHILE loop. i = 2
Testing the WHILE loop. i = 3
Testing the WHILE loop. i = 4
Testing the WHILE loop. i = 5

End of WHILE loop. i=5.
```

Now the value of *i* has the same value when it exits the loop as it had during the last **FPRINTF** function inside the loop. However, we see that this loop counts to 5 rather than 4. This is because when *i* is set to 4, the condition $i \leq 4$ is still true, so the loop takes another pass through. To make this loop count to 4, we must change the condition to $i < 4$.

```
fprintf('Testing the WHILE loop structure.\n\n');
i=0;
while i < 4
   i=i+1;
   fprintf('Testing the WHILE loop. i = %g\n',i);
end
fprintf('\nEnd of WHILE loop. i=%g.\n', i);
Testing the WHILE loop structure.

Testing the WHILE loop. i = 1
Testing the WHILE loop. i = 2
Testing the WHILE loop. i = 3
Testing the WHILE loop. i = 4

End of WHILE loop. i=4.
```

Let's look at a few example programs that use a **WHILE** loop. For the first example, we modify the guessing game of Program 2-16 to ask users if they wish to play the game again. If the user types **y**, then a new game starts. If the user types anything else, the game terminates.

Program 2-26 Guessing game using a **WHILE** loop

```
%
% This is file guess0.m
%

again = 'y';
```

```
while again == 'y'
   real_number=rand(1); % Generate a number between 0 and 1
   real_number=10*real_number; % Scale the number between 0 and 10
   number=ceil(real_number); % Round the number to the next highest integer.

   % Ask the user to guess the number:

   fprintf('I am thinking of a number between 1 and 10.\n');
   fprintf('You have three guesses.\n');
   correct=0;

   for i = 1:3
      if ~correct
         guess = input('Guess a number: ');

         if guess == number
            fprintf('Congratulations - You guessed the correct number.\n');
            correct=1;
         else
            if guess > number
                fprintf('Your guess was too high.\n');
            else
                fprintf('Your guess was too low.\n');
            end
         end
      end
   end

   if ~ correct
      fprintf('Sorry - The number I was thinking of was %g.\n\n', number);
   end

   again=input('Do you wish to play again? ','s');
end %of while loop
```

The entire guessing program of Program 2-16 is now enclosed within the **WHILE** loop:

```
%
% This is file guess0.m
%

again = 'y';
while again == 'y'
      .

      .

   Program statements from Program 2-16

      .

      .

   again=input('Do you wish to play again? ','s');
end %of while loop
```

The program initializes the variable *again* to the value that makes the condition of the **WHILE** loop true. The program then executes the statements of the game. When one session of the game is finished, the computer asks users if they wish to play again, requests input, and then jumps to the **WHILE** statement at the beginning of the program where it tests the condition again. Let's test the program.

```
EDU» guess0
I am thinking of a number between 1 and 10.
You have three guesses.

Guess a number: 5
Your guess was too low.

Guess a number: 7
Your guess was too high.

Guess a number: 6
Congratulations - You guessed the correct number.

Do you wish to play again? y
I am thinking of a number between 1 and 10.
You have three guesses.

Guess a number: 5
Your guess was too low.

Guess a number: 8
Your guess was too low.

Guess a number: 10
Your guess was too high.
Sorry - The number I was thinking of was 9.


Do you wish to play again? n
EDU»
```

The program appears to work correctly. However, if a user makes a mistake when responding to the question, the program could terminate inadvertently:

```
EDU» guess0
I am thinking of a number between 1 and 10.
You have three guesses.

Guess a number: 5
Your guess was too high.
```

```
Guess a number: 3
Your guess was too high.

Guess a number: 2
Your guess was too high.
Sorry - The number I was thinking of was 1.

Do you wish to play again? yes
EDU» guess0
I am thinking of a number between 1 and 10.
You have three guesses.

Guess a number: 5
Your guess was too high.

Guess a number: 3
Your guess was too high.

Guess a number: 1
Congratulations - You guessed the correct number.

Do you wish to play again?
EDU»
```

A longer response terminates the program.

No response terminates the program.

As you can see, if the user enters anything other than **y**, the program terminates. Now let's modify the program to test for user inputs **y** and **n**. A **y** causes the program to start another game, and an **n** terminates the program. Any other response causes the program to emit an error message.

Program 2-27 Modified guessing program

```
%
% This is file guess1.m
%

again = 'y';
while again == 'y'
    real_number=rand(1); % Generate a number between 0 and 1
    real_number=10*real_number; % Scale the number between 0 and 10
    number=ceil(real_number); % Round the number to the next highest integer.

    % Ask the user to guess the number:

    fprintf('I am thinking of a number between 1 and 10.\n');
    fprintf('You have three guesses.\n');
    correct=0;
```

```
for i = 1:3
    if ~correct
        guess = input('Guess a number: ');

        if guess == number
            fprintf('Congratulations - You guessed the correct number.\n');
            correct=1;
        else
            if guess > number
                fprintf('Your guess was too high.\n');
            else
                fprintf('Your guess was too low.\n');
            end
        end
    end
end

if ~ correct
    fprintf('Sorry - The number I was thinking of was %g.\n\n', number);
end

correct_response = 0;
while ~correct_response
    again=input('Do you wish to play again? (y/n) ','s');
    if (again ~= 'y') & (again ~= 'n')
        fprintf('Error - Enter either ''y'' or ''n''.\n');
    else
        correct_response = 1;
    end
end  %of while loop 2
end %of while loop1
```

Look at the added statements at the end of Program 2-27. Once again, Program 2-16 is contained within a **WHILE** loop:

```
%
% Program 2-20
%

again = 'y';
while again == 'y'

        .

        .

    Program 2-16

        .

        .

correct_response = 0;
```

```
    while ~correct_response
       again=input('Do you wish to play again? (y/n) ','s');
       if (again ~= 'y') & (again ~= 'n')
          fprintf('Error - Enter either ''y'' or ''n''.\n');
       else
          correct_response = 1;
       end
    end  %of while loop 2

end %of while loop1
```

The second **WHILE** loop is added inside the main loop of the program. This loop asks for user input and loops until the user responds with a valid response. If the user enters an invalid response, an error message is generated and the question is repeated. The inner loop is only terminated when a valid response is entered.

```
EDU» guess1
I am thinking of a number between 1 and 10.
You have three guesses.

Guess a number: 5
Your guess was too high.

Guess a number: 3
Your guess was too high.

Guess a number: 2
Your guess was too high.
Sorry - The number I was thinking of was 1.


Do you wish to play again? (y/n) hjk
Error - Enter either 'y' or 'n'.

Do you wish to play again? (y/n) mlsp
Error - Enter either 'y' or 'n'.

Do you wish to play again? (y/n)
Error - Enter either 'y' or 'n'.

Do you wish to play again? (y/n) bad
Error - Enter either 'y' or 'n'.

Do you wish to play again? (y/n) n
EDU»
```

Upon further testing, we find that any string that contains a **y** or **n** is treated by the inner **WHILE** loop as a valid response:

```
EDU» guess1
I am thinking of a number between 1 and 10.
You have three guesses.

Guess a number: 5
Your guess was too low.

Guess a number: 8
Your guess was too high.

Guess a number: 7
Congratulations - You guessed the correct number.

Do you wish to play again? (y/n) xyz
EDU» guess1
I am thinking of a number between 1 and 10.
You have three guesses.

Guess a number: 5
Congratulations - You guessed the correct number.

Do you wish to play again? (y/n) name
EDU»
```

This string contains the letter **y**.

This string contains the letter **n**.

With further testing, we conclude that any string that contains a **y** or **n** is treated as a valid response by the inner **WHILE** loop. Note that even though the inner **WHILE** loop exits because one character in the string is a **y** or **n**, the main **WHILE** loop sees that a string such as **xyz** is not equal to the string **y** and terminates. Thus, the outer **WHILE** loop works correctly, but the inner **WHILE** loop does not.

The inner loop fails because of the array nature of MATLAB. When MATLAB makes the check **if again ~= 'y'**, it checks for all characters in the string. The test is only true if all characters in the string are not equal to **y**. Suppose the user enters the text string **yes**. In this case, *again*(1) = y, *again*(2) = e, and *again*(3) = s. In MATLAB, the statements **if again ~= 'y'** and **if (again(1) ~= 'y') & (again(2) ~= 'y') & (again(3) ~= 'y')** are equivalent for the example *again* = yes.

Both versions are only true if all characters in the string are not equal to **y**. Sometimes the array nature of MATLAB makes programming easier. Certainly the command **if again ~= 'y'** is easier to use than **if (again(1) ~= 'y') & (again(2) ~= 'y') & (again(3) ~= 'y')**, especially if a string is long. However, as in this example, the statement may not work as planned.

To fix the program, we'll have the inner loop check whether the response from the user is one character. If the response is longer than one character, the response could not have been **y** or **n**. Program 2-28 shows how we code this.

Program 2-28 Fix of Program 2-27

```
%
% This is file guess2.m
%

again = 'y';
while again == 'y'
    real_number=rand(1); % Generate a number between 0 and 1
    real_number=10*real_number; % Scale the number between 0 and 10
    number=ceil(real_number); % Round the number to the next highest integer.

    % Ask the user to guess the number:

    fprintf('I am thinking of a number between 1 and 10.\n');
    fprintf('You have three guesses.\n');
    correct=0;

    for i = 1:3
        if ~correct
            guess = input('Guess a number: ');

            if guess == number
                fprintf('Congratulations - You guessed the correct number.\n');
                correct=1;
            else
                if guess > number
                    fprintf('Your guess was too high.\n');
                else
                    fprintf('Your guess was too low.\n');
                end
            end
        end
    end

    if ~ correct
        fprintf('Sorry - The number I was thinking of was %g.\n\n', number);
    end

    correct_response = 0;
    while ~correct_response
        again=input('Do you wish to play again? (y/n) ','s');
        if ((again ~= 'y') & (again ~= 'n')) | (length(again) ~=  1)
            fprintf('Error - Enter either ''y'' or ''n''.\n');
        else
            correct_response = 1;
        end
    end  %of while loop 2
end %of while loop1
```

This line is modified.

One statement has been modified:

```
if ((again ~= 'y') & (again ~= 'n')) | (length(again) ~= 1)
```

This line now generates an error message if the response does not contain a **y** or an **n**, or if the response is not one character in length.

```
EDU» guess2
I am thinking of a number between 1 and 10.
You have three guesses.

Guess a number: 5
Your guess was too low.

Guess a number: 8
Congratulations - You guessed the correct number.

Do you wish to play again? (y/n)
Error - Enter either 'y' or 'n'.

Do you wish to play again? (y/n) yes
Error - Enter either 'y' or 'n'.

Do you wish to play again? (y/n) no
Error - Enter either 'y' or 'n'.

Do you wish to play again? (y/n) hgfkdhgf
Error - Enter either 'y' or 'n'.

Do you wish to play again? (y/n) hfuei
Error - Enter either 'y' or 'n'.

Do you wish to play again? (y/n) n
EDU»
```

> No response. The user only pressed the ENTER key.

The program appears to work correctly.

MATLAB Built-In Function

The MATLAB function **STRCMP** is the best way to check whether two strings are equal. This function returns a 1 if the strings are the same length and have the same characters in the string and a 0 if the strings are a different length or the characters in the strings are different. Some examples are given here:

```
EDU» strcmp('y','yes')
ans =
     0
```

```
EDU» strcmp('y','xx')
ans =
     0
EDU» strcmp('y','y')
ans =
     1
EDU» strcmp('y','n')
ans =
     0
EDU» strcmp('yes','yes')
ans =
     1
EDU» strcmp('yes','sey')
ans =
     0
EDU»
```

EXERCISE 2-7 Change Program 2-28 so that it uses the **STRCMP** function in the inner **WHILE** loop.

As another example of a **WHILE** loop, let's create a program to find the smallest integer N such that $\sum_{i=1}^{N} i$ is greater than a specified limit. For example, find N if the limit is 5. The answer is $N = 3$ since $1 + 2 + 3 = 6$. If the limit is 15, then N is 6 since $1 + 2 + 3 + 4 + 5 + 6 = 21$. Note that N is not equal to 5 because $1 + 2 + 3 + 4 + 5 = 15$, and the sum must be greater than the limit.

We'll use two **WHILE** loops in this program. The first **WHILE** loop asks the user for the value of the limit and checks to see that a valid response is entered. The second **WHILE** loop finds the sum.

Program 2-29 Find the sum < limit

```
% This is file limit1.m
%
% This program finds the largest value of N such that the sum of
% 1, 2, 3, ..., N-1, N is greater than a specified limit. The user
% specifies the limit. The limit must be a real number greater than or equal
% to 1.
%

% Request a number from the user and check to see if it is a valid number.
good_input = 0;
```

```
while ~good_input
    limit = input('Specify the limit for the sum: ');
    if limit < 1
        fprintf('Error, you must specify a real number greater than 1.\n');
    elseif ~isreal(limit)
        fprintf('Error, you must specify a real number.\n');
    else
        good_input = 1;
    end
end

% Now that we have a good number from the user, find N.

N = 0;
sum = 0;

while sum <= limit
    sum=0;
    N=N+1;
    for i = 1:N
        sum=sum+i;
    end
end

fprintf('N = %g, The sum of 1...N is %g\n\n',N,sum);
```

Program 2-29 has two parts. The first part asks the user for a number. This part is implemented as a loop that repeats indefinitely until the user enters a valid number for the limit. This code segment is shown here:

```
good_input = 0;
while ~good_input
    limit = input('Specify the limit for the sum: ');
    if limit < 1
        fprintf('Error, you must specify a real number greater than 1.\n');
    elseif ~isreal(limit)
        fprintf('Error, you must specify a real number.\n');
    else
        good_input = 1;
    end
end
```

This loop repeats until the user inputs a real number that is greater than or equal to 1. Here is an example run of the code segment:

```
Specify the limit for the sum: -5
Error, you must specify a real number greater than 1.
```

```
Specify the limit for the sum: 3+4i
Error, you must specify a real number.
Specify the limit for the sum: 5.1
EDU» limit
limit =
    5.1000
EDU»
```

The code for finding *N* is

```
N = 0;
sum = 0;

while sum <= limit
    sum=0;
    N=N+1;
    for i = 1:N
        sum=sum+i;
    end
end
```

The algorithm for finding *N* is fairly straightforward. First, *N* is set to 1, and the sum of 1 to 1 is calculated. If the sum is less than or equal to the limit, *N* is set to 2 and the sum from 1 to 2 is calculated. If the sum is less than or equal to the limit, *N* is set to 3 and the sum from 1 to 3 is calculated. This process is repeated until a value of *N* is found where the sum is greater than the limit. To illustrate the process, we'll display *N* and the sum for each value of *N*.

Let's modify Program 2-29 slightly to display *N* and the sum of 1 to *N* for each value of *N* it tests.

Program 2-30 Modification of Program 2-29 to display intermediate results

```
% This is file limit2.m
%
% This program finds the largest value of N such that the sum of
% 1, 2, 3, ..., N-1, N is greater than a specified limit. The user
% specifies the limit. The limit must be a real number greater than or equal
% to 1.
%

good_input = 0;
while ~good_input
    limit = input('Specify the limit for the sum: ');
    if limit < 1
        fprintf('Error, you must specify a real number greater than 1.\n');
```

```
        elseif ~isreal(limit)
            fprintf('Error, you must specify a real number.\n');
        else
            good_input = 1;
        end
end


% Find N

N = 0;
sum = 0;

fprintf('\n\n    N          Sum of 1...N\n');
fprintf('    —          ————\n');
while sum <= limit
    sum=0;
    N=N+1;
    for i = 1:N
        sum=sum+i;
    end
    fprintf('   %4.0f            %4.0f\n', N, sum);
end
fprintf('\n\nN = %g, The sum of 1...N is %g\n\n',N,sum);
```

Program 2-30 is the same as Program 2-29 except for a few additions to display intermediate results. Let's run the program. We won't test the error detection code of the first **WHILE** loop since we tested that code segment earlier.

```
EDU» limit2

Specify the limit for the sum: 5


            N          Sum of 1...N
        ----           --------------
            1               1
            2               3
            3               6



        N = 3, The sum of 1...N is 6
EDU» limit2

Specify the limit for the sum: 10
```

```
        N              Sum of 1...N
      ----            --------------
        1                   1
        2                   3
        3                   6
        4                  10
        5                  15


N = 5, The sum of 1...N is 15

EDU» limit2

Specify the limit for the sum: 25


        N              Sum of 1..N
      ----            --------------
        1                   1
        2                   3
        3                   6
        4                  10
        5                  15
        6                  21
        7                  28


N = 7, The sum of 1...N is 28

EDU»
```

As a final example of a **WHILE** loop, we'll modify the command center of Program 2-9 so that it continues asking for commands until the user enters the command to exit the command center. The program has been modified so that valid responses include **E** for exit.

Program 2-31 Command center modified with **WHILE** loop

```
%
% This is file command4.m
%
% This program asks for a number between 1 and 5 inclusive.
% The program then performs the tasks assigned to the number.
% The tasks are not specified for this program.

exit=0;
```

```
while ~exit
    command=-99;
    str_cmd=input('Specify your command [1-5,E]: ', 's');

    if strcmp(str_cmd,'E')
        exit=1;
    else
        command=str2num(str_cmd);
    end

    if isempty(command)
        command = -99;
    end

    if command==1
        fprintf('You chose option 1.\n');
        % Command for option 1 entered here.
    elseif command==2
        fprintf('You chose option 2.\n');
        % Command for option 2 entered here.
    elseif command==3
        fprintf('You chose option 3.\n');
        % Command for option 3 entered here.
    elseif command==4
        fprintf('You chose option 4.\n');
        % Command for option 4 entered here.
    elseif command==5
        fprintf('You chose option 5.\n');
        % Command for option 5 entered here.
    else
        if exit ~= 1
            fprintf('Error - You must enter an integer from 1 to 5, or ➡
the letter E.\n');
        end
    end
end
fprintf('Good Bye.\n');
```

Let's first look at the input statement, **str_cmd=input('Specify your command [1-5,E]: ', 's')**. This statement requests input as a string because the response of **E** is a valid input. Thus, to allow responses of **1**, **2**, **3**, **4**, **5**, and **E**, we must request input as a string, and then convert the string to a numerical value when needed. To convert a text string such as **'123'** to a number 123, we use the **STR2NUM** function.

Before we discuss the operation of Program 2-31, let's examine the MATLAB **STR2NUM** function. This function converts a text string into a number. We need to examine a few special cases so that we can handle erroneous input.

```
EDU» str2num('5')
ans =
     5
EDU» str2num('555')
ans =
   555
EDU» str2num('55.55')
ans =
   55.5500
EDU» str2num('55.5e5')
ans =
     5550000
EDU» str2num('3+5i')
ans =
   3.0000 + 5.0000i
```

As you can see, **STR2NUM** handles integers, complex numbers, numbers with decimal places, and numbers in scientific notation. The next question is, What does **STR2NUM** do for an invalid string?

```
EDU» str2num('123ab')
ans =
     []
EDU» str2num('')
ans =
     []
```

Entering an invalid string or no string at all results in **STR2NUM** returning a null array.
    The code segment

```
str_cmd=input('Specify your command [1-5,E]: ', 's');
if strcmp(str_cmd,'E')
        exit=1;
else
        command=str2num(str_cmd);
end
```

requests input as a string. It then tests that input using the **STRCMP** function to see if the input string was **E**. If so, variable *exit* is set to 1. That is, *exit* is the flag that signals that the user wishes to exit the program. If the input string is not **E**, then the string is converted to a number using the **STR2NUM** function.
    The code segment

```
if isempty(command)
    command = -99;
end
```

checks the value returned by the **STR2NUM** function. When the **STR2NUM** function en-counters an invalid text string, it returns the empty set or null set **[ ]** as a value. Thus, in some cases, the value of *command* will be **[ ]**. MATLAB does not like making com-parisons using the null set:

```
EDU» x=[ ];
EDU» if x==[ ] fprintf('What happened here.\n');end
Warning: X == [] is technically incorrect. Use isempty(X) in-
stead.
What happened here.
EDU»
```

Note that MATLAB performs the **IF** statement properly but it also generates an error mes-sage. To check whether a number is equal to the empty set, use the **ISEMPTY**(*x*) com-mand. This command returns a 1 if *x* is equal to the empty set, and returns a 0 if *x* is not equal to the empty set.

```
EDU» x=[ ];
EDU» if isempty(x) fprintf('What happened here.\n');end
What happened here.
EDU»
```

MATLAB does not generate a warning message here. Since the **STR2NUM** function re-turns an empty set as a value, the code segment

```
    if isempty(command)
        command = -99;
    end
```

is added to the program to check whether *command* is equal to the empty set and to change its value to –99. Note that the **IF-ELSEIF-ELSE-END** command used later in the program will generate warning messages if the condition in the **IF** or **ELSEIF** statement uses a variable whose value is the empty set. Changing the value to –99 avoids this warning message.

Now let's examine the remainder of Program 2-31. As you can see, a large **WHILE** loop encloses all of the statements inside the **WHILE** loop.

```
%
% This is file command4.m
%
% This program asks for a number between 1 and 5 inclusive.
% The program then performs the tasks assigned to the number.
% The tasks are not specified for this program.

exit=0;
while ~exit
    ..
```

```
Commands to execute.
  ..
end
fprintf('Good Bye.\n');
```

This **WHILE** loop will loop indefinitely until variable *exit* is set to 1. Thus, the command center continues to request commands until the user enters the command that causes variable *exit* to be set to 1.

The first few lines inside the main **WHILE** loop request user input and check to see whether the user entered the letter **E** or a null string (nothing).

```
exit=0;
while ~exit
  command=-99;
  str_cmd=input('Specify your command [1-5,E]: ', 's');
  if strcmp(str_cmd,'E')
        exit=1;
  else
        command=str2num(str_cmd);
  end
      if isempty(command)
        command = -99;
  end


  ...
  ... Remainder of Program.
  ...
end
fprintf('Good Bye.\n');
```

This portion of the program shows the **WHILE** loop that repeats until the user enters the command to exit.

The remainder of the program either executes the specified command (if the user entered **1**, **2**, **3**, **4**, or **5**), does nothing if the user entered **E**, or generates an error message:

```
if command==1
      fprintf('You chose option 1.\n');
      % Command for option 1 entered here.
elseif command==2
      fprintf('You chose option 2.\n');
      % Command for option 2 entered here.
elseif command==3
      fprintf('You chose option 3.\n');
      % Command for option 3 entered here.
elseif command==4
      fprintf('You chose option 4.\n');
```

```
        % Command for option 4 entered here.
elseif command==5
        fprintf('You chose option 5.\n');
        % Command for option 5 entered here.
else
        if exit ~= 1
        fprintf('Error - You must enter an integer from 1 to 5, or the ➡
letter E.\n');
        end
end
```

Let's run the program.

EDU» **command4.m**

```
Specify your command [1-5,E]: HFJJH
Error - You must enter an integer from 1 to 5, or the letter E.

Specify your command [1-5,E]:  ◄
Error - You must enter an integer from 1 to 5, or the letter E.

Specify your command [1-5,E]: 1
You chose option 1.
```

> User pressed the ENTER key as a response.

```
Specify your command [1-5,E]: 2
You chose option 2.

Specify your command [1-5,E]: 3
You chose option 3.

Specify your command [1-5,E]: 4
You chose option 4.

Specify your command [1-5,E]: 5
You chose option 5.

Specify your command [1-5,E]: e
Error - You must enter an integer from 1 to 5, or the letter E.

Specify your command [1-5,E]: E
Good Bye.
EDU»
```

EXERCISE 2-8   Modify Program 2-31 so that it exits the program if the user enters **E**, **e**, **Exit**, **exit**, or **EXIT**.

2.7 Problems

Problem 2-1   Write a quiz program that asks `What is the unit of power?` The program should display `Your answer is correct` if the user inputs `watt` and should display `No, a watt` otherwise. Call this program "xxx_3_1.m" where xxx are your initials.

Problem 2-2   A particle, initially at rest, is subjected to an applied force F whose direction is constant but whose magnitude F_m changes according to the relation

$$F = F_m \left(1 - \frac{t-T}{T}\right)^2$$

for time t between 0 and T. For $t \geq T$, $F = F_m$. Write a script file that requests the values of F_m, T, and t as inputs, and displays the value of F.

Problem 2-3   Write a program that calculates how much to tip a waitperson at a restaurant. The tip should be 18% of the check with a minimum of $2. Your script file should request the amount of the check from the user.

Problem 2-4   Write a program to calculate how much to tip a waitperson based on quality of service. The script file should ask for the amount of the check and whether the service was good, fair, or poor. If the service was good, the tip should be 15% with a minimum tip of $2. If the service was fair, the tip should be 10% with a minimum of $1. If the tip was poor, the tip should be 1 cent.

Problem 2-5   Write a script file that requests a word as input and translates the word into pig Latin. The rules for pig Latin are

- If the word begins with a consonant, move the first letter to the end of the word and add the letters *ay*. Thus, *frog* becomes *rogfay*.
- If the word begins with a vowel, add the text *way* to the end of the word. Thus, *add* becomes *addway*.

Problem 2-6   The National Weather Service uses the following amounts of cloud cover to specify weather conditions. 0%–28% is "clear"; 29%–71% is "partly cloudy"; 72%–99% is "cloudy"; 100% is "overcast." Write a script file that requests the percentage of cloud cover and displays the National Weather Service's weather conditions.

Problem 2-7   Write a program that requests three exam scores as input and displays the average of the two highest scores. Call this program "xxx_3_2.m" where xxx are your initials.

Problem 2-8   Write a program to find the first odd integer whose cube is greater than 3000.

Problem 2-9   Write a program that asks the user to enter positive real numbers into the computer until the product of the numbers is greater than 500. The program should check for invalid input: The user should not be able to enter strings, negative numbers, complex numbers, or arrays. When a total of 500 is exceeded, the program should display the product of the numbers, how many numbers were entered, and list the numbers. Use **FPRINTF** commands to display results and make the output look nice.

Problem 2-10 Write a program that generates three random numbers, each between 0 and 9. The three numbers are displayed, and the message "Lucky 7" is displayed if two or more of the digits are 7. Call this program "xxx_3_3.m" where xxx are your initials.

Problem 2-11 Run the program you wrote in Problem 2-10 10,000 times and calculate the odds of winning.

Problem 2-12 Labor law requires employees who are paid hourly to be paid "time-and-a-half" for hours worked in over 40 hours in a week. For example, if a person's hourly wage is $8.50 and they work 63 hours in one week, their gross pay should be

$$(40 \cdot 8.5) + [1.5 \cdot 8.5 \cdot (63 - 40)] = \$633.25$$

Write a program that requests an employee's hours and hourly wage, and then displays their gross pay.

Problem 2-13 Write a program that requests a number from 1 to 80 with the **INPUT** function and displays a row of that many asterisks.

Problem 2-14 Write a program that requests a number from 1 to 80 with the **INPUT** function and displays a filled square having that number of asterisks on each side.

Problem 2-15 Write a program that requests a number from 1 to 80 with the **INPUT** function and displays the outline of a square having that number of asterisks on each side.

Problem 2-16 Write a program that requests two positive integers greater than 3 (with error checking). The numbers represent the x- and y-dimensions of a rectangle. Your program should then display a rectangle of asterisks (*) with the specified dimensions. The rectangle with dimensions $x = 5$ and $y = 7$ would be displayed as

```
* * * * *
*       *
*       *
*       *
*       *
*       *
* * * * *
```

Problem 2-17 Write a program that requests a text string from the user. Count the number of times the letter t appears in the text string and display the result.

Problem 2-18 Request a number θ with the **INPUT** function. Check that the number is $0 < \theta < 1$. Find $\sum_{i=0}^{1000} \theta^i$. Verify that this sum is equal to $\dfrac{1}{(1-\theta)}$ where θ is the number entered by the user. Display the results of both calculation methods.

Problem 2-19   Write a program that asks the user to input a distance in miles and then converts the distance to kilometers. The program should display the distance in kilometers. The process should be repeated until the user enters the text string 'quit'.

Problem 2-20   Write a program that requests a positive integer as input and then follows this procedure: If the number is even, divide it by 2. If not, multiply the number by 3 and then add 1. Repeat this process with the resulting number and continue repeating until the number 1 is reached. After the number 1 is reached, the program should display the number of iterations required to reach the number 1. When the program requests input, it should check that the user inputs a positive integer greater than 1. If not, an error message should be generated.

Problem 2-21   Write a program that finds all decimal numbers between 1 and 10,000 that are palindromes in both decimal and binary notations. Examples are

Decimal	Binary
33	100001
99	1100011
313	100111001
585	1001001001
717	1011001101
7447	1110100010111

You can use the MATLAB functions **NUM2STR** and **DEC2BIN**.

Problem 2-22   Write a program that requests a list of scores (number of scores is unknown) and then displays the two highest scores.

Problem 2-23   Write a program that calculates the fewest bills needed to make up any sum of money for amounts of $1 to $1000. Available bills are $1, $5, $10, $20, $50, and $100. Your program should display the list of bills used to make up the sum.

Problem 2-24   Write a program that reads a text string such as +20-4+8-9 and evaluates the expression numerically. You may use the **STR2NUM** function, but do not use the **EVAL** function.

Problem 2-25   Write a program that reads a text string such as +20-4+8-9 and evaluates the expression numerically. Solve the problem using the **EVAL** function.

Problem 2-26   Using a Taylor series expansion, the function e^x can be approximated by the series

$$\sum_{n=0}^{\infty} \frac{x^n}{n!}$$ for values of x close to zero. Write a program that determines how many terms in the sum are required so that the difference between e^x and the sum is less than 10^{-6}, for values of x equal to 0.1, 0.01, and 0.001. Your program should display a table of x, e^x, the sum, and the difference between e^x and the sum.

for   sum = 0
       j = 1: n
        sum = sum + fact (j)

Problem 2-27 Write a program that requests a real number as input and then requests an integer called *i*. The program then tells the user what the *i*th digit of the number is. If the number does not have *i* digits, the program should emit an error message.

Problem 2-28 Write a program that requests a real number as input and displays the number of digits in the number and whether or not the number contains a decimal point. Examples are given below. The name of the script file is meh_hw.m.

```
EDU>> meh_hw
Specify a real number: 278.789
The number you specified has 6 digits and a decimal point.
EDU>> meh_hw
Specify a real number: 27
The number you specified has 2 digits with no decimal point.
EDU>>
```

Problem 2-29 Write a program that reads and prints the average of numbers input by a user. You do not know beforehand how many numbers will be entered. Your program should follow this procedure.

    a. Print the statement "Specify a number."

    b. Request input from the user.

    c. If the user specifies a number other than **NaN**, add the number to the array storing the numbers, and then go to step (a) to request another number. If the user enters **NaN**, print the statement "End of Data Input." and go to step (d).

    d. Calculate the average of the numbers.

    e. Print out the following statements:

```
You entered n numbers.
The numbers you entered were: _, _, _, ..., _.
The average of the numbers is x.
```

An example session follows. The name of the script file that contains the example solution is called prob2.m.

```
EDU» prob2
Marc E. Herniter
```

[ ] combin string

```
Specify a number:
5
Specify a number:
6
Specify a number:
7
Specify a number:
NaN
```

```
End of Data Input.

You entered 3 numbers.
The numbers you entered were: 5, 6, 7.
The average of the numbers is 6.


Good bye.

EDU»
```

Your program should work for the special case when the user does not enter any numbers.

```
EDU» prob2
EGR 222 Homework # 3 Solution.



Specify a number:
NaN

End of Data Input.

You entered 0 numbers.
No numbers were entered. Average cannot be calculated.


Good bye.

EDU»
```

Problem 2-30 Write a script file in MATLAB to solve the following problem. Name the file xxx_230.m, where xxx are your initials.

Write a script file that requests a number as input and displays the following information:

a. Displays the number of numeric characters in the number.
b. States whether the number is positive, negative, or complex.
c. Displays the number of decimal points in the number.
d. If the number is not complex, states whether the number is an integer.
e. If the number is complex, displays the number as a magnitude and a phase in degrees.
f. If the number is complex, determines whether the user used the letter i or j in the number. If the user used the letter i, the program should emit the statement, You used the letter i to write a complex number. You must be a geek from the

physics department. If the user used the letter j, the program should emit the statement, `You used the letter j to write a complex number. You must be a cool dude from Engineering.`

You should read the number as a string and convert it to a number if necessary. Complex numbers can only be entered in rectangular coordinates. Your script should not check for invalid input from the user. Some examples are shown here.

```
EDU>> meh_224
Specify a number: 278.789
The number of numeric characters is 6.
The number is a positive real number.
The number of decimal points in the number is 1.
The number is not an integer.
EDU>> meh_224
Specify a real number: -27
The number of numeric characters is 2.
The number is a negative real number.
The number of decimal points in the number is 0.
The number is an integer.
EDU>> meh_224
Specify a real number: -27.62+71.3i
The number of numeric characters is 7.
The number is complex.
The number of decimal points in the number is 2.
The magnitude is 76.4628 the phase in degrees is 111.1752.
You used the letter i to write a complex number. You must be a
geek from the physics department.
EDU>>
```

Problem 2-31 Write a program that reads and prints the average of numbers input by a user. You do not know beforehand how many numbers will be entered. Your program should follow the procedure outlined below:

a. Print the statement `Do you wish to enter a number?`
b. If the user enters **n** or **no**, go to step (f).
c. Print the statement `Specify a number.`
d. Request input from the user.
e. If the user enters a real number, add the number to the array storing the numbers and then go to step (a). If the user enters anything other than a real number, ignore the entry, display the statement `Invalid Number Specified - Entry Ignored`, and then go to step (a).
f. Find the average, minimum, and maximum of the numbers entered.
g. Print out the following statements:
    `You entered n numbers.`
    `The numbers you entered were: _, _, _, ..., _.`

```
            The average of the numbers is x.
            The maximum number you entered was y.
            The minimum number you entered was z.
```

An example session is shown here:

```
EDU» meh_hw3


Do you wish to enter a number? n
No numbers were entered.



Good bye.

EDU» meh_hw3


Do you wish to enter a number? y
Specify a number: 5
Do you wish to enter a number? n

You entered 1 number.
The number you entered was: 5.
The average of the numbers is 5.
The maximum number you entered was 5.
The minimum number you entered was 5.



Good bye.

EDU» meh_hw3


Do you wish to enter a number? y
Specify a number: 1
Do you wish to enter a number? y
Specify a number: 3
Do you wish to enter a number? y
Specify a number: 5
Do you wish to enter a number? y
Specify a number: 7
Do you wish to enter a number? y
Specify a number: 9
Do you wish to enter a number? y
Specify a number: 11
Do you wish to enter a number? y
Specify a number: 13
Do you wish to enter a number? no
```

```
You entered 7 numbers.
The numbers you entered were: 1, 3, 5, 7, 9, 11, 13.
The average of the numbers is 7.
The maximum number you entered was 13.
The minimum number you entered was 1.
Good bye.
EDU»
```

Problem 2-32   Write a program that requests a numerical value of resistance from the user and then requests the unit for the value just input (for example, ohms, kilo-ohms, mega-ohms, and
so on). The program then converts the resistance to ohms and displays the value followed by the unit (which is now ohms). Your program should work for all possible vari-
ants of the multiplier. For example, **K, k, kohms, Kohms, kilo-ohms**, should be
treated the same. Your program should treat upper- and lowercase letters the same. Use
the **SWITCH** statement.

Problem 2-33   Write a program that finds all perfect numbers between 2 and n, where n is an integer
entered by a user. A perfect number is one whose prime factors add to the number. The
number itself is not included in the sum. An example is 6, whose factors are 1, 2, and 3.

Functions

Functions are subprograms that perform tasks that you use frequently. An example would be a function that asks the user a question, waits for a response, checks the validity of the response, emits an error message if necessary, and returns the user response if a valid response is entered. In a large program, you may want to ask the user several questions. Instead of repeating the same block of code over and over, you can write a function to perform the task and then call the function whenever you need to ask a question. This has several benefits: (1) You only write the function once. (2) When you debug the function and it works correctly, it works correctly no matter how many times you use it. (3) You can easily use the function in other programs, and other programmers can also use your function. MATLAB has a plethora of built-in functions, and we will show you how to use these functions and how to create your own functions.

3.1 General Structure of a Function

MATLAB Predefined
LOOKFOR
command
HELP function
INPUT function
WHILE loop
FOR loop
STRCMP function

Functions are MATLAB subprograms that can be used to

1. Break up a large program into several smaller programs.
2. Perform a task that is used frequently. Instead of repeating the same code in several places, the function is simply called.

   The general form of a function is

```
FUNCTION y = function_name ( input arguments )
% Comments for help using the function.
%
       .
       .
       .
    statements
       .
       .
       .
y = something
```
◄——————————————————⎤ Set the return value of *y* here. ⎦

The word **function** is a MATLAB keyword and must be used as shown. The variable *y* is the value the function returns. The output *y* can have a single value (which may even be a "don't care") or several values. The function name is the name by which you call the function in MATLAB, and the file name is where you store the function. The function and file names by convention are named the same. Thus,

```
function y = fact(n)
```

is a function designed to calculate the factorial of an integer *n*. The name of the function is fact. The function is saved in a file called fact.m.

A function can have no inputs, a single input, or as many inputs as you like. Function fact(n) has a single input. An example of a function with three inputs is

```
function q=zeta(a,b,c)
```

The name of the function is zeta. It returns a single variable *q*. It has three input variables, *a*, *b*, and *c*. Note that *a*, *b*, and *c* can be individual numbers (scalars) or arrays. Thus, even though this function only has three input variables (*a*, *b*, and *c*), the variables themselves could contain several values in the form of arrays.

Function **zeta(a,b,c)** returns a single variable *q*. Note that *q* can be a scalar or an array. In general, functions can return many variables. For example,

```
function [x,y,z]=beta(a,b,c,d)
```

This function has four input variables and returns three variables. Again, variables *x*, *y*, and *z* can be scalars (a single value), or they can be arrays. Thus, even though this function returns only three variables, the variables can each contain many numerical values in the form of arrays.

Note in the description of functions above that the last line of the function sets the value of *y*.

The first line of the function declares *y* as the variable it will return. The last line sets the value of *y*. If you omit the last line and never specify a value for *y*, the function returns a null value, which may not be what you want. It is a general programming practice to set the value of the return variables on the last line. This is not required, but it does make understanding the function a little easier.

The last thing to note about the function definition is the first few comment lines of the function. If you type **help** followed by the name of the function, the MATLAB help facility will echo the comment lines immediately after the line **function y = ...**). Thus, the beginning comment lines should be written with the MATLAB help facility in mind and should provide information for using the function and what the function does. Also note that the MATLAB **LOOKFOR** command looks at the first comment line of each function when it is performing a search. Thus, the first comment line should contain some keywords that one might use when looking for a function that performs a task similar to your function.

Let's look at some examples that make functions a little clearer. For a simple example, we'll convert the factorial program of Program 2-19 to a function in Function 3-1.

Function 3-1 Factorial program converted to a function

```
function x=fact(n)
% function x = fact(n) - This function calculates the factorial of an integer n.
% The function checks to see if n is a positive integer.
% The function returns the following values depending on
% the input variable n.
%
% x = n! - if n is a valid input
% x = -1 - if n is not an integer.
% x = -2 - if n is not positive.
% x = -3 - if n is not an integer and n was negative.

error_flag=0;
if floor(n) ~= n % Not an integer
   error_flag=1;
   x=-1;
end

if n < 0
   if error_flag
        x = -3;    % Not positive and not an integer
   else
        x=-2;
        error_flag=1; %Not positive
   end
end
```

```
if error_flag==0 % The number entered was a good number
   if n == 0
        factorial=1;
   else
        factorial=1;
        for i = n:-1:2
            factorial=factorial*i;
        end
   end
   x = factorial;
end
```

Note that the function does some error checking on the input *n*, and then calculates the factorial of *n* if it is an integer ≥ 0. The comments at the beginning of the function describe what the user should expect for return values. If *n* is a valid number for which a factorial can be calculated, the value of the factorial will be returned. If *n* is not a positive integer, then the function will return one of the error codes (-1, -2, or -3) that describe the nature of the error. These commands sufficiently describe how the function works and what the user should expect for return values.

First, let's see how the help facility displays our introductory comments.

EDU» **help fact**

```
 function x = fact(n)- This function calculates the factorial ➥
 of an integer n.
  The function checks to see if n is a positive integer.
  The function returns the following values depending on
  the input variable n.

  x = n! - if n is a valid input
  x = -1 - if n is not an integer.
  x = -2 - if n is not positive.
  x = -3 - if n is not an integer and n is negative.
```

EDU»

As you can see, the **HELP** function echoes the beginning comments of the function. Keep this in mind when writing functions that will be used later. Next, we will see what happens when we use the **LOOKFOR** command and specify the keyword **factorial**:

EDU» **lookfor factorial**
```
fact.m: % function x = fact(n)- This function calculates the
factorial of an integer n.
EDU»
```

The **LOOKFOR** command finds our function. Thus, our comments at the beginning allow someone to find our function and also instruct them on how to use the function.

Next, we'll test the operation of the function:

```
EDU» fact(5)
ans =
 120
EDU» fact(2)
ans =
 2
EDU» fact(0)
ans =
 1
EDU» fact(3.1)
ans =
 -1
EDU» fact(-5)
ans =
 -2
EDU» fact(-3.1)
ans =
 -3
```

Now that we have a function that computes the factorial, we can use it in other programs. Using the function allows us to simplify other programs and make them more readable. As an example, in Program 3-1 we create a function that asks for a number and then calculates the factorial if the specified number is an integer greater than or equal to zero.

Program 3-1 Example using factorial function

```
% This is file fact_ex.m
% This program calculates the factorial of a number.
%

n=input('Specify a non-negative integer: ');

x=fact(n);

if x == -1
   fprintf('The number you entered is not an integer.\n')
elseif x == -2
   fprintf('The number you entered is not positive.\n')
elseif x == -3
   fprintf('The number you entered is not positive.\n')
   fprintf('The number you entered is not an integer.\n')
```

```
else
   str=int2str(x);
   fprintf('%g factorial equals ',n);
   fprintf(str);
   fprintf('.\n');
end
```

Program 3-1 uses the fact function to find the factorial of a number. Depending on the result returned by the function, the program displays either an error message or the value of the factorial. Since the function returns an error code rather than displaying an error message, the calling program chooses how to handle the error. In this case, the calling program just emits error messages.

Let's test the program:

```
EDU» fact_ex

Specify a non-negative integer: 5
5 factorial equals 120.
EDU» fact_ex

Specify a non-negative integer: 12
12 factorial equals 479001600.
EDU» fact_ex

Specify a non-negative integer: 0
0 factorial equals 1.
EDU» fact_ex

Specify a non-negative integer: 4.5
The number you entered is not an integer.
EDU» fact_ex

Specify a non-negative integer: -3
The number you entered is not positive.
EDU» fact_ex

Specify a non-negative integer: -3.5
The number you entered is not positive.
The number you entered is not an integer.
EDU»
```

As another example, we will rewrite Program 2-22 to use the factorial function. Recall that Program 2-22 calculated the sum of factorials

$$\sum_{i=1}^{n} i! = 1! + 2! + 3! + \cdots + (n-1)! + n! = 1 + (2 \cdot 1) + (3 \cdot 2 \cdot 1) + (4 \cdot 3 \cdot 2 \cdot 1) + \cdots + (n-1)! + n!$$

The nature of summation lends itself easily to using the factorial function since the sum can be rewritten as

$$\sum_{i=1}^{n} fact(i)$$

We just need to write a single **FOR** loop to do the sum, which we do in Program 3-2.

Program 3-2 Sum of factorials–version 2

```
% This is file sum_fact2
% This program calculates the sum from i = 1 to n of n!
%

n=input('Specify a non-negative integer: ');

error_flag=0;
if floor(n) ~= n
   fprintf('The number you entered is not an integer.\n')
   error_flag=1;
end

if n < 1
   fprintf('The number you entered is not greater than 1.\n')
   error_flag=1;
end


if error_flag==0 % The number entered was a good number
   sum=0;
   for j=1:n
        sum=sum+fact(j);
   end
   str=int2str(sum);
   fprintf('The sum equals ');
   fprintf(str);
   fprintf('.\n');
end
```

Most of the program is for error checking to see if the number specified by the user is a valid number. The lines that actually calculate the sum are

```
sum=0;
for j=1:n
        sum=sum+fact(j);
end
```

Very few lines are needed. Compare these lines to those needed in Program 2-22 to accomplish the same task:

```
sum=0;
for j=1:n
        fact=1;
        for i = j:-1:2
                fact=fact*i;
        end
        sum=sum+fact;
  end
```

Programs 2-22 and 3-2 are identical except for these two sections of code. Although both methods are fairly easy to read and understand, the use of functions is considered better programming practice because they break larger tasks into smaller easy-to-understand sections. Now we will show that Programs 2-22 and 3-2 yield the same results:

```
EDU» sum_fact2

Specify a non-negative integer: 5
The sum equals 153.
EDU» sum_fact2

Specify a non-negative integer: 8
The sum equals 46233.
EDU» sum_fact2

Specify a non-negative integer: 0
The number you entered is not greater than 1.
EDU» sum_fact2

Specify a non-negative integer: -1
The number you entered is not greater than 1.
EDU»
```

Running Program 2-22 yields the same results:

```
EDU» sum_fact1

Specify a non-negative integer: 5
The sum equals 153.
EDU» sum_fact1

Specify a non-negative integer: 8
The sum equals 46233.
EDU» sum_fact1
```

```
Specify a non-negative integer: 0
The number you entered is not greater than 1.
EDU» sum_fact1

Specify a non-negative integer: -1
The number you entered is not greater than 1.
EDU»
```

As a second example, let's write a function that calculates the resistance value of two resistors in parallel. The form of the function is

$$\texttt{Function Req = fpar(R1,R2)}$$

The function must check to see that R_1 and R_2 are not negative, and if either R_1 or R_2 is equal to zero. If you remember, the equation for the equivalent resistance of two resistors in parallel is

$$\frac{1}{R_{EQ}} = \frac{1}{R_1} + \frac{1}{R_2}$$

This equation can be rewritten in several forms:

$$R_{EQ} = \frac{1}{\dfrac{1}{R_1} + \dfrac{1}{R_2}} \quad \text{or} \quad R_{EQ} = \frac{R_1 R_2}{R_1 + R_2}$$

We can use any of these forms. However, we must check for the special case of R_1 or R_2 equal to zero. If either or both of the resistors has a value of zero, the equivalent resistance is zero. Function 3-2 shows how this works.

Function 3-2 Resistors in parallel

```
function Req = fpar(R1,R2)
% Function Req = fpar(R1,R2) - This function calculates the equivalent resis-
tance of two resistors in parallel.
%
% R1 and R2 are the resistances in ohms.
% Req is the equivalent parallel resistance in ohms.
%
% Return Values
% Req = R1//R2 if R1 and R2 have values >= 0.
% Req = -1 if R1 is negative.
% Req = -2 if R2 is negative.


if R1 < 0
  Req = -1;
elseif R2 < 0
  Req = -2;
```

```
elseif (R1 == 0) | (R2 == 0)
 Req = 0;
else
 Req = (R1*R2)/(R1+R2);
end
```

EXERCISE 3-1 Change Function 3-2 so that it calculates the equivalent resistance of three resistors in parallel.

As another example, look at Function 3-3, which asks the user for a response. If the user gives an invalid response, the function announces the error, specifies a list of valid responses, displays the question again, and then waits for the user to enter another response. When a correct response is entered, the function returns a numerical value that indicates which correct response was entered.

Function 3-3 Question-and-answer function

```
function answer=ask_q(valid_answers, output_string);
% function answer=ask_q(valid_answers, output_string);
%
% This function asks a question of the user and loops indefinitely
% until the user enters one of the correct responses. The question asked
% is contained in the string variable output_string.
% The valid responses are contained in string valid_answers. All
% correct responses can only be 1 character in length.
% The return value, answer, is the position of the correct response in the
% string of valid answers. For example, if we use the command:
%
% xx=ask_q('abcde', 'Enter your command:')
%
% If the user types the letter a, function ask_q will return the numerical ➡
value 1.
% If the user types the letter b, function ask_q will return the numerical ➡
value 2.
% If the user types the letter c, function ask_q will return the numerical ➡
value 3.
% If the user types the letter d, function ask_q will return the numerical ➡
value 4.
% If the user types the letter e, function ask_q will return the numerical ➡
value 5.


good_input=0;
```

```
while ~good_input
  in_string=input(output_string,'s');
  for i = 1:length(valid_answers)
      if strcmp(in_string,valid_answers(i))
          good_input=1;
          answer=i;
      end
  end
  if ~good_input
      fprintf('Error - Input one of the following: ');
      for i = 1:(length(valid_answers)-1)
          fprintf(valid_answers(i));
          fprintf(',');
      end
      fprintf([valid_answers(length(valid_answers)),'.\n']);
  end
end
```

Function 3-3 has two input arguments: *valid_answers* and *output_string*. Both variables contain text strings. *Output_string* is the message printed when the function requests input from the user. *Valid_answers* contains a string of characters. Each character in the string is a valid response. For example, if *valid_answers* = 'abc123', the user is allowed to enter any of the characters **a**, **b**, **c**, **1**, **2**, or **3**. Note that this function does not allow the user to input a response of more than one character. That is, a valid response cannot be **yes**.

The return value of Function 3-3 is the numerical position of the valid response in string *valid_answers*. For example, if *valid_answers* = 'abc123' and the user response is c, the function will return a numerical value of 3. If the user response is 2, the function returns a numerical value of 5 because 2 is in the fifth position of the string 'abc123'.

Before we look at the details of the function, let's examine a few examples using Function 3-3, the **ask-q** function, so that we can better understand its operation.

```
EDU» ask_q('abcde','Specify your response:\n')
Specify your response:
1
Error - Input one of the following: a,b,c,d,e.
Specify your response:
abc
Error - Input one of the following: a,b,c,d,e.
Specify your response:
q
Error - Input one of the following: a,b,c,d,e.
Specify your response:
d
answer =
    4
EDU»
```

Valid responses are **a**, **b**, **c**, **d**, and **e**.

Not a one-letter response.

Not on the list of valid responses.

d is the fourth item in the list of valid responses, so the function returns a code of 4.

Note that the function only allows single-letter responses and loops indefinitely until the user enters a single-character response that is on the list of valid responses.

```
EDU» ask_q('12345E','Specify your command number, E to exit.');
Specify your command number, E to exit.
vf
Error - Input one of the following: 1,2,3,4,5,E.
Specify your command number, E to exit.
hfdsk
Error - Input one of the following: 1,2,3,4,5,E.
Specify your command number, E to exit.
3
answer =
     3
EDU»
```

Using the \n text in the output string enables us to have an output message several lines long. Each \n produces a new line when displayed with the **INPUT** command.

```
EDU» out_str='Specify your command:\n 1=format disk\n ➡
2=directory\n 3=run MATLAB\n E=exit\n\n'

out_str =

Specify your command:\n 1=format disk\n 2=directory\n 3=run ➡
MATLAB\n E=exit\n\n

EDU» ask_q('123E',out_str);
Specify your command:
 1=format disk
 2=directory
 3=run MATLAB
 E=exit


6
Error - Input one of the following: 1,2,3,E.
Specify your command:
 1=format disk
 2=directory
 3=run MATLAB
 E=exit


e
Error - Input one of the following: 1,2,3,E.
Specify your command:
```

```
1=format disk
2=directory
3=run MATLAB
E=exit


E
answer =
  4
EDU»
```

The only real limitation in this function is that it doesn't allow responses such as **yes** or **no**. See Section 4.7.1 on page 276 for a modification to this function that allows responses such as **yes** and **no**. Now let's examine the operation of the program.

```
good_input=0;
while ~good_input
    ...
    ...remainder of program...
    ...
end
```

Note that a large **WHILE** loop encloses most of the program. The first line initializes *good_input* to 0. The **WHILE** loop will loop indefinitely until something inside the loop sets *good_input* to 1. The code inside the loop is split into two portions. The first portion prints the question, asks for user input, and compares the user response to the list of correct responses. The second portion prints an error message and lists the correct responses, in the event the user does not respond with a correct response.

We will look at the first portion:

```
in_string=input(output_string,'s');
for i = 1:length(valid_answers)
      if strcmp(in_string,valid_answers(i))
            good_input=1;
            answer=i;
      end
end
```

The first line prints the question (variable *output_string* is displayed) and then requests input from the user in the form of a text string. The user can type in anything and the MATLAB **INPUT** function will not generate an error message because input is read as a string. A **FOR** loop is then used to increment *i* from 1 to the number of characters in variable *valid_answers*. The **IF** statement uses the **STRCMP** function to compare the user input (variable *in_string*) to the *i*th character of *valid_answers*. If the condition is true, variable *good_input* is set to 1 and *answer* is set to *i*. We see that the code segment compares the user input to each character in *valid_answers*. When a match is found, *answer* is set to the index of the character in *valid_answers* that matches the user input, and *good_input* is set to 1 when a match occurs to allow the outer **WHILE** loop to exit.

The second portion of the program displays an error message and lists the valid responses. The code segment is only executed if the user response does not match a character in *valid_response*.

```
if ~good_input
        fprintf('Error - Input one of the following: ');
        for i = 1:(length(valid_answers)-1)
                fprintf(valid_answers(i));
                fprintf(',');
        end
        fprintf([valid_answers(length(valid_answers)),'.\n']);
end
```

If variable *valid_answers* has *n* characters, the **FOR** loop prints out the first *n*–1 characters, each separated by a comma. The line

```
fprintf([valid_answers(length(valid_answers)),'.\n']);
```

prints out the last character followed by a period.

The **ask_q** function would be very useful in many of our previous programs that required input from the user. For example, let's modify Program 2-27 on page 132. The original program is repeated here, and the code shown in larger type will be modified when we use the **ask_q** function.

Program 2-27

```
%
% This is file guess3.m
%
again = 'y';
while again == 'y'
   real_number=rand(1); % Generate a number between 0 and 1
   real_number=10*real_number; % Scale the number between 0 and 10
   number=ceil(real_number); % Round the number to the next highest integer.

   % Ask the user to guess the number:

   fprintf('I am thinking of a number between 1 and 10.\n');
   fprintf('You have three guesses.\n');
   correct=0;

   for i = 1:3
        if ~correct
            guess = input('Guess a number: ');
```

```
            if guess == number
                fprintf('Congratulations - You guessed the correct number.\n');
                correct=1;
            else
                if guess > number
                    fprintf('Your guess was too high.\n');
                else
                    fprintf('Your guess was too low.\n');
                end
            end
        end
    end

    if ~ correct
        fprintf('Sorry - The number I was thinking of was %g.\n\n', number);
    end

    correct_response = 0;
    while ~correct_response
        again=input('Do you wish to play again? (y/n) ','s');
        if (again ~= 'y') & (again ~= 'n')
            fprintf('Error - Enter either ''y'' or ''n''.\n');
        else
            correct_response = 1;
        end
    end %of while loop 2
end %of while loop1
```

The lines shown in larger type in the preceding program ask the user if they want to play another game and check for a correct response. The entire **WHILE** loop at the end of the program can thus be replaced by the **ask_q** function. Program 3-3 shows the modifications, which are in larger type.

Program 3-3 Guessing game with **ask-q** function

```
%
% This is file guess4.m
%

again = 1;
while again == 1
    real_number=rand(1); % Generate a number between 0 and 1
    real_number=10*real_number; % Scale the number between 0 and 10
    number=ceil(real_number); % Round the number to the next highest integer.

    % Ask the user to guess the number:

    fprintf('I am thinking of a number between 1 and 10.\n');
```

```
fprintf('You have three guesses.\n');
correct=0;

for i = 1:3
    if ~correct
        guess = input('Guess a number: ');

        if guess == number
            fprintf('Congratulations - You guessed the correct number.\n');
            correct=1;
        else
            if guess > number
                fprintf('Your guess was too high.\n');
            else
                fprintf('Your guess was too low.\n');
            end
        end
    end
end

if ~ correct
    fprintf('Sorry - The number I was thinking of was %g.\n\n', number);
end

again=ask_q('yn','Do you wish to play again? (y/n) ');

end %of while loop1
```

As you can see, the entire **WHILE** loop at the bottom of Program 2-27 has been replaced by the **ask_q** function. This makes Program 3-3 easier to understand. Also note that the lines

```
again = 'y';
while again == 'y'
```

at the top of Program 2-27 have been replaced by

```
again = 1;
while again == 1
```

in Program 3-3. This is because, with the **ask_q** function, a response of **y** is returned as the numerical value 1, and a response of **n** is returned as the numerical value 2.

An example session of Program 3-3 is shown here.

```
EDU» guess4
I am thinking of a number between 1 and 10.
You have three guesses.
```

```
Guess a number: 5
Your guess was too low.

Guess a number: 8
Your guess was too high.

Guess a number: 7
Congratulations - You guessed the correct number.
Do you wish to play again? (y/n)
y
I am thinking of a number between 1 and 10.
You have three guesses.

Guess a number: 5
Your guess was too low.

Guess a number: 9
Your guess was too high.

Guess a number: 7
Congratulations - You guessed the correct number.
Do you wish to play again? (y/n)
f
Error - Input one of the following: y,n.
Do you wish to play again? (y/n)
1
Error - Input one of the following: y,n.
Do you wish to play again? (y/n)
n
EDU»
```

EXERCISE 3-2 Modify Program 3-3 to accept responses **y**, **Y**, **n**, and **N**. Upper- and lowercase letters should have the same function.

For another example, in Program 3-4 we modify the command center of Program 2-31 so that it uses the **ask_q** function.

Program 3-4 Command center with the **ask_q** function

```
%
% this if file command3.m
%
% This program asks for a number between 1 and 5 inclusive.
% The program then performs the tasks assigned to the number.
```

```
% The tasks are not specified for this program.

str1='\n\nSpecify your option: [1-5,E]:\n 1=Execute Command 1\n 2=Execute ➡
Command 2\n';
str2=' 3=Execute Command 3\n 4=Execute Command 4\n';
str3=' 5=Execute Command 5\n E=Exit\n\n';
out_str=[str1, str2, str3];

command=-99;
while command ~= 6

   command=ask_q('12345E', out_str);

   if command == 1
       fprintf('You chose option 1.\n');
       % Command for option 1 entered here.
   elseif command == 2
       fprintf('You chose option 2.\n');
       % Command for option 2 entered here.
   elseif command ==3
       fprintf('You chose option 3.\n');
       % Command for option 3 entered here.
   elseif command ==4
       fprintf('You chose option 4.\n');
       % Command for option 4 entered here.
   elseif command ==5
       fprintf('You chose option 5.\n');
       % Command for option 5 entered here.
   end
end
fprintf('Good Bye.\n');
```

Note that variable *out_str* contains a long string that is the output message in the **ask_q** function. The string we want to use is too long to fit on a single line, so we created *out_str* by concatenating several smaller strings together. (See Section 1.1.2, Concatenating Strings, for more information.) Note that the lines

```
str1='\n\nSpecify your option: [1-5,E]:\n 1=Execute Command 1\n 2=Execute Command 2\n';
str2=' 3=Execute Command 3\n 4=Execute Command 4\n';
str3=' 5=Execute Command 5\n E=Exit\n\n';
out_str=[str1, str2, str3];
```

could be replaced by

```
out_str='\n\nSpecify your option: [1-5,E]:\n 1=Execute Command 1\n 2=Execute Command 2\n';
out_str=[out_str,' 3=Execute Command 3\n 4=Execute Command 4\n'];
out_str=[out_str,' 5=Execute Command 5\n E=Exit\n\n'];
```

The two methods produce the same results, but the first one is a bit easier to understand.

Note that the **IF-ELSEIF** statement in Program 3-4 is slightly different than the one in Program 2-31 because the **ask_q** function does all of the necessary error checking. This eliminates the need for the error message emitted by the **else** clause at the bottom of Program 2-31.

```
EDU» command3


Specify your option: [1-5,E]:
 1=Execute Command 1
 2=Execute Command 2
 3=Execute Command 3
 4=Execute Command 4
 5=Execute Command 5
 E=Exit


1
You chose option 1.


Specify your option: [1-5,E]:
 1=Execute Command 1
 2=Execute Command 2
 3=Execute Command 3
 4=Execute Command 4
 5=Execute Command 5
 E=Exit


2
You chose option 2.


Specify your option: [1-5,E]:
 1=Execute Command 1
 2=Execute Command 2
 3=Execute Command 3
 4=Execute Command 4
 5=Execute Command 5
 E=Exit
```

```
gfs
Error - Input one of the following: 1,2,3,4,5,E.


Specify your option: [1-5,E]:
 1=Execute Command 1
 2=Execute Command 2
 3=Execute Command 3
 4=Execute Command 4
 5=Execute Command 5
 E=Exit


e
Error - Input one of the following: 1,2,3,4,5,E.


Specify your option: [1-5,E]:
 1=Execute Command 1
 2=Execute Command 2
 3=Execute Command 3
 4=Execute Command 4
 5=Execute Command 5
 E=Exit


E
Good Bye.
EDU»
```

EXERCISE 3-3   Modify Program 3-4 to accept responses **e** and **E** in addition to the other valid responses. Both **e** and **E** should cause the program to exit.

3.2 Scope of Variables

A variable's scope is where in a program the variable is defined. Suppose we have this function:

```
function z=example(x)
fprintf('Inside function example.\n');
a=5;
b=6;
```

```
c=7;
z=10;
fprintf('a = %g, b=%g, c=%g, z=%g\n',a, b, c, z);
fprintf('Leaving function example.\n');
```

This function uses variables *a*, *b*, *c*, and *z* inside the function. Now let's use this function inside the following program:

```
a=27;
b=53;
c=100;
p=example(5);
fprintf('Back in the main program.\n');
fprintf('a = %g, b=%g, c=%g.\n',a, b, c);
```

When we run the program, we see this dialog:

```
Inside function example.
a = 5, b=6, c=7, z=10
Leaving function example.
Back in the main program.
a = 27, b=53, c=100.
```

Note inside the function, variables *a*, *b*, and *c* have the values 5, 6, and 7. In the main program, *a*, *b*, and *c* have the values 27, 53, and 100. Although we use variables with the same name in both the main program and the function, MATLAB treats variables in the main program as different from those in the function, even though they have the same name. That is, *a*, *b*, and *c* in the main program are different from *a*, *b*, and *c* in the function. Thus, any calculations in which we use *a*, *b*, and *c* in the function will not affect the values of *a*, and *b*, and *c* in the main program.

As another example, let's define function example2 as follows:

```
function z=example2(x)
fprintf('Inside function example2.\n');
fprintf('a = %g, b=%g, c=%g\n',a, b, c);
fprintf('Leaving function example2.\n');
```

This function never sets the values of *a*, *b*, and *c* inside the function. However, it does attempt to print out their values. We will call this function with the following code segment:

```
a=27;
b=53;
c=100;
p=example2(5);
fprintf('Back in the main program.\n');
fprintf('a = %g, b=%g, c=%g.\n',a, b, c);
```

When we run the code segment, this output is generated:

```
Inside function example2.
??? Undefined function or variable a.

Error in ==> c:\matlab\toolbox\examples\example2.m
On line 3 ==> fprintf('a = %g, b=%g, c=%g\n',a, b, c);
```

Note that the code segment sets values for *a*, *b*, and *c* and then calls function example2. MATLAB enters the function and prints out the message:

```
Inside function example2.
```

When we attempt to print out the values of *a*, *b*, and *c*, we get the error messages:

```
??? Undefined function or variable a.
```

and

```
Error in ==> c:\matlab\toolbox\examples\example2.m.
```

This means that inside function example2, variable *a* is not known. Even though we define values for *a*, *b*, and *c* in the main program, their values are not known inside the function.

Next, we'll show that variables defined inside a function are not known to the main program. Let's use function example here:

```
function z=example(x)
fprintf('Inside function example.\n');
a=5;
b=6;
c=7;
z=10;
fprintf('a = %g, b=%g, c=%g, z=%g\n',a, b, c, z);
fprintf('Leaving function example.\n');
```

Note that we define variables *a*, *b*, and *c* inside the function; that is, they are known inside the function. We will now call the function from the following code segment:

```
p=example(5);
fprintf('Back in the main program.\n');
fprintf('a = %g, b=%g, c=%g.\n',a, b, c);
```

When we run this code segment, we see this output:

```
Inside function example.
a = 5, b=6, c=7, z=10
Leaving function example.
Back in the main program.
??? Undefined function or variable a.
```

Values for a, b, and c are not set in the code segment. When we call function example, the values of a, b, and c (5, 6, and 7, respectively) are known inside the function. However, when we leave the function and attempt to print out the values of a, b, and c, we get the error message: ??? Undefined function or variable a. This means that in the main program variable a is unknown.

From these examples, you can see that variables defined in the main program are known to the main program, but not to the function. Conversely, variables defined inside the function are known to the function but not to the main program. And variables with the same name that are defined in both the main program and a function are treated as different variables.

This property of variables being defined in one location and not another is referred to as scope. The scope of a variable is where in a program and in what functions it is defined. Suppose we have function example2,

```
function z=example2(x)
fprintf('Inside function example2.\n');
a=5;
b=6;
c=7;
z=10;
fprintf('a = %g, b=%g, c=%g, z=%g\n',a, b, c, z);
fprintf('Leaving function example2.\n');
```

and we use function example2 in the follwing program:

```
w=10;
q=99;
y='This is a test';
p=example2(w)
y=w/q;
```

Variables w, q, y, and p are known only in the main program; that is, they are undefined inside function example2. So, the scope of variables w, q, y, and p is the main program only. Variables a, b, c, and z are defined only inside function example2. Thus, the main program does not have access to the variables and does not know their value, so the scope of variables a, b, c, and z is the function example2 only. Also note that variables a, b, c, and z are referred to as local variables since they are only defined within the function.

3.3 Passing Parameters

Suppose we are using function example3:

```
function z=example3(a, b, c)
fprintf('Inside function example3.\n');
fprintf('Upon Entering function example3, ');
fprintf('a = %g, b=%g, c=%g.\n\n',a, b, c);
```

```
a=a/10;
b=b/10;
c=c/10;
z=a+b+c;
fprintf('After performing a few calculations, ');
fprintf('a = %g, b=%g, c=%g, z=%g\n',a, b, c, z);
fprintf('Leaving function example3.\n\n');
```

We pass three values a, b, and c, to this function. When we first enter the function, we display the values of a, b, and c. We then modify their values and print out their values again, and the function ends. Let's use this function in the following section of code:

```
x=1;
y=2;
w=3;


fprintf('In the main program, ');
fprintf('x = %g, y=%g, w=%g.\n\n',x, y, w);
answer=example3(x,y,w);
fprintf('Back in the main program, ');
fprintf('x = %g, y=%g, w=%g.\n\n',x, y, w);
```

When the code segment runs, this output is generated:

```
In the main program, x = 1, y=2, w=3.
Inside function example3.
Upon Entering function example3, a = 1, b=2, c=3.
After performing a few calculations, a = 0.1, b=0.2, c=0.3, z=0.6
Leaving function example3.
Back in the main program, x = 1, y=2, w=3.
```

This section of code defines values for w, x, and y. The values are then printed to show their values: In the main program, x = 1, y = 2, w = 3. MATLAB then executes the statement **answer=example3(x,y,w)**. This line calls function example3 and passes the values of x, y, and w to a, b, and c in function example3. The definition of function example3 is **function z=example3(a, b, c)**. The order in which we pass variables determines their values. When you compare the function call (**answer=example3(x,y,w)**) to the function definition (**function z=example3(a, b, c)**) we know that the value of x is placed into variable a in the function, the value of y is placed into variable b in the function, and the value of w is placed into variable c in the function. This is evident from the output of the program and the function. The main program output is:

```
In the main program, x = 1, y=2, w=3
```

and the output of the function is:

```
Upon Entering function example3, a = 1, b=2, c=3.
```

Thus, you can see that x and a have the same value, y and b have the same value, and w and c have the same value. The variables are different variables, but upon entering the function they have the same value. This is how values are passed to functions.

The next thing the function does is divide variables a, b, c, and z by 10 and then print out the new values: `After performing a few calculations, a = 0.1, b=0.2, c=0.3, z=0.6`. Note that the values of a, b, c, and z have changed inside the function. MATLAB then returns to the main program and displays the values of x, y, and w: `Back in the main program, x = 1, y=2, w=3` and you can see that x, y, and w are unchanged in the main program. Thus, when we call a function with a line like: `answer=example3(x,y,w)`, the values of x, y, and w are passed to the function, but the variable's value doesn't change even though the value of their placeholders is modified. This method of passing parameters is referred to as **call by value.** With this method, the values of the variables pass to the function but the variables themselves are unchanged. In our example above, the values contained in x, y, and w passed to the function. When the function finished processing and control returned to the main program, the values of x, y, and w remained the same, as if the function had not been called.

3.4 Global Variables

The following function is repeated from Section 3.2.

```
function z=example(x)
fprintf('Inside function example.\n');
a=5;
b=6;
c=7;
z=10;
fprintf('a = %g, b=%g, c=%g, z=%g\n',a, b, c, z);
fprintf('Leaving function example.\n');
```

The function was used in this code segment:

```
a=27;
b=53;
c=100;
p=example(5);
fprintf('Back in the main program.\n');
fprintf('a = %g, b=%g, c=%g.\n',a, b, c);
```

We learned from Section 3.2 that variables a, b, and c in the function are different from a, b, and c in the code segment. These variables are not shared between the code segment and function; that is, they are treated as completely different variables by MATLAB. Local variables are desirable because in different functions, variables with the same name are independent. Suppose we have functions 1, 2, and 3. We can use variables x, y, and z in function 1 *and* in functions 2 and 3, and MATLAB will treat the variables as if they all have unique names. The calculations we perform on x, y, and z in function 1 will not affect calculations in functions 2 or 3. Thus, we can use variables with the same name in as many different functions as we want. The variables can have different values and meanings in each function, and a variable in one function will not affect variables with the same name in other functions.

Local variables are desirable because we use many common variables like i, j, and x frequently. It would be confusing if every function needed unique variable names. If the variables weren't local, major problems occur if function A used variable x and function B also used variable x, but for a different purpose. If the variables weren't local, then using variable x in one function would affect how other functions worked. To avoid this problem, MATLAB makes all variables local.

On occasion, it is desirable for a variable to be used in the same way in more than one function. That is, we have a variable x in function A, and a variable x in function B, and we want x to be the same variable in both functions. A variable of this type is called a global variable because it's defined in more than one function. Let's look at an example of a local variable, but first we need to define function xxx1:

```
function z=xxx1(in)
fprintf('The value of a is %g\n', a);
```

The only thing this function does is print the value of a. Next we write a code segment that assigns a value to a and then calls function xxx1:

```
a=10;
xxx1(25);
??? Undefined function or variable a.

Error in ==> c:\MATLAB\toolbox\examples\xxx1.m
On line 2 ==>
```

An error message is generated inside function xxx1 because variable a is local to the code segment and is not defined inside function xxx1. We can share variables between the code segment and the function by defining variable a as **GLOBAL** in both the code segment and the function:

```
function z=xxx2(in)
global a;
fprintf('The value of a is %g\n', a);
```

Function xxx2 defines variable a as **GLOBAL** but never sets its value. This will be done in the code segment. Both the code segment and the function must define a as **GLOBAL**:

```
global a;
a = 5;
xxx2(2);
```

When we run the code segment, we see this output:

```
The value of a is 5
```

An error message is not generated when we run the code segment this time because variable *a* is global and known to both the code segment and the function. Since we define *a* in the code segment and give it a value of 5, and because both the code segment and function xxx2 define *a* as global, variable *a* is also defined inside the function with a value of 5.

The next example shows that when we modify variable *a* in the function, its value is also modified in the code segment.

```
function z=xxx3(in)
global a;
fprintf('Inside the function the value of a is %g\n', a);
a=1;
```

This function first prints the current value of *a*, changes the value to 1, and then exits. Let's test it with the following code segment:

```
global a;
a = 5;
xxx3(7);
fprintf('In the code segment, a = %g.\n', a);
Inside the function the value of a is 5

In the code segment, a = 1.
```

As you can see, setting the value of *a* to 5 in the code segment sets *a* to 5 in the function as well. In the function, when we change the value of *a* to 1, *a* also becomes 1 in the code segment. This shows that *a* is the same variable in both the code segment and the function. Thus, the **GLOBAL** statement allows us to share variables when we specifically declare the variables as **GLOBAL**.

Variables are shared only between functions that declare the variable as **GLOBAL**. We've created the three functions below to demonstrate this concept:

```
function z=yyy1(in)
global a;
fprintf('Inside function yyy1 the value of a is %g\n', a);

function z=yyy2(in)
global a;
fprintf('Inside function yyy2 the value of a is %g\n', a);

function z=yyy3(in)
fprintf('Inside function yyy3 the value of a is %g\n', a);
```

Note that functions yyy1 and yyy2 declare variable *a* as **GLOBAL**, but function yyy3 does not. When we call these functions, variable *a* will be known in functions yyy1 and yyy2, but will generate an error message in function yyy3 because it is not defined as **GLOBAL** in yyy3. Let's test the functions:

```
global a;
a=27;
yyy1(5);
yyy2(5);
yyy3(5);
Inside function yyy1 the value of a is 27
Inside function yyy2 the value of a is 27
??? Undefined function or variable a.

Error in ==> c:\MATLAB\toolbox\examples\yyy3.m
On line 2 ==> fprintf('Inside function yyy3 the value of a is %g\n', a);
```

As you can see, *a* is defined in functions yyy1 and yyy2, but not in yyy3. This is because functions yyy1 and yyy2 define *a* as global, but function yyy3 does not. Thus, a **GLOBAL** variable is only shared between functions that declare it as **GLOBAL**.

3.5   The RETURN Statement

Normally, a function returns to its caller after it executes the last line of the function. However, a function can be made to return to its caller at any point by using the MATLAB **RETURN** statement, as we show in Function 3-4.

Function 3-4 The return statement

```
function y=ret_ex(choice)
%
fprintf('Entering function 3_4.\n');
if choice==1
   fprintf('Exiting early.\n');
   y=0;
   return
end

fprintf('After The IF Statement.\n');
y=1;
fprintf('Executing the last line of the function.\n');
```

If the function is called with an input argument of 1, the commands inside the **IF** statement will execute and the program will exit early by using the **RETURN** statement. If the function is called with any other argument, the function will return after executing the last statement of the function.

```
ret_ex(1)

Entering function 3_4.
Exiting early.
ans =
      0
```

We see that the function never reaches the statement **fprintf('After The IF Statement.\n')**. Instead, it reaches the **RETURN** statement and returns immediately to its calling routine.

```
ret_ex(0)
Entering function 3_4.
After The IF Statement.
Executing the last line of the function.
ans =
    1
```

In this example, the function reaches the last line of the function and returns.

3.6 *nargin* and *nargout*

MATLAB Predefined
RESIDUE
function

You may have noticed in the MATLAB help files, and in some of the examples in this book, that you can call MATLAB functions with a different number of input variables, and functions can return a different number of output variables. A good example is the MATLAB **RESIDUE** function. If you read the help information for the function, you will see that the function can be used in two forms:

[r,p,k] = residue(b,a)   or   **[b,a] = residue(r,p,k)**

In the first case, the function has two input arguments and three output variables. In the second case, it has three input arguments and two output variables. The same function handles both cases.

The question is, how does the function *know* how to handle both cases? It must know the number of input and output arguments inside the function. The MATLAB variables *nargin* and *nargout* contain this information. We will use Function 3-5 to demonstrate these variables.

Function 3-5 Function with an arbitrary number of inputs

```
function [a,b,c,d,e,f]=arb_in(s,t,u,v,w,x,y,z);
fprintf('The number of input arguments was %g.\n', nargin);
fprintf('The number of output variables was %g.\n', nargout);
```

Let's test Function 3-5 with different numbers of input and output arguments:

```
arb_in(1,2,5)
The number of input arguments was 3.

The number of output variables was 0.

xs=arb_in(1,2,5,7,4);
The number of input arguments was 5.
The number of output variables was 1.
```

```
[b,t,w]=arb_in;
The number of input arguments was 0.

The number of output variables was 3.
```

As you can see, the function knows the number of input and output arguments. With this information, the function reacts differently depending on how it is used.

If you look at the file residue.m in Function 3-6, you will see that the function checks the value of *nargin* to determine the action to take. Note that the first thing the function does is check the number of input arguments. This is how the function determines how to react based on the number of input arguments.

Function 3-6 The RESIDUE function

```
function [coeffs,poles,k] = residue(u,v,k)

%RESIDUE Partial-fraction expansion or residue computation.
%  [R,P,K] = RESIDUE(B,A) finds the residues, poles and direct term of
%  a partial fraction expansion of the ratio of two polynomials,
%  B(s) and A(s). If there are no multiple roots,
%     B(s)        R(1)        R(2)              R(n)
%     ----  =  --------  +  --------  + ... + --------  + K(s)
%     A(s)      s - P(1)     s - P(2)          s - P(n)
%  Vectors B and A specify the coefficients of the polynomials in
%  descending powers of s. The residues are returned in the column
%  vector R, the pole locations in column vector P, and the direct
%  terms in row vector K. The number of poles is
%    n = length(A)-1 = length(R) = length(P)
%  The direct term coefficient vector is empty if length(B) < length(A);
%  otherwise
%    length(K) = length(B)-length(A)+1
%
%  If P(j) = ... = P(j+m-1) is a pole of multiplicity m, then the
%  expansion includes terms of the form
%       R(j)         R(j+1)                R(j+m-1)
%     --------  +  -----------  + ... + -----------
%     s - P(j)     (s - P(j))^2         (s - P(j))^m
%  [B,A] = RESIDUE(R,P,K), with 3 input arguments and 2 output arguments,
%  converts the partial-fraction expansion back to the polynomials with
%  coefficients in B and A.
%
%  Warning:
%  Numerically, the partial-fraction expansion of a ratio of polynomials
%  represents an ill-posed problem. If the denominator polynomial, A(s),
%  is near a polynomial with multiple roots, then small changes in the
%  data, including roundoff errors, can make arbitrarily large changes
%  in the resulting poles and residues. Problem formulations making use
%  of state-space or zero-pole representations are preferable.
```

```
% Reference: A.V. Oppenheim and R.W. Schafer, Digital
% Signal Processing, Prentice-Hall, 1975, p. 56-58.
% C.R. Denham and J.N. Little, MathWorks, 1986, 1989.
% Copyright (c) 1984-94 by The MathWorks, Inc.
tol = 0.001; % Repeated-root tolerance; adjust as needed.
% This section rebuilds the U/V polynomials from residues,
% poles, and remainder. Multiple roots are those that
% differ by less than tol.

if nargin > 2
  [p,i] = sort(-abs(v)); % Sort poles by magnitud e.
  p = v(i); r = u(i); % Rearrange residues similarly.
  n = length(p);
  q = [p(:).' ; ones(1,n)]; % Poles and multiplicities.
  for j = 2:n
    if abs(p(j) - p(j-1)) < tol
      q(2,j) = q(2,j-1) + 1; % Multiplicity of pole.
    end
  .
  .(This is an incomplete listing of the file.)
```

> The function checks the value of *nargin*.

As an example of using *nargin*, let's create a function that calculates the equivalent resistance of several resistors in parallel. Recall that if you have *N* resistors in parallel, the equivalent resistance of the *N* resistors is

$$\frac{1}{R_{EQ}} = \sum_{i=1}^{N} R_i$$

Our function will calculate this sum and and will work if the user specifies anywhere from 1 to 10 resistors. When a user uses this function, it will be called as **R=parallel(R1,R2,R3,...,RN)** where the number of resistors passed to the function can be between 1 and 10. Here are some examples:

```
R5 = parallel(10000,15000);
R23 = parallel(1000,1200,680,270);
Rx = 15000 + parallel(1200,10000,2700,33000,560,15000);
```

For the function parallel to work, it must know how many arguments are passed to it. An example implementation of the function is shown in Function 3-7.

Function 3-7 Function that calculates the equivalent resistance of resistors in parallel

```
function Req=parallel(R1,R2,R3,R4,R5,R6,R7,R8,R9,R10)
% Function Parallel - Calculates the equivalent
% resistance of up to 10 parallel resistors.
% Req = parallel resistance if 1 through 10 resistors are specified.
% = -1 if no resistors were specified.
% = NaN - if a resistor has a value of zero.
```

```
if nargin == 0
  Req = -1;
elseif nargin==1
  Req = R1;
elseif nargin == 2
  Geq = (1/R1) + (1/R2);
  Req = 1/Geq;
elseif nargin == 3
  Geq = (1/R1) + (1/R2) + (1/R3);
  Req = 1/Geq;
elseif nargin == 4
  Geq = (1/R1) + (1/R2) + (1/R3) + (1/R4);
  Req = 1/Geq;
elseif nargin == 5
  Geq = (1/R1) + (1/R2) + (1/R3) + (1/R4) + (1/R5);
  Req = 1/Geq;
elseif nargin == 6
  Geq = (1/R1) + (1/R2) + (1/R3) + (1/R4) + (1/R5) + (1/R6);
  Req = 1/Geq;
elseif nargin == 7
  Geq = (1/R1) + (1/R2) + (1/R3) + (1/R4) + (1/R5) + (1/R6) + (1/R7);
  Req = 1/Geq;
elseif nargin == 8
  Geq = (1/R1) + (1/R2) + (1/R3) + (1/R4) + (1/R5) + (1/R6) + (1/R7) + (1/R8);
  Req = 1/Geq;
elseif nargin == 9
  Geq = (1/R1) + (1/R2) + (1/R3) + (1/R4) + (1/R5) + (1/R6) + (1/R7) + (1/R8) ➡
+ (1/R9);
  Req = 1/Geq;
elseif nargin == 10
  Geq = (1/R1) + (1/R2) + (1/R3) + (1/R4) + (1/R5) + (1/R6) + (1/R7) + (1/R8) ➡
+ (1/R9) + (1/R10);
  Req = 1/Geq;
end
```

It's easy to see from Function 3-7 that *nargin* directly affects how this function works. Function 6-1 (page 347) shows a second implementation of the parallel function using the **EVAL** function.

3.7 Recursive Functions

MATLAB Predefined
CLOCK function
ETIME function

A recursive function is a function that calls itself. A classic example is the factorial function; that is, 5! can be calculated as $5 \times 4!$; 4! as $4 \times 3!$, 3! as $3 \times 2!$, and 2! as $2 \times 1!$ Finally, 1! is equal to 1. Thus, for any positive integer *n, n*! can be calculated as $n \times (n-1)!$ The process stops when we reach a known value of the factorial. In this case, 1! is equal to 1 and we don't need to calculate 1! as $1 \times (1-1)!$

A recursive version of the factorial function is shown in Function 3-8.

Function 3-8 Recursive factorial function

```
function x=fact2(n)

if n ~= 1
   x=n*fact2(n-1);
else
   x=1;
end
```

This function has been simplified by removing all of the error-checking code at the beginning of the previous factorial function. Note that this function will cause problems if the number passed to it is not a positive integer.

As you can see, Function 3-8 returns a value of 1 if n is equal to 1. If n is greater than 1, it returns a value of $n \times (n-1)!$ It calls itself to find the value of $n-1$. Thus, this function calls itself repeatedly, each time finding the factorial of a number closer to 1. Eventually it asks for the factorial of 1, which is immediately returned as 1. In Function 3-9, we modify the function so that we can see how it calls itself.

Function 3-9 Recursive factorial function, modified

```
function x=fact3(n)

fprintf('\n\nEntering function fact3, n = %g\n', n);
if n ~= 1
   fprintf('Calling function fact3 for n= %g\n', n-1);
   x=n*fact3(n-1);
else
   x=1;
end
fprintf('Returning from fact3 function, n = %g, n! = %g.\n', n, x);
```

Whenever we call function fact3, it displays a message that it was called. If it needs to call itself, it displays a message that it is about to call itself, and then displays the value it is going to pass to itself. Before the function returns, it displays the value of the factorial it has just calculated.

```
fact3(5)

Entering function fact3, n = 5
Calling function fact3 for n= 4


Entering function fact3, n = 4
Calling function fact3 for n= 3
```

```
Entering function fact3, n = 3
Calling function fact3 for n= 2


Entering function fact3, n = 2
Calling function fact3 for n= 1


Entering function fact3, n = 1
Returning from fact3 function, n = 1, n! = 1.
Returning from fact3 function, n = 2, n! = 2.
Returning from fact3 function, n = 3, n! = 6.
Returning from fact3 function, n = 4, n! = 24.
Returning from fact3 function, n = 5, n! = 120.
ans =
   120
```

Note that the function continues to call itself. It never returns a value until it is asked to find 1! The function knows the value of 1! and immediately returns a result. The function then calculates 2! as 2×1! It returns the number 2 and goes on to calculate 3! as 3×2! The function continues to unwind until it calculates the factorial of the original number.

```
EDU» fact3(10)

Entering function fact3, n = 10
Calling function fact3 for n= 9


Entering function fact3, n = 9
Calling function fact3 for n= 8


Entering function fact3, n = 8
Calling function fact3 for n= 7


Entering function fact3, n = 7
Calling function fact3 for n= 6


Entering function fact3, n = 6
Calling function fact3 for n= 5
```

```
Entering function fact3, n = 5
Calling function fact3 for n= 4


Entering function fact3, n = 4
Calling function fact3 for n= 3


Entering function fact3, n = 3
Calling function fact3 for n= 2


Entering function fact3, n = 2
Calling function fact3 for n= 1


Entering function fact3, n = 1
Returning from fact3 function, n = 1, n! = 1.
Returning from fact3 function, n = 2, n! = 2.
Returning from fact3 function, n = 3, n! = 6.
Returning from fact3 function, n = 4, n! = 24.
Returning from fact3 function, n = 5, n! = 120.
Returning from fact3 function, n = 6, n! = 720.
Returning from fact3 function, n = 7, n! = 5040.
Returning from fact3 function, n = 8, n! = 40320.
Returning from fact3 function, n = 9, n! = 362880.
Returning from fact3 function, n = 10, n! = 3.6288e+006.
ans =
 3628800
```

To make a complete function, we should add the error checking of our original factorial function, which we do in Funciton 3-10.

Function 3-10 Recursive factorial function with error checking

```
function x=fact4(n)
% function x = fact4(n)
% This function calculates the factorial of an integer n
% using recursion.
% The function checks to see if n is a positive integer.
% The function returns the following values depending on
% the input variable n.
%
% y = n! - n was a valid input
% y = -1 - n was not an integer.
% y = -2 - n was not positive.
```

```
% y = -3 - n was not an integer and n was negative.

error_flag=0;
if floor(n) ~= n % Not an integer
   error_flag=1;
   x=-1;
end

if n < 0
   if error_flag
        x = -3;   % Not positive and not an integer
   else
        x=-2;
        error_flag=1; %Not positive
   end
end


if error_flag==0 % The number entered was a good number
   if (n == 0) | (n == 1)
        factorial=1;
   else
        factorial=n*fact4(n-1);
   end
   x = factorial;
end
```

Functions fact and fact4 perform exactly the same function using different methods. Recursion gives us a concise way to solve some problems, but calling functions recursively is very "expensive" in that it takes a lot of computer time compared to a nonrecursive solution. Let's calculate 15! using functions fact and fact4 and compare the execution times.

```
t1=clock;
fact(15)
t2=clock;
etime(t2,t1)
ans =
 1.3077e+012
ans =
 0.1700
```

The built-in **CLOCK** function gives you the current time. The **ETIME** function calculates the time in seconds between two different values returned by the **CLOCK** function. From the preceding example, you can see that 15! is equal to 1.3077e+012 and that it took 0.1700 seconds to calculate the result using the nonrecursive function fact.

Next, let's calculate 15! using our recursive function:

```
t1=clock;
fact4(15)
t2=clock;
etime(t2,t1)
ans =
  1.3077e+012
ans =
  0.7700
```

The functions produce the same numerical results, but function fact4 took 0.7700 seconds to calculate the result. This is more than 4.5 times longer than the nonrecursive method in function fact. In general, recursive functions are much slower than nonrecursive functions.

3.8 Problems

Problem 3-1   Write a MATLAB function called **isodd(a)** that checks whether a number is an odd integer. The definition of the function is

```
function x = isodd(a)
```

$$\text{where } x = \begin{cases} 1 & \text{if } a \text{ is an odd integer} \\ 0 & \text{if } a \text{ is an even integer} \\ -1 & \text{if } a \text{ is not an integer} \end{cases}$$

The value of x should be set to -1 if a is a real number, a complex number, an array, a *NaN*, or a string.

Problem 3-2   Write a MATLAB function called **isoddm(a)** that checks whether the elements of a are odd integers; a may be an array of numbers or a scalar. The definition of the function is

```
function x = isoddm(a)
```

$$\text{where the elements of } x = \begin{cases} 1 & \text{if the corresponding element in } a \text{ is an odd integer} \\ 0 & \text{if the corresponding element in } a \text{ is an even integer} \\ -1 & \text{if the corresponding element in } a \text{ is not an integer} \end{cases}$$

This program is the same as Problem 3-1, except that the input variable a may be a scalar or an array. The dimensions of x should be the same as the dimensions of a. The elements of x should be set to 1, 0, -1 according to the above equation and the individual elements of a. An individual element of x should be set to -1 if the corresponding element in a is a real number, a complex number, a *NaN*, or a string. Here are several examples:

$$a = 5 \qquad\qquad x = 1$$
$$a = 3.7 \qquad\qquad x = -1$$
$$a = 44 \qquad\qquad x = 0$$

$$a = [1\ 2\ 0\ 4\ 5\ 88.6\ 97\ 0\ 3.14] \qquad x = [1\ 0\ 0\ 0\ 1\ -1\ 1\ 0\ -1]$$

$$a = \begin{bmatrix} 1 & 3 & 4.4 \\ NaN & 4 & -6 \\ 27 & 33.1 & 5 \end{bmatrix} \qquad x = \begin{bmatrix} 1 & 1 & -1 \\ -1 & 0 & 0 \\ 1 & -1 & 1 \end{bmatrix}$$

Problem 3-3   Rewrite your solution of Program 2-20 so that it uses your **isodd** function of Problem 3-1.

Problem 3-4   Modify function **ask_q** discussed on page 167 so that the calling program can pass an error message as well as the question. The syntax of the function should be

```
function answer=ask_q(valid_answers, output_string, error_msg).
```

If the error message equals a null string, function **ask_q** should emit its standard error message. If the error message is not null, the function should display the user-specified error message and not display any portion of the standard error message.

Problem 3-5   Write a function called **powr_xxx (x,n)**, where **xxx** are your initials and n must be an integer. The function calculates x^n using recursion. Note that x^n can be calculated as $x*x^{n-1}$. Remember that a recursive function calls itself.

Problem 3-6   Write a function called **pow_xxx (x,n)**, where **xxx** are your initials and n must be an integer. The function calculates x^n using a **FOR** loop. Do not use the MATLAB built-in functions for calculating powers.

Problem 3-7   Write a script file that calculates 2^{32} three times, once using function **powr_xxx**, once using function **pow_xxx**, and once using the MATLAB power operator **2^32**. The script should use the MATLAB functions **TIC** and **TOC** to check how long each method takes to calculate the result.

Problem 3-8   Write a recursive MATLAB function called **pow_xxx(k, R)** that performs the sum specified below. Note that **xxx** in the function name are your initials. The function definition is

```
function x=pow_xxx(k, R)
```

where R is a real number such that $|R| < 1$ and k is an integer such that $0 \le k \le 90$. The value the function returns is

$$x = \sum_{i=k}^{90} R^i$$

If we want to write a recursive function to compute the sum, we rewrite the sum as

$$\sum_{i=k}^{90} R^i = \begin{cases} R^k + \sum_{i=k+1}^{90} R^i & \text{if } k < 90 \\ R^{90} & \text{if } k = 90 \end{cases}$$

Your function should do the appropriate error checking on k and R to see that the function is used correctly. If R or k are not valid values, your function should return a value of *NaN* for x.

Problem 3-9   Write a function similar to the **ask_q** function covered on page 167. This function will be called **ask_num_xxx**, where **xxx** are your initials. The format of the function is

```
function num_ans=ask_num(code, question, min, max).
```

The function only returns a valid numerical value. Variable *question* is a text string that contains the question that the function will ask. Variable *code* contains a numeric value that specifies the types of number requested. Variables *min* and *max* specify a range of acceptable numbers.

Code	Numeric type
0	any integer
1	any integer between *min* and *max* inclusive
2	any real number
3	any real number between *min* and *max* inclusive
4	any complex number in rectangular coordinates

- If the calling function specifies an invalid code, the **ask_num** function should emit an error message and return a numerical value of *NaN* to the calling function.
- If the calling function specifies a *min* value greater than the *max* value, the **ask_num** function should emit an error message and return a numerical value of *NaN* to the calling function.
- Your function should only return to the caller when a valid number has been entered. If an invalid answer is entered, the function should display an error message, state what the valid responses are, and then repeat the question.
- You must request the input as a text string and then use the **str2num** function to convert it to a numeric value. Your function should handle invalid inputs such as strings that do not represent numeric values.

Structured Data Types

4

Thus far, we have focused on simple data types that hold a single value. For example, x has had only single values, say the value 5. By using arrays, x can hold several values, say 5, 6, 78,432, 6743, 77, 99. One strength programming solutions to problems gives us is that if we can write a program to solve a problem for a single value, we can usually generalize the program to solve the problem many times. So if we solve the equation $y = 5 * \sin(\log(x))$ for a single value of x and y, we can solve the equation for many values of x and y if we use an array. Arrays also allow us to store large amounts of data in a single variable. Instead of using $x1$ to store one value, $x2$ another value, and $x3$ yet another value, we can store all the values in a single variable named x.

Structured data types allow us to store large amounts of data and different types of data in a single variable. We could, for example, designate one variable to hold address information for an address book. Not only does the variable hold data for several entries in the address book, it also holds information for each entry, such as name, phone number, zip code, and e-mail address. With structured data types, we can easily hold all of this information in a single variable.

In this chapter we'll use arrays and examine a few of their many applications.

4.1 Introduction

MATLAB Predefined
SIN function
`linspace` function
`plot` function

Arrays are variables that can store many values. Almost all programming languages support arrays. The power of programming is not that you can write a program to solve a problem for a single value but that, by using arrays, you can easily modify the program to solve for many values. For example, we can use the MATLAB **SIN** function to find the value $\sin(x)$ for a single value of x:

```
EDU» sin(0)
ans =
0
```

Here, the **SIN** function returns a single value equal to the sin(0). We can also pass the **SIN** function an array of values:

```
EDU» sin([0, pi/4, pi/2, (3*pi/4), pi])
ans =
0     0.7071     1.0000     0.7071     0.0000
```

Here we passed the **SIN** function 5 values, 0, $\dfrac{\pi}{4}$, $\dfrac{\pi}{2}$, $\dfrac{3\pi}{4}$, π, and the **SIN** function returned 5 values, sin(0), $\sin\left(\dfrac{\pi}{4}\right)$, $\sin\left(\dfrac{\pi}{2}\right)$, $\sin\left(\dfrac{3\pi}{4}\right)$, and sin($\pi$). We could just have easily asked for 1000 values. Let's plot the **SIN** function from 0 to 2π using 5, 10, 50, and 1000 points. But first we'll use the **LINSPACE** function to create the array of input values. This function creates a row array of values from a beginning number to an ending one using a specified number of points. The syntax is **LINSPACE** (*start_number*, *end_number*, *number_of_points*). So, to create a row array with numbers from 1 to 5 with a total of 5 points, we use

```
EDU» linspace(1,5,5)
ans =
1     2     3     4     5
```

To create a row array from 1 to 5 with 9 points, we use

```
EDU» linspace(1,5,9)
ans =
  Columns 1 through 7
1.0000     1.5000     2.0000     2.5000     3.0000     3.5000     4.0000
Columns 8 through 9
4.5000     5.0000
```

And to create a row array from 1 to 5 with 51 points, we would use

```
EDU» linspace(1,5,51)
ans =
  Columns 1 through 7
1.0000     1.0800     1.1600     1.2400     1.3200     1.4000     1.4800
Columns 8 through 14
1.5600     1.6400     1.7200     1.8000     1.8800     1.9600     2.0400
Columns 15 through 21
2.1200     2.2000     2.2800     2.3600     2.4400     2.5200     2.6000
Columns 22 through 28
2.6800     2.7600     2.8400     2.9200     3.0000     3.0800     3.1600
```

```
Columns 29 through 35
3.2400    3.3200    3.4000    3.4800    3.5600    3.6400    3.7200
Columns 36 through 42
3.8000    3.8800    3.9600    4.0400    4.1200    4.2000    4.2800
Columns 43 through 49
4.3600    4.4400    4.5200    4.6000    4.6800    4.7600    4.8400
Columns 50 through 51
   4.9200    5.0000
```

Let's use the **PLOT** function to plot the function sin(x). The syntax of the **PLOT** function is plot(x, y) where x and y are arrays containing the x- and y-coordinates of the points to be plotted. First we plot 5 points.

```
EDU» x=linspace(0,2*pi,5);
EDU» y=sin(x);
EDU» plot(x,y);
```

Here are the results of this code segment:

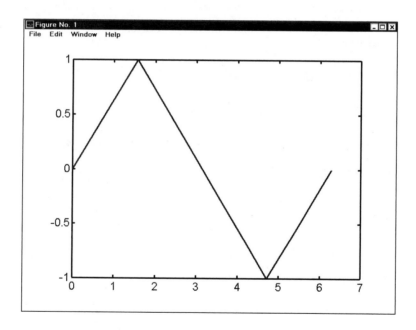

This is not a great plot because there are too few points in the plot, 5 in this case.
Let's try it with 10 points:

```
EDU» x=linspace(0,2*pi,10);
EDU» y=sin(x);
EDU» plot(x,y);
```

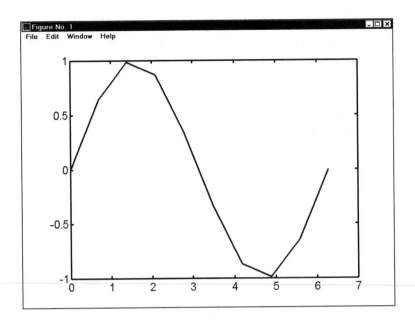

This is a little better, but the plot still doesn't look like a smooth sinusoidal curve, so we try 50 points:

```
EDU» x=linspace(0,2*pi,50);
EDU» y=sin(x);
EDU» plot(x,y);
```

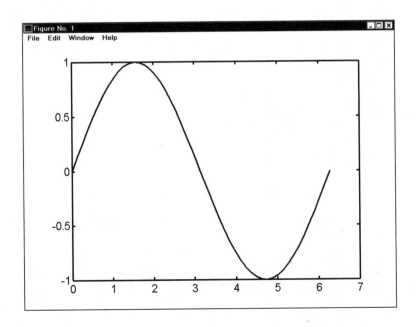

This plot is much nicer, and it was just as easy to plot 50 points as it was 5. Next, we plot 1000 points:

```
EDU» x=linspace(0,2*pi,1000);
EDU» y=sin(x);
EDU» plot(x,y);
```

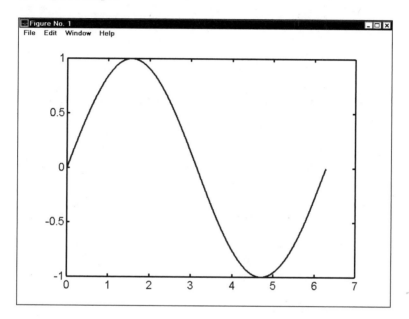

There's not a lot of difference between this graph and the previous one, but because we used arrays it was just as easy to construct. The power of arrays is that if you can write a program for a single value, using arrays enables you to modify the program to do the same thing for many values.

4.2 Arrays in MATLAB

In most programming languages, variables that store multiple values are called arrays. Arrays can have many dimensions. Here we will look at one-, two-, and three-dimensional arrays.

4.2.1 One-Dimensional Arrays

A one-dimensional array can be thought of as a table with one row. Here variable x contains 10 values:

$x(1)$	$x(2)$	$x(3)$	$x(4)$	$x(5)$	$x(6)$	$x(7)$	$x(8)$	$x(9)$	$x(10)$
5	3.4	6	27	19.5	3.9	213	21.87	0.9	2

Each element of the array can be individually addressed. For example, the fourth element of x is referenced as $x(4)$; its value is 27. The eighth element of x is referenced as $x(8)$ and its value is 21.87. This one-dimensional array has a single row. We'll refer to this structure as a "row array."

A one-dimensional array can also be thought of as a table with a single column:

$x(1)$	5
$x(2)$	3.4
$x(3)$	6
$x(4)$	27
$x(5)$	19.5
$x(6)$	3.9
$x(7)$	213
$x(8)$	21.87
$x(9)$	0.9
$x(10)$	2

The variables are still addressed in the same way; that is, the ninth variable is referenced as $x(9)$ and its value is 0.9. This one-dimensional array has a single column and we'll refer to this structure as a "column array."

To define x as a row array, we use this command:

```
EDU» x=[5, 3.4, 6, 27, 19.5, 3.9, 213, 21.87, 0.9, 2];
```

If we ask for the value of x, it is displayed as a row array:

```
EDU» x
x =
  Columns 1 through 7
5.0000    3.4000    6.0000    27.0000    19.5000    3.9000  213.0000
Columns 8 through 10
21.8700    0.9000    2.0000
```

We can ask for individual elements of the array:

```
EDU» x(5)
ans =
19.5000
EDU» x(7)
ans =
213
```

To define x as a column array, we use this command:

```
EDU» x=[5; 3.4; 6; 27; 19.5; 3.9; 213; 21.87; 0.9; 2];
```

Note that each element of the array is separated by a semicolon (;). Now the value of x is displayed as a column array:

```
EDU» x
x =
    5.0000
    3.4000
    6.0000
   27.0000
   19.5000
    3.9000
  213.0000
   21.8700
    0.9000
    2.0000
```

Each element of the array is addressed the same way whether it is a row or a column array:

```
EDU» x(1)
ans =
5
EDU» x(3)
ans =
6
EDU» x(9)
ans =
0.9000
```

One advantage of arrays is that we can use a variable to address array elements. Instead of referencing a cell by a fixed number—$x(5)$, for example—we reference the cell by an index i—for example, $x(i)$.

```
EDU» i=5;
EDU» x(i)
ans =
19.5000
```

Since the value of i can change, we can address different cells in the array rather than a specific cell:

```
for i = 1:10
fprintf('The value of x(%1.0f) is %g.\n', i, x(i));
end
The value of x(1) is 5.
The value of x(2) is 3.4.
The value of x(3) is 6.
The value of x(4) is 27.
The value of x(5) is 19.5.
```

```
The value of x(6) is 3.9.
The value of x(7) is 213.
The value of x(8) is 21.87.
The value of x(9) is 0.9.
The value of x(10) is 2.
```

In the code segment above, i takes on several values and we can thus access all elements of x with a few simple lines of code.

In older student editions of MATLAB, array sizes are limited. In MATLAB's Student version 5, the number of elements in an array is limited to 16,384. The student version 4 is limited to 8192 elements, and the student version 5.3 doesn't have array size limitations.

To show you how to use arrays, we'll write a MATLAB function with a single input argument that is the value of a resistor. The function returns the nearest standard 5% resistor. There are two sequences of standard 5% resistors. We will use the series 1, 1.2, 1.5, 1.8, 2.2, 2.7, 3.3, 3.9, 4.7, 5.6, 6.8, 8.2. Every decade of resistor values follows this sequence. A decade is a factor of 10 in numerical values. For example, 1 to 10 is a decade, 10 to 100 is a decade, as is 100 to 1,000, 1,000 to 10,000, and so on. Resistor values in a decade will follow the sequence above. For example, the available resistor values from 1000 to 10,000 are 1000, 1200, 1500, 1800, 2200, 2700, 3300, 3900, 4700, 5600, 6800, 8200, and 10,000. Function 4-1 finds the nearest standard value of resistance to a nonstandard resistance value. The function works for resistor values from 0.1 Ω to 10 MΩ.

Function 4-1 Finding the nearest standard 5% resistor for a given resistance value

```
function R_std = std_res(Rin)
% This is m-file std_res.m
% This function finds the closest standard 5% resistor
% to the specified resistor Rin.
% The allowed resistors are standard 5% resistors in the series
% 1,1.2, 1.5, 1.8, 2.2, 2.7, 3.3, 3.9, 4.7, 5.6, 6.8, and 8.2
% If the input resistance is not in the range of 0.1 Ohm to
% 10 MOhm, the function returns a value of [ ].

std = [1,1.2, 1.5, 1.8, 2.2, 2.7, 3.3, 3.9, 4.7, 5.6, 6.8, 8.2];
Resistors = [0.1*std,std, 10*std, 100*std, 1000*std, 10000*std, ➥
100000*std,1000000*std];
Resistors = [Resistors, 10e6];

if Rin < Resistors(1) | Rin > Resistors(length(Resistors))
    R_std = [ ];
    return
end

j=length(Resistors);
for i = 1:(length(Resistors)-1)
    if (Rin >= Resistors(i)) & (Rin < Resistors(i+1))
        j = i;
        break;
```

```
        end
end


if j == length(Resistors)
    R_std = Resistors(length(Resistors));
else
    if (Rin-Resistors(j))< (Resistors(j+1) - Rin)
        R_std = Resistors(j);
    else
        R_std = Resistors(j+1);
    end
end
```

The first line, `std = [1,1.2, 1.5, 1.8, 2.2, 2.7, 3.3, 3.9, 4.7, 5.6, 6.8, 8.2]`, creates an array with the sequence of basic standard values. The next line uses concatenation to create a large array with all of the standard values from 0.1 to 8.2×10^6. A shortened example would be `R=[std, 10*std]` where *std* is the variable holding the values listed above, and `10*std` is the array [10, 12, 15, 18, 22, 27, 33, 39, 47, 56, 68, 82]. The line `R=[std, 10*std]` concatenates these two arrays into a single array that has the standard resistor values for 1 Ω to 82 Ω.

```
EDU» R=[std, 10*std]
R =
  Columns 1 through 7
    1.0000    1.2000    1.5000    1.8000    2.2000    2.7000    3.3000
  Columns 8 through 14
    3.9000    4.7000    5.6000    6.8000    8.2000   10.0000   12.0000
  Columns 15 through 21
   15.0000   18.0000   22.0000   27.0000   33.0000   39.0000   47.0000
  Columns 22 through 24
   56.0000   68.0000   82.0000
EDU»
```

Another example is `R=[0.1*std, std, 10*std]`, where `0.1*std` creates a row array with the resistance values from 0.1 to 0.82, *std* is the array the contains values from 1 to 8.2, and the array `10*std` contains the standard values from 10 to 82. The line `R=[0.1*std, std, 10*std]` concatenates these three arrays into a single one that contains the standard resistor values from 0.1 Ω to 82 Ω. The line `Resistors = [Resistors, 10e6]` adds the value of 10 MΩ to the end of the array so that array Resistors contains the standard values from 0.1 Ω to 10 MΩ.

The code segment

```
for i = 1:(length(Resistors)-1)
    if (Rin >= Resistors(i)) & (Rin < Resistors(i+1))
        j = i;
        break;
    end
end
```

searches through the array of standard values and compares the input resistor value to the ith element and the $(i + 1)$st element. If Rin is greater than or equal to the ith element and less than the $(i + 1)$st element, j is set to i and the loop exits. When the loop exits, we know that the value of Rin is between the jth and the $(j + 1)$st element.

The statement

```
if (Rin-Resistors(j))< (Resistors(j+1) - Rin)
    R_std = Resistors(j);
else
    R_std = Resistors(j+1);
end
```

looks at the two values the surround Rin and picks the one closest to Rin.

Text Strings in MATLAB

A text string in MATLAB is a row array in which each element holds a single character of the string. For example, if $x =$ **this is a test**, then $x(1) =$ **t**, $x(4) =$ **s**, $x(7) =$ **s**, and so on.

```
EDU» x='this is a test';
EDU» x(1)
ans =
t
EDU» x(4)
ans =
s
EDU» x(7)
ans =
s
```

Representing strings as row arrays allows all of the MATLAB facilities available for manipulating arrays to be used with strings.

4.2.2 Two-Dimensional Arrays

Two-dimensional arrays can be thought of as a table. Here is an array with 4 rows and 5 columns:

x	Column 1	Column 2	Column 3	Column 4	Column 5
Row 1	25.3	26.7	18.4	2	17.9
Row 2	−33	−27.9	13.6	3.1415	88
Row 3	−2.7	3.9	1.9	7.8	99
Row 4	13.9	−12.2	17	26.9	−199

In this array, x contains 20 values. An individual cell is referenced by its row and column numbers, x(row, column). Note that the row number comes first. Thus, $x(1, 2) = 26.7$, $x(3, 1) = -2.7$, and $x(2, 5) = 88$. To enter this array in MATLAB, we use the command below:

```
EDU» x=[25.3, 26.7, 18.4, 2, 17.9; -33, -27.9, 13.6, 3.1415, 88;
-2.7, 3.9, 1.9, 7.8, 99; 13.9, -12.2, 17, 26.9, -199];
```

Note that elements in the same row are separated by a comma or a space. Rows are separated by a semicolon. When we display x, it is displayed as an array.

```
EDU» x
x =
      25.3000    26.7000    18.4000     2.0000    17.9000
     -33.0000   -27.9000    13.6000     3.1415    88.0000
      -2.7000     3.9000     1.9000     7.8000    99.0000
      13.9000   -12.2000    17.0000    26.9000  -199.0000
EDU»
```

We can now reference individual elements of the array by specifying its row and column numbers:

```
EDU» x(1,2)
ans =
26.7000
EDU» x(3,1)
ans =
   -2.7000
EDU» x(2,5)
ans =
88
```

Arrays of Text Strings

In Section 4.2.1, we saw that a text string is a one-dimensional array whose elements are the individual characters in the string. Suppose we want to create an array whose elements are the days of the week. That is, we want array *days* to contain the text strings `'Monday'`, `'Tuesday'`, `'Wednesday'`, `'Thursday'`, `'Friday'`, `'Saturday'`, and `'Sunday'`. A first attempt might be to use this command:

```
EDU» days=['Monday', 'Tuesday', 'Wednesday', 'Thursday', ➡
'Friday', 'Saturday', 'Sunday']
days =
MondayTuesdayWednesdayThursdayFridaySaturdaySunday
```

The result is the concatenation of all of the strings. If you review Section 1.9.2, you will see that the command we used above is the command for concatenating strings. Thus, we can't create a row array whose elements are strings unless we want a single long string. Now let's try to create a column array where each element of the array is one of the strings. With a column array, a new row is specified with a semicolon (;).

```
EDU» days=['Monday'; 'Tuesday'; 'Wednesday'; 'Thursday'; ➡
'Friday'; 'Saturday'; 'Sunday']
??? All rows in the bracketed expression must have the same
number of columns.
```

This command generates an error because each element in the column array is a string, and a string is a row array. We are creating a column array whose *elements* are row arrays—that is, a two-dimensional array. The problem is our strings are not the same length. In an array, each row must have the same number of columns, but 'Monday' has 6 columns (one for each letter) while 'Wednesday' has 9 columns. So, to make a column array with these text strings, we must add spaces to the strings so that they all have the same number of characters. The maximum number of characters in these strings is 9, and we add spaces so that each string contains 9 characters:

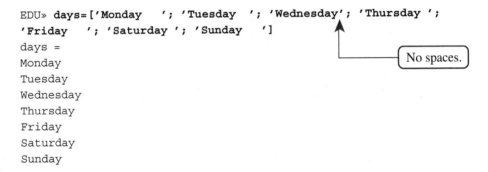

```
EDU» days=['Monday   '; 'Tuesday  '; 'Wednesday'; 'Thursday ';
'Friday   '; 'Saturday '; 'Sunday   ']
days =
Monday
Tuesday
Wednesday
Thursday
Friday
Saturday
Sunday
```

No spaces.

We now have an array of characters. This array can be viewed as a table:

	Col 1	Col 2	Col 3	Col 4	Col 5	Col 6	Col 7	Col 8	Col 9
Row 1	M	o	n	d	a	y			
Row 2	T	u	e	s	d	a	y		
Row 3	W	e	d	n	e	s	d	a	y
Row 4	T	h	u	r	s	d	a	y	
Row 5	F	r	i	d	a	y			
Row 6	S	a	t	u	r	d	a	y	
Row 7	S	u	n	d	a	y			

We address individual characters in the array by specifying their row and column numbers:

```
EDU» days(3,1)
ans =
W
```

```
EDU» days(4,7)
ans =
a
EDU» days(7,2)
ans =
u
```

To ask for an entire row, we use a colon (:) for the column address:

```
EDU» days(1,:)
ans =
Monday
EDU» days(4,:)
ans =
Thursday
```

It's difficult to make an array of text strings using a standard array, so MATLAB has a data structure called a cell array that allows us to make such arrays easily. Cell arrays are covered in more detail in Section 4.7. To make a cell array, we use curly brackets instead of square brackets:

```
EDU» days = {'Monday'; 'Tuesday'; 'Wednesday'; 'Thursday'; ➡
'Friday'; 'Saturday'; 'Sunday'}
days =
    'Monday'
    'Tuesday'
    'Wednesday'
    'Thursday'
    'Friday'
    'Saturday'
    'Sunday'
EDU»
```

Each text string is now an element of the cell array. We can use many of the addressing methods that we used with standard arrays, but instead of using parentheses, we use curly brackets:

```
EDU» days{2}
ans =
Tuesday
EDU» days{5}
ans =
Friday
EDU» days{1:2}
ans =
Monday
```

```
ans =
Tuesday
EDU»
```

To address a single character of a text string, we address the cell first, and then the element of the array. For example, to find the third letter of the fifth cell, we would use

```
EDU» days{5}(3)
ans =
i
EDU»
```

4.2.3   Multidimensional Arrays

MATLAB version 5 also handles arrays of three or more dimensions. If we picture a one-dimensional array as a row of cells with one address specifying the address of the cell, then a two-dimensional array is a table of cells with two addresses specifying the row and column numbers of the cell, or the x- and y-coordinates of the cell, and a three-dimensional array is a cube of cells. To address a cell in the cube, we specify three addresses for the x-, y-, and z-coordinates of the cell.

As an example, let's specify a 2×2×2 array of zeros in MATLAB:

```
EDU» x=zeros(2,2,2)
x(:,:,1)  =
       0      0
       0      0

x(:,:,2)  =
       0      0
       0      0

EDU»
```

The cube can be viewed as two tables of data placed one behind the other. The front table is `x(:,:,1)` and the back table is `x(:,:,2)`.

Another example is

```
EDU» y=ones(2,2,4)
y(:,:,1)  =
       1      1
       1      1

y(:,:,2)  =
       1      1
       1      1
```

```
y(:,:,3) =
       1       1
       1       1


y(:,:,4) =
       1       1
       1       1


EDU»
```

This cube of data contains four 2×2 tables of data.

To address an element, all we do is specify the three coordinates of the cell. Let's change the cell in the first row, second column, and third table to 5:

```
EDU» y(1,2,3)=5
y(:,:,1) =
       1       1
       1       1


y(:,:,2) =
       1       1
       1       1


y(:,:,3) =
       1       5
       1       1


y(:,:,4) =
       1       1
       1       1


EDU»
```

4.3 Addressing Arrays

MATLAB Predefined
For loops
length function

MATLAB has several facilities for addressing arrays. These include specifying a single element to specifying a small array within a larger array. We'll start with one-dimensional arrays and then extend the techniques to two-dimensional arrays.

4.3.1 Addressing One-Dimensional Arrays

First let's define a string variable.

```
EDU» a='1234567890abcdefghijklmnopqrstuvwxyzABCDEFGHIJKLMNOPQRSTUVWXYZ';
```

We address individual elements by specifying the position in the array. To see what the 10th element is, we type **a(10)**.

```
EDU» a(10)
ans =
0
```

The 10th element in string *a* is the character zero. The 36th element is

```
EDU» a(36)
ans =
z
```

The number in the parentheses is called the index. So in **a(10)** the index is 10, and in **a(36)** the index is 36.

If we want to *change* the value of an element, we also use the index. Let's change the value of the 10th element to the character A.

```
EDU» a(10)='A';
a
a =
123456789AabcdefghijklmnopqrstuvwxyzABCDEFGHIJKLMNOPQRSTUVWXYZ
```

The 10th element of string *a* is now A. We can use indexed arrays in equations in the same way we use scalar variables. For example, the equation

average=(x+y+z)/3

could also be written with indexed variables as

average=(x(1)+x(2)+x(3))/3

Both are valid MATLAB statements.

Next, suppose we want to use a subarray of array *a*. For example, string *a* contains the characters 0–9, a–z, and A–Z. From array *a*, we want to create three new arrays; the first array contains only numbers, the second one only lowercase letters, and the third array only uppercase letters. We could easily create these smaller arrays using **FOR** loops, but MATLAB has built-in addressing functions that make the task even easier.

To create the numbers-only subarray, we copy the first 10 elements of array *a* to a new array. This can be done with the command

```
EDU» a='1234567890abcdefghijklmnopqrstuvwxyzABCDEFGHIJKLMNOPQRSTUVWXYZ';
EDU» numbers=a(1:10);
```

The term **a(1:10)** means return a subarray whose elements are $a(1)$, $a(2)$, $a(3)$, $a(4)$, $a(5)$, $a(6)$, $a(7)$, $a(8)$, $a(9)$, and $a(10)$. The MATLAB statement **a(1:10)** is equivalent to the statement [**a(1)**, **a(2)**, **a(3)**, **a(4)**, **a(5)**, **a(6)**, **a(7)**, **a(8)**, **a(9)**, **a(10)**]. The result of the preceding command is

```
EDU» numbers
numbers =
1234567890
```

As you can see, text string *numbers* contains 10 elements and is a subarray of the original array *a*. To create the lowercase subarray, we specify the indices for the lowercase elements in array *a*:

```
EDU» lower_case=a(11:36)
lower_case =
abcdefghijklmnopqrstuvwxyz
```

To create the uppercase array, we specify the indices for the uppercase elements:

```
EDU» upper_case=a(37:62)
upper_case =
ABCDEFGHIJKLMNOPQRSTUVWXYZ
```

We can also create this last example using the MATLAB **LENGTH** function. The **LENGTH** function returns the number of elements in an array:

```
EDU» length(a)
ans =
62
```

Thus, the two statements **a(37:62)** and **a(37:length(a))** are equivalent.

```
EDU» upper_case=a(37:length(a))
upper_case =
ABCDEFGHIJKLMNOPQRSTUVWXYZ
```

MATLAB Example

Create a substring containing the lowercase letters using a **FOR** loop rather than MATLAB indexing such as **a(11:36)**.

```
a='1234567890abcdefghijklmnopqrstuvwxyzABCDEFGHIJKLMNOPQRSTUVWXYZ';
lower_case = '                          ';
for i = 11:36
  lower_case(i-10)=a(i);
end
lower_case
```

```
lower_case =
abcdefghijklmnopqrstuvwxyz
```

Exercise 4-1   Modify the code segment in the box above so that one substring contains only upper-case letters.

So now you may be asking, How can I create a substring that contains several substrings of the original array? It's pretty easy, actually. We simply specify substrings and concatenate them. We can do this several ways. Let's create an array that contains both the numbers and the uppercase letters. The string is 1234567890ABCDEFGHIJKLM NOPQRSTUVWXYZ. Here is one way:

```
EDU» x=a(1:10);
EDU» y=a(37:62);
EDU» num_caps=[x, y]
num_caps =
1234567890ABCDEFGHIJKLMNOPQRSTUVWXYZ
```

We could also use

```
EDU» num_caps=[a(1:10), a(37:62)]
NUM_CAPS =
1234567890ABCDEFGHIJKLMNOPQRSTUVWXYZ
```

The two methods are equivalent.

Note the similarity of the statement **37:62** used in addressing arrays and in **FOR** loops. We could use the statements

```
for i = 37:62
    x=x(i-36)=a(i)
end
```

or we could use

```
EDU» x=a(37:62)
```

In both cases, **37:62** means generate numbers 37–62 in unit steps—that is, 37, 38, 39, ..., 61, 62. In **FOR** loops, we saw a much more general use of this counting method. For example, **1:10** means generate numbers from 1 to 10 incrementing by 1—that is, 1, 2, 3, 4, 5, 6, 7, 8, 9, 10. We can specify other increments. Thus, **0:2:10** means count from 0 to 10 stepping by 2. This generates the numbers 0, 2, 4, 6, 8, 10. To count backwards, we specify a negative increment. The statement **10:-1:1** generates numbers 10, 9, 8, 7, 6, 5, 4, 3, 2, 1.

Only one counting method does not work for indexing arrays—specifying an increment that is not an integer. For example, `1:0.25:3` generates numbers from 1 to 3 with 0.25 increments—that is, 1, 1.25, 1.5, 1.75, 2, 2.25, 2.5, 2.75, 3. These numbers are not valid indices for arrays.

With these new counting methods, we can manipulate the arrays with a few different techniques. To obtain every other character in string *a,* we could use

```
EDU» a_half=a(1:2:62)
a_half =
13579acegikmoqsuwyACEGIKMOQSUWY
```

We can get every third character:

```
EDU» a_three=a(1:3:62)
a_three =
1470cfiloruxADGJMPSVY
```

We can reverse the order of the characters:

```
EDU» a_reverse=a(62:-1:1)
a_reverse =
ZYXWVUTSRQPONMLKJIHGFEDCBAzyxwvutsrqponmlkjihgfedcba0987654321
```

We can reverse the characters and skip every other character:

```
EDU» a_rev_two=a(62:-2:1)
a_rev_two =
ZXVTRPNLJHFDBzxvtrpnljhfdb08642
```

The only time you will generate errors or erroneous results is if you generate an index that is not a positive integer greater than 0, or the index is not within the range of indices for a particular variable.

```
EDU» bad_num=a(1:0.25:3)
bad_num =
112222333
```

Here the noninteger index generates a result, but it is not the result we intended.

```
EDU» bad_num=a(10:-1:-10)
???  Index into matrix is negative or zero.
```

This example generates indices from 10 to –10. The indices 0, –1, –2, …, *n* generate the error message, `???  Index into matrix is negative or zero.`

4.3.2 Addressing Multidimensional Arrays

To address multidimensional arrays, we use the same techniques as with one-dimensional arrays, except that we specify more than one index. For the first few examples, we will use the 4×7 array defined as follows:

```
EDU» a=[1 2 3 4 5 6 7; 21 22 23 24 25 26 27; 31 32 33 34 35 36 37; ➡
41 42 43 44 45 46 47]
a =
        1     2     3     4     5     6     7
       21    22    23    24    25    26    27
       31    32    33    34    35    36    37
       41    42    43    44    45    46    47
```

To address an individual element of the array, we specify the row number and column number of the element:

```
EDU» a(2,3)
ans =
23
EDU» a(1,1)
ans =
1
EDU» a(4,7)
ans =
47
```

Note that the row number must be specified first. For this example, **a(7,4)** generates an error because there are not 7 rows in this array:

```
EDU» a(7,4)
???  Index exceeds matrix dimensions.
```

Suppose we want to create a row array that has the same elements as the second row of array *a*. We do this by specifying a range of numbers for the column index.

```
EDU» row2=a(2,1:7)
row2 =
21    22    23    24    25    26    27
```

Note that the term **1:7** generates the numbers 1, 2, 3, 4, 5, 6, 7. For this array, **1:7** specifies the entire row. A shortcut for specifying the entire row uses the colon by itself. The following statement is equivalent to the one above:

```
EDU» row2a=a(2,:)
row2a =
21    22    23    24    25    26    27
```

In this case, a colon specifies the entire row, independent of the length of the row.

We can create a row array with fewer elements:

```
EDU» row2_short=a(2,1:4)         or        EDU» row2_shorty=a(2,3:5)
row2_short =                               row2_shorty =
21    22    23    24                       23    24    25
```

To make a column of the table, we specify a range of numbers for the row index. To get the fifth column of the array and place it in a column array, we use

```
EDU» col5=a(1:4, 5)
col5 =
       5
      25
      35
      45
```

As a shortcut, we can use the colon by itself for the row index. The command below is equivalent to the previous example:

```
EDU» col5a=a(:, 5)
col5a =
       5
      25
      35
      45
```

In this case, the colon means the entire column, independent of how long the column is.

By specifying a range for both the row and column indices, we can create sub-arrays of the original array. To create a 3×3 subarray, we use the statement

```
EDU» sub1=a(1:3, 1:3)
sub1 =
       1     2     3
      21    22    23
      31    32    33
```

We can use any of the variations on indices for two-dimensional arrays that we used for one-dimensional arrays. In the following example, the subarray, the first two rows of the original array but the order of the row elements is reversed.

```
EDU» sub2=a(1:2, 7:-1:1)
sub2 =
       7     6     5     4     3     2     1
      27    26    25    24    23    22    21
```

In the next example, the subarray uses rows 2–4 and columns 4–7 of the original array. The row order is reversed.

```
EDU» sub3=a(4:-1:2, 4:7)
sub3 =
        44      45      46      47
        34      35      36      37
        24      25      26      27
```

The last example takes all rows and columns of the original array and reverses the row order and the column order.

```
EDU» sub5=a(4:-1:1, 7:-1:1)
sub5 =
        47      46      45      44      43      42      41
        37      36      35      34      33      32      31
        27      26      25      24      23      22      21
         7       6       5       4       3       2       1
```

4.4 Examples of Using Arrays

Let's look at some applications of arrays.

4.4.1 Generating a Histogram

MATLAB Predefined
abs function
setstr function
upper function
lower function
fscanf function
for loop
sort function
randn function
hist function

As a first example, we'll generate a histogram. We have a text document and we want to know how many times the letter a appears in the document, and also the letters b, c, and so on. Let's treat capital and lowercase letters as the same letter, and we'll have a category called "other" for nonalphabetic characters. Spaces will be ignored.

Before we do this, let's look at how the **ABS** function works with characters.

```
EDU» abs('a')
ans =
97
EDU» abs('b')
ans =
98
EDU» abs('c')
ans =
99
EDU» abs('d')
ans =
100
EDU» abs('z')
ans =
122
```

As you can see, the lowercase letters are represented by a number between 97 and 122, and the numbers are sequential. Look at the capital letters—they are represented by the numbers 65–90:

```
EDU» abs('A')
ans =
65
EDU» abs('B')
ans =
66
EDU» abs('C')
ans =
67
EDU» abs('D')
ans =
68
EDU» abs('Z')
ans =
90
```

Also note that the numeric difference between a and A is the same as between b and B, c and C, and z and Z.

```
EDU» abs('a')-abs('A')
ans =
32
EDU» abs('b')-abs('B')
ans =
32
EDU» abs('c')-abs('C')
ans =
32
EDU» abs('z')-abs('Z')
ans =
32
```

To convert uppercase letters to lowercase, all we do is add 32 to the lowercase letters.

MATLAB function **SETSTR** is the inverse of the **ABS** function for characters. It converts a number into a character:

```
EDU» setstr(97)
ans =
a
EDU» setstr(65)
ans =
A
```

```
EDU» setstr(90)
ans =
Z
```

With the **ABS** and **SETSTR** functions, we can perform numerical calculations on characters to manipulate them. For example, to convert an uppercase letter to lowercase, we convert the character to a number, add 32 to the number, and then convert the result back to a character.

```
EDU» x='K';
EDU» xnum=abs(x);
EDU» lc=xnum+32;
EDU» x=setstr(lc)
x =
k
```

This can be done on a single line:

```
EDU» x='Q';
EDU» x=setstr(abs(x)+32)
x =
q
```

On different systems, the difference between **'a'** and **'A'** may not be 32; however, the differences between **'a'** and **'A'**, **'b'** and **'B'**, and **'z'** and **'Z'** will be the same. To make the operation a bit more general, we define a constant for the difference between the two:

```
EDU» diff=abs('a')-abs('A');
EDU» x='M';
EDU» x=setstr(abs(x)+diff)
x =
m
```

These examples show how we check if a letter is a capital letter, convert it to lowercase, and then determine what the letter is. Now we can build the histogram function. Part 1 of the process is shown in Function 4-2.

Function 4-2 Part 1 of the histogram function

```
function z=hist1(filename)
fid=fopen(filename,'r');  %Open a file.
in_string=fscanf(fid, '%s');       %Read the file as a character string.
fclose(fid);
% Display the string.
in_string
```

The first few lines in Function 4-2 open a file for reading, read the information into a text string, and then close the file. The input to the function is variable *filename*. This text string contains the name of the file you want to open. The line **fid=fopen(filename,'r')** opens the file specified in variable *filename*. The 'r' specifies that the file is read-only. That is, we can't write to this file and alter its contents. Variable *fid* holds a code that is used to refer to the file just opened. The **FSCANF** function reads a specified file using a specified format. The '%s' tells the function to read the information as a text string. The line **in_string=fscanf(fid, '%s')** reads the entire contents of the file and stores all the information in a single text string named *in_string*. Note that the %s format removes spaces and new line characters as it reads the file. If you want *in_string* to contain the spaces and new line characters that are in the data file, use the %c format. The line **fclose(fid)** closes the file. It's always good practice to close all files that you have opened before exiting any program.

The last line of Function 4-2 displays the value of variable *in_string*. We will run these few lines to see if we've read the information correctly. We'll test the function on two files, using a file called data.txt that contains two lines:

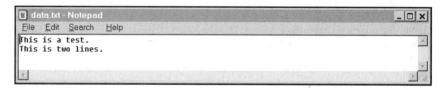

Let's see if we can read this file with Function 4-2:

```
EDU» hist1('data.txt');
in_string =
Thisisatest.Thisistwolines.
```

The function reads the entire file, removes spaces and new line characters from the text, and the entire file is contained in a single string. For a second test, let's read a file called data2.txt:

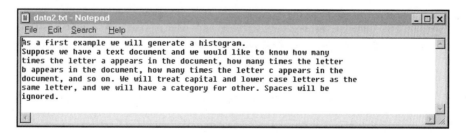

```
EDU» hist1('data2.txt');
in_string =
```

Asafirstexamplewewillgenerateahistogram.Supposewehaveatextdocumen
tandwewouldliketoknowhowhowmanytimestheletteraappearsinthedocument,h
owmanytimestheletterbappearsinthedocument,howmanytimestheletterca
ppearsinthedocument,andsoon.Wewilltreatcapitalandlowercaselettersast
hesameletter,andwewillhaveacategoryforother.Spaceswillbeignored.

It appears to work as expected. In Function 4-3, we add lines to change capital letters to lowercase.

Function 4-3 Part 2 of the histogram function

```
function z=hist2(filename)
fid=fopen(filename,'r');
in_string=fscanf(fid, '%s');
fclose(fid);

% Convert Uppercase letters to lowercase
Low_bound = abs('A');
Up_bound= abs('Z');
Diff = abs('a') - abs('A');

for i = 1:length(in_string)
  if (abs(in_string(i))>= Low_bound) & (abs(in_string(i))<= Up_bound)
      in_string(i) = setstr(abs(in_string(i)) + Diff);  % Convert to lower-
case.
  end
end
in_string
```

Function 4-3 checks each letter in the string. If the **ABS** of the letter indicates that the letter is a capital, then it converts the function to a lowercase letter. Let's test this function on our two data files:

```
EDU» hist2('data.txt')
in_string =
thisisatest.thisistwolines.
EDU» hist2('data2.txt')
in_string =
```

asafirstexamplewewillgenerateahistogram.supposewehaveatextdocumen
tandwewouldliketoknowhowhowmanytimestheletteraappearsinthedocument,h
owmanytimestheletterbappearsinthedocument,howmanytimestheletterca
ppearsinthedocument,andsoon.wewilltreatcapitalandlowercaselettersast
hesameletter,andwewillhaveacategoryforother.spaceswillbeignored.

Before we continue, note that MATLAB has a built-in function for converting uppercase letters to lowercase letters:

EDU» **help lower**

```
LOWER  Convert string to lowercase.
   B = LOWER(A) converts any uppercase characters in A to
   the corresponding lowercase character and leaves all
   other characters unchanged.

   Character sets supported:
      Mac  : Standard Roman
      PC   : Windows Latin-1
      Other: ISO Latin-1 (ISO 8859-1)

   See also UPPER.
```

EDU»

The **LOWER** function can be used to simplify Function 4-3, as Function 4-4 shows.

Function 4-4 Alternate method for part 2 of the histogram function

```
function z=hist2a(filename)
fid=fopen(filename,'r');
in_string=fscanf(fid, '%s');
fclose(fid);

% Convert Uppercase letters to lowercase

in_string=lower(in_string);

in_string
```

Let's test the new method with the second data file:

EDU» **hist2a('data2.txt')**

in_string =

asafirstexamplewewillgenerateahistogram.supposewehaveatextdocumen
tandwewouldliketoknowhowmanytimestheletteraappearsinthedocument,h
owmanytimestheletterbappearsinthedocument,howmanytimestheletterca
ppearsinthedocument,andsoon.wewilltreatcapitalandlowercaselettersast
hesameletter,andwewillhaveacategoryforother.spaceswillbeignored.

EDU»

You can use either method; however, the MATLAB **LOWER** function is a built-in function, so it's more efficient than the method presented in Function 4-3. In this example, we are explaining the inner workings of function **LOWER**, so we'll continue to use the method shown in Function 4-3.

Now let's convert the entire array to numeric values and then scan the array for characters that are not lowercase letters. All characters outside this range are given the code **abs('z')+1**, the code for the character after the letter z. We don't really care what this letter is because we never print it except for debugging purposes.

Function 4-5 Part 3 of the histogram function

```
function z=hist3(filename)
fid=fopen(filename,'r');
in_string=fscanf(fid, '%s');
fclose(fid);

% Convert Uppercase letters to lowercase
Low_bound = abs('A');
Up_bound= abs('Z');
Diff = abs('a') - abs('A');

for i = 1:length(in_string)
    if (abs(in_string(i))>= Low_bound) & (abs(in_string(i))<= Up_bound)
        in_string(i) = setstr(abs(in_string(i)) + Diff);
    end
end
% Scan for characters that are not lowercase letters.
bad_char=abs('z')+1;
Low_bound = abs('a');
Up_bound= abs('z');
num_string=abs(in_string); %Convert the text string to numbers.

for i = 1:length(num_string);
    if (num_string(i) < Low_bound) | (num_string(i) > Up_bound)
        num_string(i) = bad_char;
    end
end

setstr(num_string) %Print out the data as text for demonstration purposes.
```

The **FOR** loop sweeps through each character in the string. The **IF** statement

```
if (num_string(i) < Low_bound) | (num_string(i) > Up_bound)
        num_string(i) = bad_char;
    end
```

checks each character to see if it is below the lower bound (*Low_bound,* the code for **'a'**) or above the upper bound (*Up_bound,* the code for **'z'**). If so, that character is replaced by the value of *bad_char*, the code for the character after the letter z (remember that **bad_char=abs('z')+1**).

What is *bad_char*? It doesn't really matter, but let's see if it can be displayed:

```
EDU» setstr(bad_char)
ans =
{
```

It appears to be a left curly bracket. Now let's test our function. All commas and periods should be replaced by a left curly bracket.

```
EDU» hist3('data.txt');
ans =
thisisatest{thisistwolines{
EDU» hist3('data2.txt');
ans =
asafirstexamplewewillgenerateahistogram{supposewehaveatextdocument
andwewouldliketoknowhowmanytimestheletteraappearsinthedocument{how
manytimestheletterbappearsinthedocument{howmanytimesthelettercappe
arsinthedocument{andsoon{wewilltreatcapitalandlowercaselettersasth
esameletter{andwewillhaveacategoryforother{spaceswillbeignored{
```

Variable *num_string* is an array of numbers between 97 and 123. The number 97 represents the letter a and the number 122 represents the letter z. A 123 represents all other characters and symbols. The next thing we want to do is change the numbers so that a 1 represents the letter a, a 26 represents the letter z, and a 27 represents all other characters. To do this, we subtract 96 from all letters. Let's create a sample string to demonstrate:

```
EDU» a_str='abcxyz'
EDU» a_num=abs(a_str)
a_str =
abcxyz
a_num =
97    98    99    120   121   122
```

We could use a **FOR** loop to subtract 96 from every element of *a_num:*

```
for i = 1:length(a_num)
   a_num2(i)=a_num(i)-96;
end
a_num2
a_num2 =
1    2    3    24    25    26
```

Using the matrix properties of MATLAB, we don't need a **FOR** loop:

```
EDU» a_num3=a_num-96
a_num3 =
    1     2     3    24    25    26
```

Since 96 is a scalar, MATLAB automatically subtracts 96 from each element. Thus, the two methods are equivalent.

The next improvement we want to make is to generate the number 96 using MATLAB functions. Instead of using 96, we can use **abs('a')-1**. If we use an equation like **abs(x)-96** in our program and, in a future version of MATLAB, **abs('a')** changes to a number other than 97, then **abs(x)-96** will no longer work. If we use **abs(x)-(abs('a')-1)** instead of **abs(x)-96**, then if the computer representation of a to z changes, our code will still work. Let's make the following changes:

```
EDU» a_num4=a_num-(abs('a')-1)
a_num4 =
    1     2     3    24    25    26
```

In Function 4-6, we modify the program to shift the numerical values contained in the array from 97 down to 1.

Function 4-6   Part 4 of the histogram function

```
function z=hist4(filename)
fid=fopen(filename,'r');
in_string=fscanf(fid, '%s');
fclose(fid);

% Convert uppercase letters to lowercase
Low_bound = abs('A');
Up_bound= abs('Z');
Diff = abs('a') - abs('A');

for i = 1:length(in_string)
   if (abs(in_string(i))>= Low_bound) & (abs(in_string(i))<= Up_bound)
       in_string(i) = setstr(abs(in_string(i)) + Diff);
   end
end

% Scan for characters that are not lowercase letters.
bad_char=abs('z')+1;
Low_bound = abs('a');
Up_bound= abs('z');
num_string=abs(in_string); %Convert the text string to numbers.
for i = 1:length(num_string);
   if (num_string(i) < Low_bound) | (num_string(i) > Up_bound)
       num_string(i) = bad_char;
   end
end
```

```
%Shift the numerical codes for 'a' through 'z' down to 1 through 26.
shift=abs('a')-1;
num_string = num_string - shift;


num_string %Print out the data as numbers.
```

Let's test Function 4-6; it should display an array of numbers whose values are between 1 and 27.

```
EDU» hist4('data.txt');
num_string =
  Columns 1 through 12
    20    8    9   19    9   19    1   20    5   19   20   27
  Columns 13 through 24
    20    8    9   19    9   19   20   23   15   12    9   14
  Columns 25 through 27
     5   19   27
```

The arrays are now modified sufficiently to count the characters. We need to count the number of times the number 1 occurs, the number of times 2 occurs, the number of times 3 occurs, and so on. To do this, we'll create an array with 27 elements. Variable *count*(1) will count how many times 1 occurs, *count*(2) how many times 2 occurs, *count*(3) how many times 3 occurs, and so on. Note that array *num_string* contains elements whose values are between 1 and 27. Thus, we can use these values as the index to the count array. We'll show two ways to count the elements. First, let's create a test array *x* that contains numbers between 1 and 5:

```
x=[1 2 3 4 5 1 3 2 4 5 1 2 4 1 2 3 1 2 3 1 2 4 1 3 4 2 2 4 2 1 4 ➡
3 2 5 4 2 3 4 3];
```

We'll also initialize the count array to contain a row array of 5 zeros:

```
count=zeros(1,5);
```

Now let's count how many times each number occurs in array *x* using this method:

```
for i = 1:length(x)
    if x(i) == 1
        count(1) = count(1) +1;
    elseif x(i) == 2
        count(2) = count(2) + 1;
    elseif x(i) == 3
        count(3) = count(3) +1;
```

```
    elseif x(i) == 4
            count(4) = count(4) + 1;
    else
            count(5) = count(5) + 1;
    end
end
count
count =
8    11    8    9    3
```

The results show that 1 occurs 8 times, 2 occurs 11 times, 3 occurs 8 times, and so on. What we notice from the preceding example are lines like

```
elseif x(i) == 3
        count(3) = count(3) +1;
```

Note the repeated use of the number 3. Here $x(i) = 3$ and the index into the count array is 3. In the code

```
if x(i) == 1
        count(1) = count(1) +1;
```

you can also see that $x(i) = 1$ and that the index into the count array is 1. This occurs in all sections of the **IF** statement. Instead of the **IF** statement, we could use this code:

```
count(x(i)) = count(x(i)) +1
```

We can now rewrite the counter:

```
x=[1 2 3 4 5 1 3 2 4 5 1 2 4 1 2 3 1 2 3 1 2 4 1 3 4 2 2 4 2 1 4 ➡
3 2 5 4 2 3 4 3];
count=zeros(1,5);
for i = 1:length(x)
    count(x(i)) = count(x(i))+1;
end
count
count =
8    11    8    9    3
```

This method is much more concise, so we'll now add this code in Function 4-7.

Function 4-7 Part 5 of the histogram function

```
function z=hist5(filename)
fid=fopen(filename,'r');
in_string=fscanf(fid, '%s');
```

```
fclose(fid);
% Convert Uppercase letters to lowercase
Low_bound = abs('A');
Up_bound= abs('Z');
Diff = abs('a') - abs('A');

for i = 1:length(in_string)
   if (abs(in_string(i))>= Low_bound) & (abs(in_string(i))<= Up_bound)
        in_string(i) = setstr(abs(in_string(i)) + Diff);
   end
end

bad_char=abs('z')+1;
Low_bound = abs('a');
Up_bound= abs('z');
num_string=abs(in_string); %Convert the text string to numbers.
for i = 1:length(num_string);
   if (num_string(i) < Low_bound) | (num_string(i) > Up_bound)
        num_string(i) = bad_char;
   end
end

shift=abs('a')-1;
num_string = num_string - shift;


count=zeros(1,27);
for i = 1:length(num_string)
   count(num_string(i)) = count(num_string(i))+1;
end

count
```

This function should now count the occurrences of each letter:

```
EDU» hist5('data.txt');
count =
  Columns 1 through 12
     1      0      0      0      2      0      0      2      5      0      0      1
  Columns 13 through 24
     0      1      1      0      0      0      6      5      0      0      1      0
  Columns 25 through 27
     0      0      2
EDU» hist5('data2.txt');
count =
  Columns 1 through 12
    32      2      9     10     51      2      4     14     15      0      2     18
```

```
Columns 13 through 24
  13    18    19    11     0    17    18    35     6     2    15     2
Columns 25 through 27
   4     0     8
```

The last thing we need to do is to display the histogram. We accomplish this with a simple **FOR** loop in Function 4-8.

Function 4-8 Final histogram function

```
function z=hist6(filename)
fid=fopen(filename,'r');
in_string=fscanf(fid, '%s');
fclose(fid);

% Convert Uppercase letters to lowercase
Low_bound = abs('A');
Up_bound= abs('Z');
Diff = abs('a') - abs('A');

for i = 1:length(in_string)
   if (abs(in_string(i))>= Low_bound) & (abs(in_string(i))<= Up_bound)
       in_string(i) = setstr(abs(in_string(i)) + Diff);
   end
end

bad_char=abs('z')+1;
Low_bound = abs('a');
Up_bound= abs('z');
num_string=abs(in_string); %Convert the text string to numbers.
for i = 1:length(num_string);
   if (num_string(i) < Low_bound) | (num_string(i) > Up_bound)
       num_string(i) = bad_char;
   end
end

shift=abs('a')-1;
num_string = num_string - shift;


count=zeros(1,27);
for i = 1:length(num_string)
   count(num_string(i)) = count(num_string(i))+1;
end

for i = 1:26
   fprintf('     '); % Note that there are 5 spaces before the colon.
   fprintf(setstr(i+shift)); %Print out the letter followed by a colon.
   fprintf(': '); % Note that there is 1 space after the colon.
```

```
    for j = 1:count(i)
        fprintf('*');  % Print out one * for each occurrence of the letter.
    end
    fprintf('\n');
end
fprintf('Other: ');
for i = 1:count(27) %Print out one * for each occurrence of a non-alphabetic
    fprintf('*');
end
fprintf('\n\n\n');
```

Let's examine our added code. Since 1 represents an a, 2 a b, 3 a c, 26 a z, and 27 all other characters, we need a way to print out an a for 1, a b for a 2, and so on. Recall that the ASCII code was 97 for a, 98 for b, and so on. We subtracted 96 from each code to shift the codes from 97 down to 1. To print out the characters, we must shift the codes 1–26 back to 97–122. We do this by adding 96 to each code. Earlier in the program, variable shift was given the value of 96, **shift=abs('a')-1**. We add this value to each code to obtain the ASCII code for the characters a–z. Next, we use the **SETSTR** function to convert an ASCII code to a character:

```
EDU» setstr(97)
ans =
a
EDU» setstr(98)
ans =
b
EDU» setstr(122)
ans =
z
EDU»
```

The **FOR** loop

```
for i = 1:26
   000
end
```

loops *i* through codes 1–26. The line **setstr(i+shift)** converts *i* to the characters a–z, and the **FPRINTF** statement prints out the character. The following code segment prints out all the characters on a single line:

```
shift=abs('a')-1;
for i = 1:26
   fprintf(setstr(i+shift))
end
fprintf('\n')
abcdefghijklmnopqrstuvwxyz
EDU»
```

In Function 4-8, we have the lines

```
for i = 1:26
    fprintf('     '); % Note that there are 5 spaces before the colon.
    fprintf(setstr(i+shift)); %Print out the letter followed by a colon.
    fprintf(': '); % Note that there is 1 space after the colon.
    ...
end
```

Here, i has values 1–26. For each value of i, we print out 5 spaces, the character corresponding to i using the line **fprintf(setstr(i+shift))**, and then a colon followed by a space, all on the same line. Note that none of the **FPRINTF** statements discussed has used a new line character (**\n**).

Next, we print out one * for each occurrence of a character with the code

```
for i = 1:count(i)
    fprintf('*');  % Print out one * for each occurrence of the letter.
end
```

Variable *count(i)* contains the number of times the character corresponding to code i occurs. For example, if $i = 1$, then *count(i)* corresponds to how many times a appears. If $i = 26$, then *count(i)* corresponds to how many times z appears. Thus, this **FOR** loop prints out one asterisk for each occurrence of the character corresponding to code i. After the asterisks print, we have a final **FPRINTF** statement, which creates a new line.

Code 27 is a special code for all characters that are not a–z. The code segment

```
fprintf('Other: ');
for i = 1:count(27) %Print out one * for each occurrence of a non-alphabetic
    fprintf('*');
end
```

prints out the text **Other:** and then one asterisk for each occurrence of non a–z characters. Note that "other" has six characters. This is the same as 5 spaces followed by a single character such as a. This causes **Other:** to use the same amount of space as **'   a:'**. Thus, the 5 spaces line up the printout.

Let's test the function on our data files:

```
hist6('data.txt');
      a: *
      b:
      c:
      d:
      e: **
      f:
      g:
      h: **
      i: *****
```

```
        j:
        k:
        l: *
        m:
        n: *
        o: *
        p:
        q:
        r:
        s: ******
        t: *****
        u:
        v:
        w: *
        x:
        y:
        z:
Other: **
```

hist6('data2.txt');

```
        a: ******************************
        b: **
        c: ********
        d: *********
        e: *****************************************************
        f: **
        g: ****
        h: *************
        i: **************
        j:
        k: **
        l: *****************
        m: ************
        n: *****************
        o: ******************
        p: **********
        q:
        r: ****************
        s: ****************
        t: *********************************
        u: ******
        v: **
        w: **************
        x: **
        y: ****
        z:
Other: ********
```

EXERCISE 4-2 Modify Function 4-8 so that the number of asterisks (*) printed depends on the most frequently occurring character. That is, if the highest-occurring character occurs ≤50 times, then print out 1 asterisk for each occurrence. If the highest-occurrence character occurs ≥51 and ≤100 times, then print out 1 asterisk for 2 occurrences. If the highest-occurrence character occurs ≥101 and ≤150 times, then print out 1 asterisk for 3 occurrences, and so on. For example, if the highest-occurring letter appeared 137 times, then an asterisk would represent 3 occurrences. For example, suppose A occurred 45 times, B appeared 65 times, and C occurred 137 times. The abbreviated table would look like this:

```
A:  ***************
B:  *********************
C:  ********************************************
```

1 asterisk = 3 occurrences

The line following the **Other** line should state what an asterisk represents. Display something like **1 asterisk = 1 occurrence** or **1 asterisk = 3 occurrences** so that a reader knows how to intrepret the results.

Note that MATLAB also has a built-in histogram function for drawing the graph. Once we have the data, we can call the **HIST** function rather than display asterisks; as done in Function 4-9.

Function 4-9 Modification of the histogram function

```
function z=hist7(filename)
fid=fopen(filename,'r');
in_string=fscanf(fid, '%s');
fclose(fid);

% Convert Uppercase letters to lowercase
Low_bound = abs('A');
Up_bound= abs('Z');
Diff = abs('a') - abs('A');

for i = 1:length(in_string)
   if (abs(in_string(i))>= Low_bound) & (abs(in_string(i))<= Up_bound)
        in_string(i) = setstr(abs(in_string(i)) + Diff);
   end
end

bad_char=abs('z')+1;
Low_bound = abs('a');
```

```
Up_bound= abs('z');
num_string=abs(in_string); %Convert the text string to numbers.
for i = 1:length(num_string);
  if (num_string(i) < Low_bound) | (num_string(i) > Up_bound)
      num_string(i) = bad_char;
  end
end

shift=abs('a')-1;
num_string = num_string - shift;


hist(num_string,27);
```

Note that the code for displaying one asterisk per character occurrence in Function 4-8 is replaced by the line **hist(num_string,27)**. This function counts the number of occurrences of each number and generates a graph. The **27** tells the **HIST** function to create a histogram plot with 27 bins.

EDU» **hist7('data.txt')**

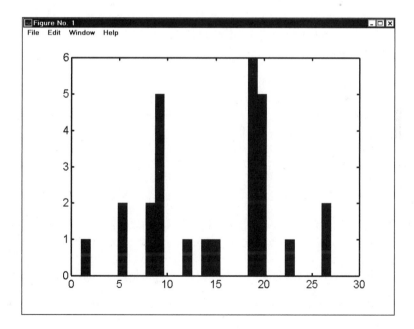

EDU» `hist7('data2.txt');`

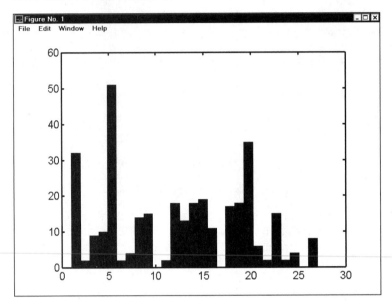

This function plots the number of occurrences of the codes 1–27; it interprets these codes as a–z and Other. The only problem with these plots is that the text **a, b, c**, . . . , and **Other** are not displayed.

4.4.2 Sorting

An application of arrays that you will use frequently involves sorting data in ascending or descending order. To see how this works, let's create a function called d_sort that sorts numbers in an array in descending order. We'll use a fairly simple algorithm that is not particularly efficient, but easy to understand. The point of this exercise is to understand arrays, not to build an efficient sorting algorithm. Normally, before creating any program or function, you should see whether MATLAB has a built-in function that solves your problem or whether an m-file on The MathWorks website solves your problem. Later in this section, we'll use a MATLAB function that solves our problem with a minimal amount of programming. Remember, the goal of this text is to develop programming skills using MATLAB as the programming language, but as good programmers, you should always check what facilities and predefined functions are available before you create new programs.

Let's sort an array that contains 5 numbers: $a = [67, 98, 22, 84, 83]$. When the sort is complete, the array should be structured as $a = [98, 84, 83, 67, 22]$. We'll start with the array $a = [67, 98, 22, 84, 83]$. For addressing purposes, note that $a(1) = 67$, $a(2) = 98$, $a(3) = 22$, $a(4) = 84$, and $a(5) = a(\text{length}(a)) = 83$. Let's start with the last element and work toward the first element.

Pass 1, Step 1. Compare $a(5)$ and $a(4)$. If $a(5)$ is greater than $a(4)$, swap the two numbers.

- $a = [67, 98, 22, 84, 83]$

After this step, we know that element 4 > element 5.

Pass 1, Step 2. Compare element 4 to element 3. If element 4 > element 3, swap the two numbers.

- $a = [67, 98, 84, 22, 83]$

After this step, we know that element 3 > elements 4 and 5.

We don't know how element 4 compares to element 5. These elements are processed later.

Pass 1, Step 3. Next, compare element 3 to element 2. If element 3 > element 2, swap the two numbers.

- $a = [67, 98, 84, 22, 83]$

After this step, we know that element 2 > elements 3, 4, and 5.

Pass 1, Step 4. Compare element 2 to element 1. If element 2 > element 1, swap the two numbers.

- $a = [98, 67, 84, 22, 83]$

After this step, we know that $a(1) \geq$ all other elements in the array

We've gone through the entire array, and the largest number is now the first element. The remaining elements, $a(2)$ through $a(5)$, are still unsorted. Now we repeat the sorting process for elements $a(2)$ through $a(5)$.

Pass 2, Step 1. Start with the last element in the array and compare element 5 to element 4. If element 5 > element 4, swap the two numbers.

- $a = [98, 67, 84, 83, 22]$

After this step, we know that element 4 > element 5.

Pass 2, Step 2. Compare element 4 to element 3. If element 4 > element 3, swap the two numbers.

- $a = [98, 67, 84, 83, 22]$

Pass 2, Step 3. Compare element 3 to element 2. If element 3 > element 2, swap the two numbers.

- $a = [98, 84, 67, 83, 22]$

Element $a(2)$ now holds the correct number. There is no need to compare elements $a(2)$ and $a(1)$ because we know from Pass 1 that element $a(1)$ is the largest number. At the end of Pass 2, we know that elements $a(1)$ and $a(2)$ contain the two largest numbers in the array, and that element $a(1) \geq$ element $a(2)$. We sort elements $a(3)$ through $a(5)$ next in Pass 3.

Pass 3, Step 1. Start with the last element in the array. Compare element $a(5)$ to element $a(4)$. If element $a(5) >$ element $a(4)$, swap the two numbers.

■ $a = [98, 84, 67, 83, 22]$

After this step, we know that element $a(4) >$ element $a(5)$.

Pass 3, Step 2. Compare element $a(4)$ to element $a(3)$. If element $a(4) >$ element $a(3)$, swap the two numbers.

■ $a = [98, 84, 83, 67, 22]$

We are done with element $a(3)$. There is no need to compare elements $a(1)$, $a(2)$, and $a(3)$ because we know from passes 1 and 2 that they are in descending order. Elements $a(1)$, $a(2)$, and $a(3)$ are sorted, all we need to do is sort the remaining elements, $a(4)$ and $a(5)$. We will call this Pass 4.

Pass 4, Step 1. Start with the last element in the array. Compare element $a(5)$ to element $a(4)$. If element $a(5) >$ element $a(4)$, swap the two numbers.

■ $a = [98, 84, 83, 67, 22]$

After Pass 4, this 5-element array is now completely sorted. Function 4-10 sorts an arbitrary row array using this method.

Function 4-10 Sort in descending order

```
function y=d_sort(x)
% This function sorts a one-dimensional array of numbers in
% descending order.
%
% Variables:
%      x - input row array of numerical values.
%      y - return row array of sorted numerical values.
% x and y can be of arbitrary length.
num_elements = length(x);
for i = 2:num_elements % (Number of passes = n-1)
   for j = num_elements:-1:i  % Step from the last element to the ith element
      if x(j) > x(j-1)
         temp=x(j-1);   %Swap two elements if necessary.
         x(j-1)=x(j);
         x(j)=temp;
      end
   end
end
y=x;
```

To test the function, let's use some fictitious data.

```
EDU» a=[1 2 3 4 5 6 7];
```

We will sort the data in descending order:

```
EDU» d_sort(a)
ans =
7    6    5    4    3    2    1
```

Let's try another test.

```
EDU» b=[678 56 838 783 927 9278 992 98 0 237 -12 99 99];
EDU» d_sort(b)
ans =
  Columns 1 through 6
9278          992          927          838          783          678
Columns 7 through 12
237           99           99           98           56           0
Column 13
          -12
```

The function appears to run properly.

EXERCISE 4-3   Modify the d_sort function so that

- If *x* is an array of real numbers, the array sorts in descending order.
- If *x* is a text string, the characters sort in reverse alphabetical order.
- If *x* is an array of complex numbers, the array sorts in descending order by the magnitude of the complex number.

Here are some examples:

```
EDU» a=[1 45 72 0 -99 28 9 0 -87];
EDU» d_sort2(a)
ans =
      72    45    28    9    1    0    0    -87    -99
EDU» b='sdgahtfgdsbnmfusdlsyfhjskyrewzmcpdyfhb';
EDU» d_sort2(b)
ans =
zyyywutssssssrpnmmlkjhhhggfffffeddddcbba
EDU» c=[1+1i, 2+2i, 3, 5, 6+7i, 10i];
EDU» d_sort2(c)
ans =
Columns 1 through 4
       0+10.0000i   6.0000+ 7.0000i   5.0000              3.0000
  Columns 5 through 6
  2.0000+ 2.0000i   1.0000+ 1.0000i
EDU»
```

As you might expect, MATLAB has a rich library of numerical functions, some of which are for sorting. In the student version, they are

```
EDU» lookfor sort
d_sort.m: % This function sorts a row vector of numbers in
d_sort2.m: % This function sorts a row vector of numbers in
CPLXPAIR Sort numbers into complex conjugate pairs.
SORT    Sort in ascending order.
SORTROWS Sort rows in ascending order.
DSORT   Sort complex discrete eigenvalues in descending order.
ESORT   Sort complex continuous eigenvalues in descending order.
EDU»
```

The function that most closely resembles our application is the **SORT** function, which sorts numbers in ascending order.

```
EDU» help sort
  SORT    Sort in ascending order.
    For vectors, SORT(X) sorts the elements of X in ascending ➡
order.
    For matrices, SORT(X) sorts each column of X in ascending ➡
order.
    For N-D arrays, SORT(X) sorts along the first non-singleton
    dimension of X.
    SORT(X,DIM) sorts along the dimension DIM.

    [Y,I] = SORT(X) also returns an index matrix I. If X is a
    vector, then Y = X(I).  If X is an m-by-n matrix, then
        for j = 1:n, Y(:,j) = X(I(:,j),j); end
    When X is complex, the elements are sorted by ABS(X). Complex
    matches are further sorted by ANGLE(X).
    Example: If X = [3 7 5
                     0 4 2]
    then sort(X,1) is [0 4 2  and sort(X,2) is [3 5 7
                       3 7 5]                   0 2 4];
    See also SORTROWS, MIN, MAX, MEAN, MEDIAN.
EDU»
```

This function is close to the one we just wrote, but it sorts in ascending order. It's usually a good idea to check MATLAB's built-in functions before you write a program. You should also check The MathWork's website (www.mathworks.com) for functions written by other engineers. Note that websites change frequently, and you should expect that The MathWorks website may differ from what we show here. However, the website always has a page that allows you to search or browse public m-files. So let's navigate The MathWorks website.

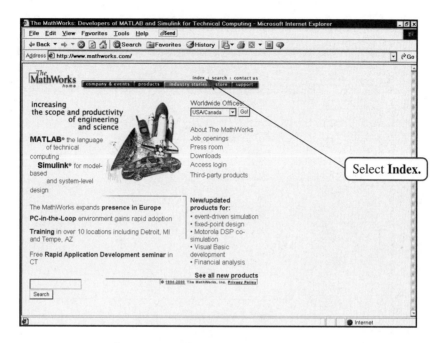

To find available m-files, select the **Index** button as shown above. This will take you to the next screen:

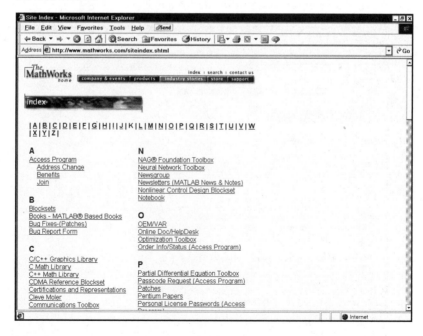

The above screen shows a list of available resources. Be sure to look at some of the other areas later, but now let's look for some m-files related to sorting.

Select the **M-Files—User Contributed** link.

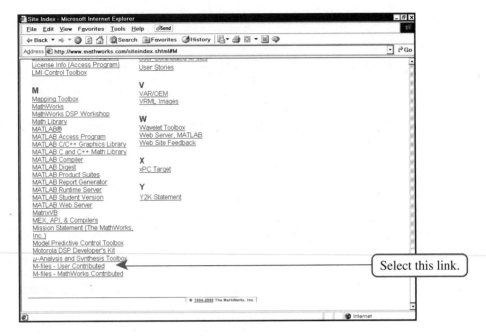

When you select the link, you should see the screen below.

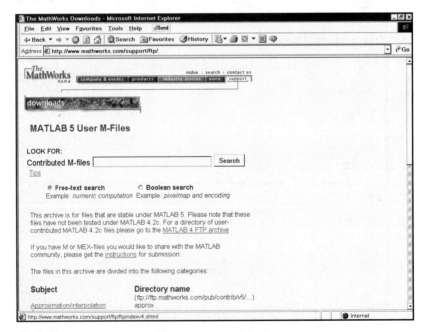

If you enter a search phrase in the **LOOK FOR: Contributed M-files** text field, you will search the MATLAB 5 archive. If you want to search the MATLAB 4 archive select the MATLAB 4 FTP archive link as shown below. Note that version 5 will run most of the MATLAB 4 m-files with little or no modification.

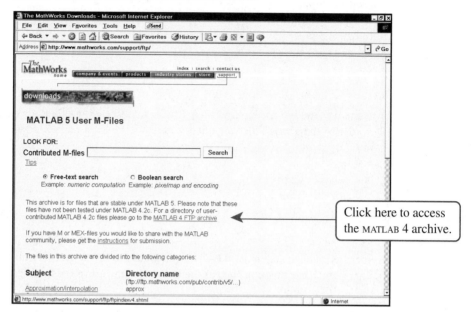

If you scroll the page down, you will see a list of libraries:

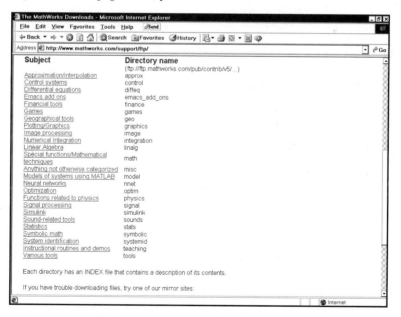

The page shows all of the libraries available for searching for m-files. We won't search through these libraries for a sorting function. However, it's a good idea to look through these archives before starting any large program.

Modifying the MATLAB **SORT** Function

It turns out that even though the **SORT** function sorts numbers in ascending order, we can easily change the sort order to descending order. First, let's demonstrate the ascending sort function:

```
EDU» a=[7589 7543 38 303 376763 3737];
EDU» ascend=sort(a)
ascend =
38        303        3737        7543        7589        376763
```

The data is sorted in ascending order. To sort in descending order, all we do is reverse the order of the array. This was done in Section 4.3.1 on page 215, which we repeat here:

```
EDU» x='abcdefgh';
EDU» x=x(8:-1:1)
x =
hgfedcba
```

Remember that the statement **8:-1:1** generates the sequence of numbers 8, 7, 6, 5, 4, 3, 2, 1. We can use this sequence as the indices for the array and reverse the order of the sort:

```
EDU» a=[7589 7543 38 303 376763 3737];
EDU» ascend=sort(a);
EDU» descend=ascend(length(ascend):-1:1)
descend =
376763        7589        7543        3737        303        38
```

The ease with which we accomplish this task is probably why MATLAB doesn't provide a separate function for sorting in descending order.

4.4.3   Monte-Carlo Analysis of a Voltage Divider

Suppose we have the following voltage divider circuit:

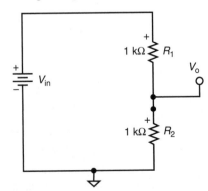

The voltage gain of the circuit is

$$\frac{V_{\text{o}}}{V_{\text{in}}} = \frac{R_2}{R_1 + R_2} = 0.5$$

The ideal voltage gain is 0.5, and this occurs when R_1 and R_2 have resistance values exactly equal to their nominal value of 1000 Ω. In reality, R_1 and R_2 are never exactly equal to their nominal values. For typical 5% resistors, a 1000-Ω resistor may have values of 1000 Ω ±5%, or values anywhere from 950 Ω to 1050 Ω. A typical distribution is a normal distribution with a standard deviation of 1.25%:

```
EDU» dev=randn(1,16384)/80;
EDU» R1=1000*(1+dev);
EDU» [n,x]=hist(R1,50);
EDU» bar(x,n);
EDU» grid;
EDU» xlabel('Resistance Value');
EDU» ylabel('Number of Occurrences')
```

The code segment produces the plot:

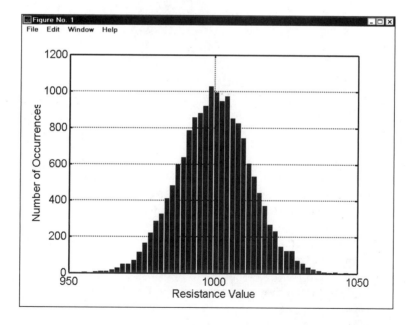

A brief explanation of the preceding code is in order. Function **RANDN**(x, y) generates an x×y array of random numbers with a normal distribution with a standard deviation of 1. The random number generator theoretically generates large positive and negative numbers, but typically the numbers are between –4 and 4. The statement **RANDN**(1, 16384) generates a row array with 16,384 random numbers between –4 and 4 with a normal distribution:

```
EDU» dev=randn(1,16384);
EDU» [n,x]=hist(dev,50);
EDU» bar(x,n);
EDU» xlabel('Random Number');
EDU» ylabel('Number of Occurrences')
EDU» grid;
```

The code segment produces the plot:

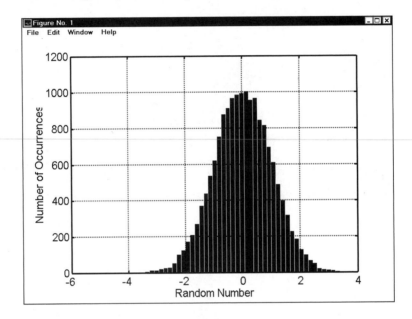

The **RANDN** function generates numbers between –4 and 4. For ±5%, we would like numbers from –0.05 to 0.05. This is done by scaling the results of the **RANDN** function by 80. The command **dev=randn(1,16384)/80** divides each element in the array by 80. The standard deviation of the 5% distribution is 1.25%.

```
EDU» dev=randn(1,16384)/80;
EDU» [n,x]=hist(dev,50);
EDU» bar(x,n);
EDU» ylabel('Number of Occurrences')
EDU» grid;
```

This code segment produces the plot:

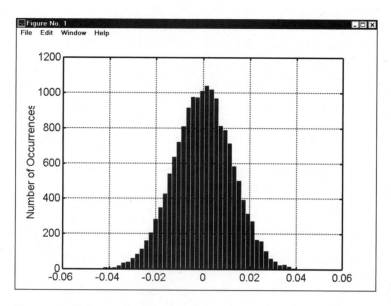

Now we add 1 to each element to generate numbers from 0.95 to 1.05. The command `dev=1 + randn(1,16384)/80` adds 1 to each element in the array. Since the number 1 is a scalar, it's added to each individual element.

```
EDU» dev=1 + randn(1,16384)/80;
EDU» [n,x]=hist(dev,50);
EDU» bar(x,n);
EDU» ylabel('Number of Occurrences')
EDU» grid;
```

This code segment produces the plot:

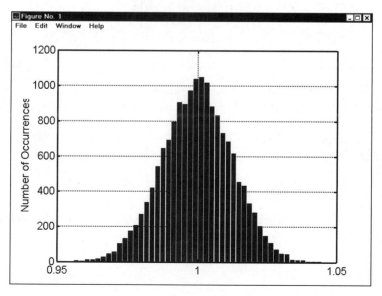

We can now use these numbers to generate random resistor values. To generate a 680-Ω resistor with a 5% tolerance, we simply multiply the distribution above by the value of the resistor, **R=680*dev**. To generate the distribution for a 680-Ω resistor, we use the following code:

```
EDU» dev=1 + randn(1,16384)/80;
EDU» R=680*dev;
EDU» [n,x]=hist(R,50);
EDU» bar(x,n);
EDU» ylabel('Number of Occurrences');
EDU» xlabel('Resistor Value');
EDU» grid;
```

This code segment generates the resistor distribution shown in the following plot:

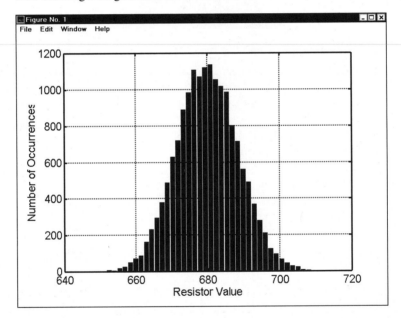

In the code segment **R=680*dev**, the number 680 is a scalar and *dev* is a row array. When we multiply an array by a scalar, every element in the array is multiplied by the scalar.

Now that we know how to generate resistors with random values centered about their nominal value, we continue with our resistor divider problem. The question is, if R_1 can have any value between 950 and 1050, and R_2 can have any value between 950 and 1050, and both have a normal distribution, what does the distribution of the voltage divider look like?

We'll attack this problem by randomly picking a value for R_1, randomly picking a value for R_2; then calculate the gain $R_2/(R_1 + R_2)$, and store the result in the gain array. We will perform this experiment 16,384 times.

```
gain=zeros(1,16384);
for i = 1:16384
    R1=1000*(1 + randn(1)/80);
    R2=1000*(1 + randn(1)/80);
    gain(i)=R2/(R1+R2);
end
[n,x]=hist(gain,50);
bar(x,n);
grid;
xlabel('Voltage Gain');
ylabel('Number of Occurrences');
```

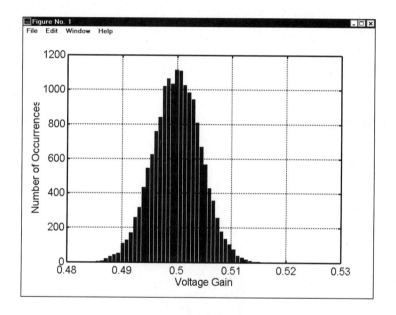

Using the array properties of MATLAB, we can simplify the code a little. Instead of using a **FOR** statement to loop 16,384 times and picking one random value for R_1 and one random value for R_2 each time through the loop, we eliminate the **FOR** loop and generate 16,384 values for R_1 and 16,384 values for R_2 in one statement each:

```
EDU» R1=1000*(1 + randn(1,16384)/80);
EDU» R2=1000*(1 + randn(1,16384)/80);
EDU» gain=R2./(R1+R2);
EDU» [n,x]=hist(gain,50);
EDU» bar(x,n);
EDU» grid;
EDU» xlabel('Voltage Gain');
EDU» ylabel('Number of Occurrences');
```

Note that R_1 and R_2 are now arrays with one row and 16,384 columns. By default, addition of arrays is performed element-by-element. That is, $R_1(1)$ adds to $R_2(1)$, $R_1(2)$ adds to $R_2(2)$, $R_1(3)$ adds to $R_2(3)$, and so on. The `/` operator (note no period) performs matrix division if possible. The `./` operator (with period) performs element-by-element division. For example, if we have the operation R1./R2, MATLAB performs element-by-element division. That is, $R_1(1)$ is divided by $R_2(1)$, $R_1(2)$ by $R_2(2)$, and $R_1(3)$ by $R_2(3)$, and so on. MATLAB executes the statement `gain=R2./(R1+R2)` element-by-element, which is equivalent to

$$gain(1) = R_2(1)/(R_1(1) + R_2(1))$$
$$gain(2) = R_2(2)/(R_1(2) + R_2(2))$$
$$gain(3) = R_2(3)/(R_1(3) + R_2(3))$$
$$gain(4) = R_2(4)/(R_1(4) + R_2(4))$$
$$\vdots$$
$$gain(16,384) = R_2(16,384)/(R_1(16,384) + R_2(16,384))$$

Since variables R_1 and R_2 have 16,384 elements, variable *gain* has 16,384 elements. The code segment above yields the following plot:

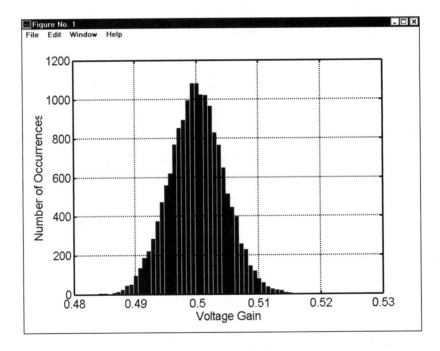

Now, how do we generate a plot that shows percentages rather than number of occurrences? Let's start by looking at the help information available with the **HIST** function:

EDU» **help hist**
HIST   Histogram.
   N = HIST(Y) bins the elements of Y into 10 equally spaced containers
   and returns the number of elements in each container. If Y is a
   matrix, HIST works down the columns.

   N = HIST(Y,M), where M is a scalar, uses M bins.

   N = HIST(Y,X), where X is a vector, returns the distribution of Y
   among bins with centers specified by X.

   [N,X] = HIST(...) also returns the position of the bin centers in X.

   HIST(...) without output arguments produces a histogram bar plot of
   the results.
EDU»

As you can see, if you use the histogram function as **[N,X]=hist (...)**, the function won't draw a graph but instead returns data in variables N and X. It isn't clear what N and X are, so let's give the function a try using this form and see if we can decipher the data. We'll use the data from the preceding example:

EDU» **R1=1000*(1 + randn(1,16384)/80);**
EDU» **R2=1000*(1 + randn(1,16384)/80);**
EDU» **gain=R2./(R1+R2);**
EDU» **[N,X]=hist(gain,25);**

Note that to keep the number of elements in N and X small, we specify 25 bins rather than 50. Next let's look at the data stored in variable N.

EDU» **N**
N =
  Columns 1 through 6
          1   2   9   25   68   113
  Columns 7 through 12
        288   504   795   1191   1614   1894
  Columns 13 through 18
       2161   2031   1777   1425   966   719
  Columns 19 through 24
        416   208   111   42   16   4
  Column 25
          4
EDU»

As you can see, N is a row array with 25 elements. These numbers look like frequency data (number of occurrences). Let's check: If the sum of the data in N equals 16,384, then it's a good guess that N contains the number of occurrences:

```
EDU» sum(N)
ans =
        16384
EDU»
```

In the example above, we ran the experiment 16,384 times. We then used the **HIST** function to sort the data into 25 bins. Variable N has 25 elements and the elements of N sum to 16384. This suggests that N contains the frequency data.

Next, let's look at the data contained in variable X:

```
EDU» X
X =
  Columns 1 through 7
    0.4824     0.4838     0.4853     0.4867     0.4881     0.4896     0.4910
  Columns 8 through 14
    0.4924     0.4939     0.4953     0.4967     0.4982     0.4996     0.5010
  Columns 15 through 21
    0.5024     0.5039     0.5053     0.5067     0.5082     0.5096     0.5110
  Columns 22 through 25
    0.5125     0.5139     0.5153     0.5168
EDU»
```

Variable X also contains 25 elements. The numbers contained in X are the bins. The first bin is from 0 to $X(1)$ or 0 to 0.4824. The number of times a result in this range occurs is $N(1)$ or 1 in the results above. The second bin is from $X(1)$ to $X(2)$ or 0.4824 to 0.4838. The number of times a result in this range occurs is $N(2)$, or 2.

As stated in the help information for the **HIST** function, we can plot histograms by using the bar graph function:

```
EDU» bar(X,N)
EDU» grid
EDU» xlabel('Voltage Divider Gain');
EDU» ylabel('Number of Occurrences')
```

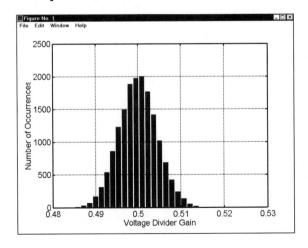

Now that we know that N is the number of occurrences, we can easily convert the data contained in N to percentages. For each bin, we take the number of occurrences, divide by the total number of trials, and then multiply by 100.

```
EDU» R1=1000*(1 + randn(1,16384)/80);
EDU» R2=1000*(1 + randn(1,16384)/80);
EDU» gain=R2./(R1+R2);
EDU» [N,X]=hist(gain,25);
EDU» Npct=100*(N/sum(N));
EDU» bar(X,Npct);
EDU» grid;
EDU» xlabel('Voltage Divider Gain');
EDU» ylabel('Percent of Trials')
```

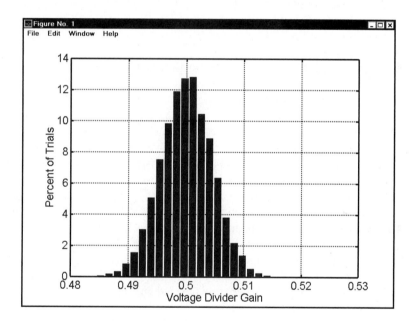

Random Numbers in MATLAB

Random numbers generated by MATLAB are actually not random; they are a sequence of numbers that appears random from one number to the next. To show this, let's start MATLAB and then ask for random numbers:

```
EDU» format long
EDU» rand(1,10)
ans =
```

```
   Columns 1 through 4
      0.95012928514718    0.23113851357429    0.60684258354179 ➡
0.48598246870930
   Columns 5 through 8
      0.89129896614890    0.76209683302739    0.45646766516834 ➡
0.01850364324822
   Columns 9 through 10
      0.82140716429525    0.44470336435319
EDU»
```

We will now exit MATLAB and then restart MATLAB. After MATLAB restarts, we ask for 10 random numbers again:

```
EDU» format long
EDU» rand(1,10)
ans =
   Columns 1 through 4
      0.95012928514718    0.23113851357429    0.60684258354179 ➡
0.48598246870930
   Columns 5 through 8
      0.89129896614890    0.76209683302739    0.45646766516834 ➡
0.01850364324822
   Columns 9 through 10
      0.82140716429525    0.44470336435319
EDU»
```

Note that the numbers are identical. This is because the random number generator produces a sequence of numbers that appears random. However, the sequence is always the same. Theoretically, the sequence will generate 2^{1492} values before repeating. The numbers generated depend on the state of the generator. When you start MATLAB, the state of the generator is initialized to zero. If you don't change the state, the sequence of numbers will always be the same because the random number generator always starts in the same state.

To start the sequence at a different location each time we use MATLAB, we start the generator in a random state. Specify the state of the generator using the command **rand('state',j)** where *j* is an integer. To generate a "random" number for *j*, use the system clock. The command **rand('state', sum(100*clock))** sets the state of the generator to a "random number." The command **sum(100*clock)** generates a number based on the system clock. The number isn't random, but it most likely will be different each time we use it.

If you are using random numbers in a program, you should always use the command **rand('state', sum(100*clock))** at the beginning of your program to initialize the random number generator and start the sequence at a new and "random" location.

4.5 Designing MATLAB Functions to Handle Array Inputs

MATLAB Predefined
sin function
plot function

Let's look at designing functions that handle a single input or an array of inputs. We saw this with the **SIN** function where we used the **SIN** function with a single input or an array of inputs.

```
EDU» sin(5)
ans =
  -0.9589
EDU» sin([1 2 3 4 5])
ans =
    0.8415    0.9093    0.1411   -0.7568   -0.9589
EDU»
```

This second case returns sin(1), sin(2), sin(3), sin(4), and sin(5). Here the elements of the data are treated individually by the **SIN** function. Since MATLAB performs both matrix operations and element-by-element operations, be sure that when you handle arrays of information, MATLAB performs the type of algebra you want. Suppose we have an array $x = [1, 2, 3]$ and we want to create a new array that contains the individual elements of x^2. That is, for each element in x we want $y = x^2$. If we input **y=x*x**, we get an error because MATLAB attempts to perform matrix multiplication:

```
EDU» x=[1, 2, 3];
EDU» y=x*x
??? Error using ==> *
Inner matrix dimensions must agree.
EDU»
```

An error message is generated because we are multiplying a row vector times a row vector: [1 2 3]*[1 2 3]. This is an invalid matrix operation, so MATLAB generates an error. What we want is $y(1) = x(1) \cdot x(1)$, $y(2) = x(2) \cdot x(2)$, and $y(3) = x(3) \cdot x(3)$. To do this in MATLAB, we use the array multiplication operator **.*** (period followed by asterisk) rather than matrix multiplication ***** (asterisk alone).

```
EDU» x=[1, 2, 3];
EDU» y=x.*x
y =
1    4    9
EDU»
```

As you can see, the **.*** operator generates the desired result. In general, any operator preceded by a period is the element-by-element array form of the operation. Instead of multiplying, we can use the power operator (**^**). This operator also has an element-by-element array form and a matrix form. The matrix form is **^**, the array form is **.^** (a period followed by a caret). Let's try the matrix form:

```
EDU» x=[1, 2, 3];
EDU» y=x^2
??? Error using ==> ^
Matrix must be square.
EDU»
```

MATLAB generates an error because we're attempting a matrix operation and the dimensions of the operands are not correct for the operation. We can square each element by using the power operator in its array form:

```
EDU» x=[1, 2, 3];
EDU» y=x.^2
y =
1    4    9
EDU»
```

The last operator we'll look at that has both a matrix and an array version is division. The matrix form is

```
EDU» x=[1, 2, 3];
EDU» y=x/x
y =
1
EDU»
```

This doesn't generate an error, but the result isn't quite what we want if we're attempting element-by-element division. The array form of division is ./.

```
EDU» x=[1, 2, 3];
EDU» y=x./x
y =
1    1    1
EDU»
```

This array form operates on individual elements of each array. That is, $y(1) = x(1)/x(1)$, $y(2) = x(2)/x(2)$, and $y(3) = x(3)/x(3)$.

By the nature of the operations, array addition and subtraction are the same for matrices and arrays:

```
EDU» x=[1, 2, 3];
EDU» y=x+x
y =
2    4    6
EDU» x=[1, 2, 3];
EDU» y=x-x
```

```
y =
      0      0      0
EDU»
```

Now we'll use the array operations to plot a few functions. Let's graph the function $y = x^3 - 3x^2 + 2x - 1$ for values of x from –5 to 5. First, we define a function (Function 4-11) that implements the equation.

Function 4-11 Quadratic function

```
function y=q1(x)
y=(x.^3) - (3*x.^2) + 2.*x -1;
```

Note that all operations in the function are the element-by-element array versions. That is, if x is an array, the function is calculated for each element of x, independent of the other values contained in x. We can use the function with a single value:

```
EDU» q1(1)
ans =
  -1
EDU» q1(-2)
ans =
  -25
EDU» q1(2)
ans =
  -1
```

We can also generate the same results with a one-dimensional array of input values:

```
EDU» q1([1 -2 2])
ans =
  -1  -25  -1
```

Note that the values returned by the function are the same as if we used the function for individual values. The function also works with an n-dimensional array of inputs. As an example, let's use the function with a two-dimensional array:

```
EDU» x=[1 2 3; 4 5 6; 7 8 9];
EDU» q1(x)
ans =
    -1    -1      5
    23    59    119
   209   335    503
```

Here we generate a plot of 100 points for x in the range of -5 to 5:

```
EDU» x=linspace(-5,5,100);
EDU» y=q1(x);
EDU» plot(x,y)
EDU» grid;
```

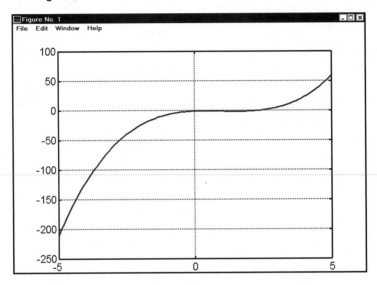

Let's narrow the range of x from -1 to 3:

```
EDU» x=linspace(-1,3,100);
EDU» y=q1(x);
EDU» plot(x,y);
EDU» grid;
```

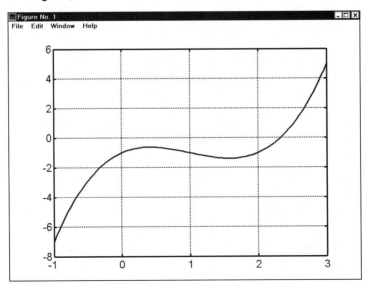

In Function 4-12, we'll plot the sinc function:

$$\text{sinc}(x) = \frac{\sin(x)}{x}$$

Function 4-12 Sinc function

```
function y=sinc1(x)
y=sin(x)./x;
```

Let's test this for a few scalar numbers:

```
EDU» sinc1(1)
ans =
0.8415
EDU» sinc1(-1)
ans =
0.8415
EDU» sinc1(pi/4)
ans =
0.9003
EDU» sinc1(0)
Warning: Divide by zero.
> In c:\MATLAB~1\examples\sinc1.m at line 2
ans =
    NaN
EDU»
```

Note that the function has a problem at $x = 0$. This is because $\text{sinc}(0) = \sin(0)/0 = 0/0$, which is an undefined mathematical operation. MATLAB assigns an undefined mathematical operation a value of *NaN*.

Next let's check if the function works for an array of inputs:

```
EDU» sinc1([1 2 3 4 5])
ans =
0.8415    0.4546    0.0470    -0.1892    -0.1918
```

The function works for a single input or an array of inputs. Now let's plot x from -10π to 10π:

```
EDU» x=linspace(-10*pi, 10*pi, 100);
EDU» y=sinc1(x);
EDU» plot(x,y);
EDU» grid
```

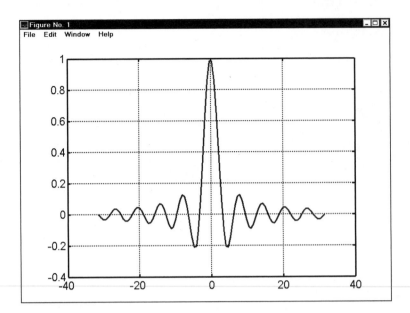

The code segment appears to work correctly, and no warning messages were generated. The question is, was there a point $x = 0$, $y = NaN$ and if there was, how did the **PLOT** function handle the *NaN* value? From the above example, we're not sure if one of the values of x was zero. Let's create some data where one of the data values is specifically set to zero:

```
EDU» x=[-3, -2, -1, 0, 1, 2, 3];
EDU» y=sinc1(x)
Warning: Divide by zero.
> In c:\MATLAB~1\examples\sinc1.m at line 2
y =
    0.0470    0.4546    0.8415       NaN    0.8415    0.4546    0.0470
EDU»
```

We do get the divide-by-zero warning message, but the calculations for all values of x are performed. That is, MATLAB didn't generate an error for $x = 0$ and then quit. It assigned the result a value of *NaN* and then continued. The warning message may or may not be OK.

The next question is, what does the **PLOT** function do when it encounters a value of *NaN*? We'll plot x and y for the seven values generated above to see what happens. Each data point is marked with a black + sign.

```
EDU» plot(x,y,'k+')
```

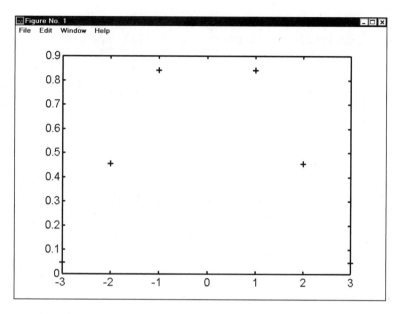

Note that a plus sign is shown for all values except the point $x = 0$, $y = NaN$. Thus, the **PLOT** function handles NaN's by ignoring the point and not plotting it. A line graph of the data appears as

EDU» **plot(x,y)**

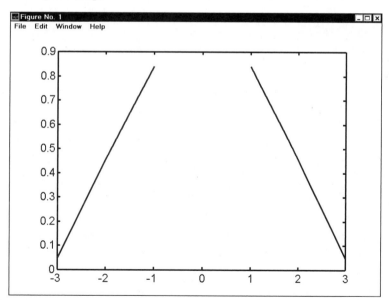

The plot function ignores the point with a y-coordinate value of NaN.

So how can we fix this problem at $x = 0$? There's no easy fix except to test if each input data point has a value of zero. Recall l'Hôpital's rule, which tells us that

$$\lim_{x \to 0} \left(\frac{\sin(x)}{x} \right) = 1$$

We'll have Function 4-13 specifically check for $x = 0$ and assign a result of 1.

Function 4-13 Revised sinc function

```
function y=sinc2(x)
len=length(x);
y=zeros(1,len);
for i = 1:len
   if x(i) == 0
        y(i)=1;
   else
        y(i)=sin(x(i))/x(i);
   end
end
```

As you can see, we have to deal with each element of x individually. For each individual element of x, we calculate a corresponding value of y. The solution is much more complicated than Function 4-12, but is required if we need to check for special conditions with individual elements. Let's test the function.

```
EDU» sinc2(0)
ans =
     1
EDU» sinc2(1)
ans =
     0.8415
EDU» sinc2(-1)
ans =
0.8415
EDU» sinc2([-3, -2, -1, 0, 1, 2, 3])
ans =
0.0470    0.4546    0.8415    1.0000    0.8415    0.4546    0.0470
```

The function now works correctly and doesn't generate any warning messages or *NaN*'s. Thus, for $x = 0$, y is assigned a value of 1.

4.6 Dynamic Arrays

MATLAB Predefined
length function
for loop
etime function

In most programming languages, array size is fixed and must be declared before the arrays are used. If you need to use an array and the number of array elements is different for different applications, the size of the array must be declared for the maximum number of elements you think you will need for all applications. An example would be an address book used in large and small companies and for personal address use. In a large business, the address book may contain several thousand entries. When used for personal records, the book might contain only 100 entries. If we're using arrays to store all of the information, the array will need enough elements to store several thousand entries even if the program is only used for personal uses. In conventional programming languages, this problem is bypassed by using pointers to create a list of elements. Every time we add a new entry to the address book, a new element is created and added to the list. With this technique, the memory used is minimal. As entries are added or deleted, the memory used grows or shrinks to accommodate the number of entries. Programming with pointers is an advanced programming topic and is covered in higher-level programming classes.

Arrays are easy to use and fast to access. Each element can be addressed individually, allowing the programmer to access an element easily. With traditional programming languages such as C or FORTRAN, the array size must be declared at the beginning of the program and cannot be changed. If someone uses the program such that the array fills up, the program will generate an error message stating that capacity has been reached. The user is then stuck unless they can access the program and change the size of the array. Pointers solve the problem, but they are more difficult to work with. To access any element in a list using pointers, the computer must walk through all elements in the list that come before the element of interest. Thus, accessing elements near the end of the list takes longer than those at the beginning of the list. All elements in the array, on the other hand, can be accessed in the same amount of time. The advantage of using pointers is that the number of elements stored in a list can change.

MATLAB has the best of both worlds. For easy access to elements, arrays can be declared at the beginning of a program. However, the size of arrays can be increased as necessary. Although adding elements to an array is slower than declaring the array at the beginning of the program, it's sometimes useful to be able to change the array size.

We'll first look at how to make an array grow. Let's start with a null row array:

```
EDU» row_vec=[ ]
row_vec =
     []
```

Presently, *row_vec* is an array with no elements. Let's add an element by concatenation. Recall that concatenation can be done with any size array:

```
EDU» row_vec=[row_vec, 0]
row_vec =
0
```

This statement means create a new row array. The beginning elements of the new array are the elements in variable *row_vec*. The element added at the end is the number 0. The resulting new row array is then stored in variable *row_vec*. The preceding statement can be repeated as many times as we wish:

```
EDU» row_vec=[row_vec, 1]
row_vec =
0    1
EDU» row_vec=[row_vec, 2]
row_vec =
0    1    2
EDU» row_vec=[row_vec, 3]
row_vec =
0    1    2    3
EDU» row_vec=[row_vec, 4]
row_vec =
0    1    2    3    4
```

Every time we repeat the statement, the size of the row array grows.

Concatenation can be used to make column arrays grow as well. Instead of separating elements by commas, the elements are separated by semicolons. Let's start with a null array:

```
EDU» col_vec = [ ]
col_vec =
     []
```

We can now add elements to the array:

```
EDU» col_vec = [col_vec; 0]
col_vec =
0
EDU» col_vec = [col_vec; 1]
col_vec =
     0
     1
EDU» col_vec = [col_vec; 2]
col_vec =
     0
     1
     2
EDU» col_vec = [col_vec; 3]
col_vec =
     0
     1
     2
     3
```

Concatenation can also be used to add columns or rows to arrays. Let's start with a null array,

```
EDU» Mat1 = [ ]
Mat1 =
     []
```

and add a column with three elements:

```
EDU» Mat1 = [Mat1, [0;0;0]]
Mat1 =
     0
     0
     0
```

Note that $[0;0;0]$ is a column array with three elements, $\begin{bmatrix} 0 \\ 0 \\ 0 \end{bmatrix}$. Now let's add a second column:

```
EDU» Mat1 = [Mat1, [1;1;1]]
Mat1 =
     0     1
     0     1
     0     1
```

Note that $[1;1;1]$ is a column array with three elements, $\begin{bmatrix} 1 \\ 1 \\ 1 \end{bmatrix}$. The process can be repeated by concatenating the array with a new column array with three elements:

```
EDU» Mat1 = [Mat1, [2;2;2]]
Mat1 =
     0     1     2
     0     1     2
     0     1     2
EDU» Mat1 = [Mat1, [3;3;3]]
Mat1 =
     0     1     2     3
     0     1     2     3
     0     1     2     3
EDU» Mat1 = [Mat1, [4;4;4]]
Mat1 =
     0     1     2     3     3     4
     0     1     2     3     3     4
     0     1     2     3     3     4
EDU» Mat1 = [Mat1, [5;5;5]]
Mat1 =
     0     1     2     3     3     4     5
     0     1     2     3     3     4     5
     0     1     2     3     3     4     5
```

```
EDU» Mat1 = [Mat1, [6;6;6]]
Mat1 =
     0     1     2     3     3     4     5     6
     0     1     2     3     3     4     5     6
     0     1     2     3     3     4     5     6
```

We can also add rows to arrays:

```
EDU» Mat2 = [ ]
Mat2 =
     []
EDU» Mat2 = [Mat2 ; [0 0 0 0]]
Mat2 =
     0     0     0     0
EDU» Mat2 = [Mat2 ; [1 1 1 1]]
Mat2 =
     0     0     0     0
     1     1     1     1
EDU» Mat2 = [Mat2 ; [2 2 2 2]]
Mat2 =
     0     0     0     0
     1     1     1     1
     2     2     2     2
EDU» Mat2 = [Mat2 ; [3 3 3 3]]
Mat2 =
     0     0     0     0
     1     1     1     1
     2     2     2     2
     3     3     3     3
EDU» Mat2 = [Mat2 ; [4 4 4 4]]
Mat2 =
     0     0     0     0
     1     1     1     1
     2     2     2     2
     3     3     3     3
     4     4     4     4
EDU» Mat2 = [Mat2 ; [5 5 5 5]]
Mat2 =
     0     0     0     0
     1     1     1     1
     2     2     2     2
     3     3     3     3
     4     4     4     4
     5     5     5     5
EDU» Mat2 = [Mat2 ; [6 6 6 6]]
```

```
Mat2 =
    0    0    0    0
    1    1    1    1
    2    2    2    2
    3    3    3    3
    4    4    4    4
    5    5    5    5
    6    6    6    6
```

Let's create a program that asks the user to input numbers and then calculates their average. The program should work if the user enters anywhere from 0 to 1000 entries. We'll first do this with fixed arrays and then with arrays that grow as needed.

Program 4-1 Finding average with a fixed array

```
% This is file avg1.m
%
% This program calculates the average of numbers entered by the user.
%

% Declare the array and fill it with zeros.
numbers = zeros(1,1000);

i = 0;
done = 0;
while ~done
  num = input('Enter a number. Type ''5i'' to exit:');
  if isreal(num)
      i = i +1;
      numbers(i) = num;
  else
      done = 1;
  end
end

if i == 0
  fprintf('No numbers entered - Average cannot be calculated\n');
else
  average = sum(numbers)/i;
  fprintf('You entered %g numbers.\n', i);
  fprintf('The average of the numbers is %g.\n', average);
end
```

You should note two points here. First, we must declare the size of the array (`numbers = zeros(1,1000)`) at the beginning of the program and specify the maximum size, 1000 elements in this case. Second, we must use a counter (*i* in this example) as an index into the array.

Let's rewrite the program using a row array that grows to accommodate each input.

Program 4-2 Finding average with a dynamic array

```
% This is file avg2.m
%
% This program calculates the average of numbers entered by the user.
%
% Initialize the array to null
numbers = [ ];

done = 0;
while ~done
   num = input('Enter a number. Type ''5i'' to exit:');
   if isreal(num)
       numbers = [numbers, num];
   else
       done = 1;
   end
end

if length(numbers) == 0
   fprintf('No numbers entered - Average cannot be calculated\n');
else
   average = sum(numbers)/length(numbers);
   fprintf('You entered %g numbers.\n', length(numbers));
   fprintf('The average of the numbers is %g.\n', average);
end
```

Numbers added to the array by concatenation.

The benefit of using dynamic arrays is that this program will work for as many numbers as the user is willing to enter. It makes use of the **LENGTH** function to determine how many numbers are entered.

4.6.1 Speed of Dynamic Arrays

As mentioned previously, dynamic arrays are slower than fixed arrays. If you create an array in MATLAB at the beginning of the program and never change its size, a fixed-size array will be much faster than a dynamic array. We'll run two examples to time the two arrays. First, let's look at Program 4-3.

Program 4-3 Fixed-array test for timing analysis

```
% This is program timer2.m
%
% Timed loop with fixed array
x = clock;
data=zeros(1,16384);
```

```
for i = 1:16384
  data(i) = i;
end
etime(clock,x)
```

This program declares an array with 16,384 elements and then uses a **FOR** loop to fill the array with numbers from 1 to 16,384. Since the program uses fixed arrays, it operates quickly. The **ETIME** function calculates the time difference between two values of the system clock. Here it displays the amount of time the program takes to execute.

```
EDU» timer2
ans =
0.3200
EDU» timer2
ans =
0.3300
EDU» timer2
ans =
0.3800
EDU» timer2
ans =
0.3900
EDU» timer2
ans =
0.3900
EDU» timer2
ans =
0.3900
EDU»
```

We've run the program several times to get an idea of its average run time. The time displayed is the real time it takes to execute the program, not the actual CPU time. The resultant times vary because the real time depends on what other programs are running at the same time. The average run time is 0.37 second. The average time required by your system will be different since it depends on the speed of your system and what other programs are running at the same time.

Program 4-4 performs the same task using dynamic arrays.

Program 4-4 Dynamic-array test for timing analysis

```
% This is file timer1.m
%
% Timed loop with dynamic array
x = clock;
data = [ ];
```

```
for i = 1:16384
   data = [data, i];
end
etime(clock,x)
```

This program creates an array with 16,384 elements and fills the elements with numbers from 1 to 16,384. In the **FOR** loop, elements add to the array by concatenation. This program uses a dynamic array that grows to accommodate each new entry, so each time we add an element to the array, a new array is created. The process of creating a new array requires a lot of time:

```
EDU» timer1
ans =
62.6100
EDU» timer1
ans =
63.0500
EDU» timer1
ans =
62.2900
EDU» timer1
ans =
63.1100
EDU» timer1
ans =
62.1200
EDU» timer1
ans =
63.1100
EDU»
```

As you can see, this program takes much longer to run—62.715 seconds on average, which is over 169 times slower than the fixed array. Obviously, for fastest program execution times, fixed arrays should be used where possible.

4.7 Cell Arrays

MATLAB Predefined **cellplot** function

Cell arrays are a data structure that allow us to store different data types in each cell. One cell can contain a multidimensional array, another cell a text string, and other cells can contain scalars. For our example, we'll create an array that contains student grade information. The cell array has the following information:

1. Name—text string
2. ID Number—integer
3. Semester—text string
4. Exam Scores—one-dimensional array (1–3 elements)

5. Homework grades—one-dimensional array (1–10 elements)
6. IQ—3×3 array of random numbers

There are six cells in this data structure, so we'll arrange it as two rows of three cells. Now let's define the cells using a method called content addressing. In this addressing method, the curly brackets are on the left side of the equals sign; they specify that variable *student₁* is a cell array, and the item on the right side of the equals sign is placed inside the cell specified by the indices.

```
EDU» student1{1,1} = 'John Q Smith';
EDU» student1{1,2} = 0296725;
EDU» student1{1,3} = 'Fall 2000';
EDU» student1{2,1} = [75 80];
EDU» student1{2,2} = [10 9 13 15 4 9 11];
EDU» student1{2,3} = rand(3,3);
```

We can view the structure graphically using the **CELLPLOT** function:

```
EDU» cellplot(student1)
```

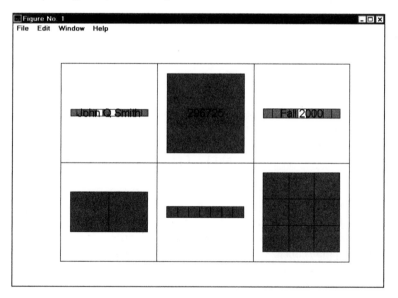

Next, we create a second cell array using a technique called cell indexing. In this method, the curly brackets are on the right side of the equals sign; they denote that the item on the right side is a cell.

```
EDU» student2(1,1) = {'Marc E. Herniter'};
EDU» student2(1,2) = {64832686};
EDU» student2(1,3) = {'Spring 2000'};
EDU» student2(2,1) = {[66 33 99]};
EDU» student2(2,2) = {[1 2 3 4 1 2 3 4 1 2]};
EDU» student2(2,3) = {rand(3,3)};
```

If you use the **CELLPLOT** function, you will see that the cell array for *student2* has the same structure as for *student₁*.

We can build cell arrays using concatenation in the same way we built standard arrays. First let's create a cell array with one cell, the student's name:

```
EDU» student3 = {'Paul J. Good'}
student3 =
'Paul J. Good'
```

Next, we concatenate a new cell onto the existing cell array. The following line creates a new cell array composed of the original cell array with a new cell tacked onto the end. Note that the curly brackets around the number indicate that it's a cell.

```
EDU» student3 = [student3, {678432}]
Student3 =
'Paul J. Good'    [678432]
```

Then we concatenate a cell containing the text string **'Fall 1999'** onto the cell array. Again the curly brackets indicate that we are concatenating a new cell onto the original cell array:

```
EDU» student3 = [student3, {'Fall 1999'}]
student3 =
'Paul J. Good'    [678432]    'Fall 1999'
```

Now we must create the second row of data for the cell array. We'll do this with another method; we'll specify the three cells at the same time:

```
EDU» bot_row = [{[55 88 88]}, {[10 10 20 20 10 10 20]}, {rand(3,3)}]
bot_row =
[1x3 double]    [1x7 double]    [3x3 double]
```

Variable *bot_row* is a cell array. The first cell is the exam scores, the second cell is the homework grades, and the third cell is a 3×3 array.

Variable *student₃* is a cell array with one row of data. Variable *bot_row* is a cell array with one row of data. We want to combine these two cell arrays into a single cell array where *student₃* is the top row and *bot_row* is the bottom row. The following method uses concatenation but adds the new element as a new row. Note that the two variables are separated by a semicolon, which means create a new row.

```
EDU» student3 = [student3;bot_row]
student3 =
'Paul J. Good'    [      678432]    'Fall 1999'
[1x3 double]    [1x7 double]    [3x3 double]
```

If you look at the *student₂* and *student₁*, you will see that they both have the same structure as *student₃*.

```
EDU» student2
student2 =
'Marc E. Herniter'    [    64832686]    'Spring 2000'
[1x3 double]    [1x10 double]    [3x3 double]
EDU» student1
student1 =
'John Q Smith'    [    296725]    'Fall 2000'
[1x2 double]    [1x7 double]    [3x3 double]
EDU»
```

We would now like to create a cell array that holds all of the data. Each cell in this array will be one of the student cell arrays:

```
EDU» data = {student1; student2; student3}
data =
{2x3 cell}
{2x3 cell}
{2x3 cell}
```

Let's view the data for individual students. Below, we ask for the data for the first student (first cell in cell array *data*).

```
EDU» data{1}
ans =
'John Q Smith'    [    296725]    'Fall 2000'
[1x2 double]    [1x7 double]    [3x3 double]
```

If we want to view the name of the first student, we specify cell 1 of cell array *data,* then cell 1, 1 of the student cell array:

```
EDU» data{1}{1,1}
ans =
John Q Smith
```

We could also do this in two steps. First, we place cell 1 of *data* into a new variable *stu,* then we ask for cell 1, 1 of variable *stu.*

```
EDU» stu=data{1}
stu =
'John Q Smith'    [    296725]    'Fall 2000'
[1x2 double]    [1x7 double]    [3x3 double]
EDU» stu{1,1}
ans =
John Q Smith
EDU»
```

It's useful to view this structure of variable *data* using the **CELLPLOT** function:

EDU» `cellplot(data)`

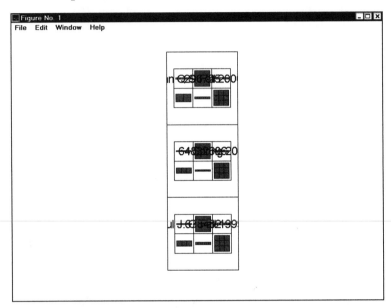

Note that the variable *data* contains detailed information on three students. This is possible because we used a cell array, which can contain many different types of data. If we have data on another student, we can easily add that student's data to variable *data*. First, we create the cell array for the student:

```
EDU» studnt{1,1} = 'Jan Fourth';
EDU» studnt{1,2} = 998822;
EDU» studnt{1,3} = 'Fall 2001';
EDU» studnt{2,1} = [88 99 89];
EDU» studnt{2,2} = [10 10 11 10 15 10];
EDU» studnt{2,3} = rand(3,3);
```

Next, we concatenate this cell onto the data cell array as a new row:

```
EDU» data = [data;{studnt}]
data =
{2x3 cell}
{2x3 cell}
{2x3 cell}
{2x3 cell}
```

The fourth cell has the same structure as any of the other cells:

EDU» `cellplot(data)`

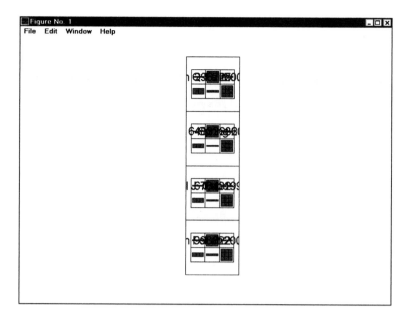

4.7.1 Cell Arrays and Text Strings

Cell arrays are very useful for holding text strings of different sizes. In Section 4.2.2, we saw that if we create an array of text strings, each text string has to be the same size. With cell arrays, it's easy to create arrays with different-size text strings. As an example, let's create a cell array that contains the days of the week:

```
EDU» days = {'Monday'; 'Tuesday'; 'Wednesday'; 'Thursday';
'Friday'; 'Saturday'; 'Sunday'}
days =
'Monday'
'Tuesday'
'Wednesday'
'Thursday'
'Friday'
'Saturday'
'Sunday'
EDU»
```

We can now reference individual cells using the cell address in curly brackets:

```
EDU» days{2}
ans =
Tuesday
EDU» days{3}
ans =
Wednesday
EDU»
```

We can use cells as we would any text string. Below are examples of the **FPRINTF** statement and concatenation:

```
EDU» fprintf('The fourth day is %s.\n',days{4})
The fourth day is Thursday.
EDU» two_days = [days{1},days{2}]
two_days =
MondayTuesday
EDU» str=[two_days,days{3}
str =
MondayTuesdayWednesday
EDU» fprintf(['Try this—',str])
Try this—MondayTuesdayWednesday
EDU»
```

We access individual characters in a text string by specifying the cell address and then the index of the character. Here we ask for the third character of the second cell, or the third character of **Tuesday**:

```
EDU» days{2}(3)
ans =
e
EDU»
```

A good application of cell arrays is to modify the **ask_q** function discussed on page 167. Recall that the **ask_q** function can only ask the user for 1-character responses. We'll modify the function to allow user responses that are several characters long. First, we rename the function ask_ques. Function 4-14 prints out a question and then waits for a user response. If the user gives an invalid response, the function announces the error, specifies a list of valid responses, displays the question again, and then waits for the user to enter another response. When a correct response is entered, the function returns a numerical value that indicates which response was entered.

Function 4-14 Question-and-answer function

```
% function answer=ask_ques(valid_answers, output_string);
% This is file ask_ques.m
% This function asks a question of the user and loops indefinitely
% until the user enters one of the correct responses. The question asked
% is contained in the string variable output_string.
% The valid responses are contained in cell array valid_answers. valid_answers
% is a cell array that contains text strings of valid user responses.
% The return value, answer, is the position of the correct response in the
% cell array of valid answers. For example, if we use the command:
%
% xx=ask_ques({'yes';'no'; 'help';'exit';'end'}, 'Enter your command:')
%
```

```
% If the user types the response yes, function ask_q will return the ➥
numerical value 1.
% If the user types the response no, function ask_q will return the ➥
numerical value 2.
% If the user types the response help, function ask_q will return the ➥
numerical value 3.
% If the user types the response exit, function ask_q will return the ➥
numerical value 4.
% If the user types the response end, function ask_q will return the ➥
numerical value 5.


good_input=0;
while ~good_input
   in_string=input(output_string,'s');
   for i = 1:length(valid_answers)
       if strcmp(in_string,valid_answers{i})
           good_input=1;
           answer=i;
       end
   end
end
if ~good_input
   fprintf('Error - Input one of the following: ');
   for i = 1:(length(valid_answers)-1)
       fprintf(valid_answers{i});
       fprintf(',');
   end
   fprintf([valid_answers{length(valid_answers)},'.\n']);
end
```

Function 4-14 has two input arguments: *valid_answers* and *output_string*. *Output_string* is the message printed when the function requests input from the user. *Valid_answers* is a cell array. Each element in the cell array is a text string. Each cell in the array is a valid user response. For example, if *valid_answers* equals {'**yes**'; '**no**';'**help**';'**exit**';'**end**'} the user is allowed to enter any of the strings **yes**, **no**, **help**, **exit**, or **end**.

The return value of Function 4-14 is the numerical position of the valid response in cell array *valid_answers*. For example, if *valid_answers* equals {'**yes**';'**no**'; '**help**';'**exit**';'**end**'} and the user response is **help**, the function returns a numerical value of 3. If the user response is **end**, the function returns a numerical value of 5 because '**end**' is in the fifth cell in the cell array {'**yes**';'**no**';'**help**'; '**exit**';'**end**'}.

If you compare Function 4-14 to Function 3-3, you will see that the only difference is the use of variable *valid_answers*. In Function 3-3, *valid_answers* is indexed with parentheses ( ); in Function 4-14 *valid_answers* is indexed with curly brackets { }. The changed lines are as follows:

Original lines in Function 3-3

```
if strcmp(in_string,valid_answers(i))
fprintf(valid_answers(i));
fprintf([valid_answers(length(valid_answers)),'.\n']);
```

In this code, when we use parentheses, the index refers to a character in text string *valid_answers*. Thus, **valid_answers(i)** refers to the *i*th character in text string *valid_answers*.

Modified lines in Function 4-14

```
if strcmp(in_string,valid_answers{i})
fprintf(valid_answers{i});
fprintf([valid_answers{length(valid_answers)},'.\n']);
```

In this code, when we use curly brackets, the index refers to a cell in cell array *valid_answers*. For example, **valid_answers{i}** refers to the *i*th cell in cell array *valid_answers,* and each cell in *valid_answers* can contain an arbitrary-length text string.

Let's look at a few examples of this function so that we can better understand its operation.

```
EDU» xx=ask_ques({'yes';'no'; 'help';'exit';'end'}, 'Enter your command:')
Enter your command:
Error - Input one of the following: yes,no,help,exit,end.
Enter your command:xhy
Error - Input one of the following: yes,no,help,exit,end.
Enter your command:yes
xx =
1
EDU» xx=ask_ques({'yes';'no'; 'help';'exit';'end'}, 'Enter your command:')
Enter your command:no
xx =
2
EDU» xx=ask_ques({'yes';'no'; 'help';'exit';'end'}, 'Enter your command:')
Enter your command:help
xx =
3
EDU» xx=ask_ques({'yes';'no'; 'help';'exit';'end'}, 'Enter your command:')
Enter your command:exit
xx =
4
EDU» xx=ask_ques({'yes';'no'; 'help';'exit';'end'}, 'Enter your command:')
Enter your command:end
xx =
5
EDU»
```

User pressed the **ENTER** key.

Program 4-5 uses the **ask_ques** function.

Program 4-5 Using the **ask_ques** function

```
% Example use of function ask_ques.
% This is file ask_examp.m
work_days = {'Monday'; 'Tuesday'; 'Wednesday'; 'Thursday'; 'Friday'; 'Satur-
day'; 'Sunday'; 'None'; 'All'};
d = ask_ques(work_days, 'Specify the day you can work this week:\n');
if d == 8
  fprintf('You cannot work any days this week.\n');
elseif d == 9
  fprintf('You can work all days this week.\n');
else
  fprintf('The day you can work this week is ');
  fprintf(work_days{d});
  fprintf('.\n');
end
```

Here is an example session using Program 4-5:

```
EDU» ask_examp
Specify the day you can work this week:
None
You cannot work any days this week.
EDU» ask_examp
Specify the day you can work this week:
All
You can work all days this week.
EDU» ask_examp
Specify the day you can work this week:
Tuesday
The day you can work this week is Tuesday.
EDU» ask_examp
Specify the day you can work this week:
hjk
Error - Input one of the following:
Monday,Tuesday,Wednesday,Thursday,Friday,Saturday,Sunday,None,All.
Specify the day you can work this week:
Wednesday
The day you can work this week is Wednesday.
EDU»
```

EXERCISE 4-4   One problem with Function 4-14 is that it distinguishes between upper- and lowercase letters. That is, if a valid response is **yes**, then responses such as **Yes**, **YES**, and **YEs** will be treated as invalid responses. Modify Function 4-14 so that responses are independent of case. For example, if a valid response is **yes**, then **YES, Yes, YeS, YEs, yEs, yES, yeS** and **yes** are all valid responses. If a valid response is **sunshine**, all possible upper- and lowercase combinations of the word are valid responses.

4.8 Structures

Structure arrays resemble cell arrays in that they store different data types in a single variable. They differ from cell arrays in that they are addressed by a field name rather than a cell address. To illustrate, we'll create a structure that contains the same student information we used in Section 4.7. Our structure is named student, and it has field names of Name, ID, Semester, Exams, HW, and IQ. The following code defines the information for the first student:

```
EDU» student.Name = 'John Q Smith';
EDU» student.ID = 0296725;
EDU» student.Semester = 'Fall 2000';
EDU» student.Exams = [75 89];
EDU» student.HW = [10 9 13 15 4 9 11];
EDU» student.IQ = rand(3,3);
```

The code segment defines the field names as well as shows us how to reference a field of data. The variable name is *student,* and it contains six fields. The following code shows how MATLAB displays variable *student:*

```
EDU» student
student =
Name: 'John Q Smith'
ID: 296725
Semester: 'Fall 2000'
Exams: [75 89]
HW: [10 9 13 15 4 9 11]
IQ: [3x3 double]
EDU»
```

If we want to look at a specific field in *student,* we specify the field using a dot:

```
EDU» student.Name
ans =
John Q Smith
EDU» student.Exams
ans =
   75   89
EDU» student.IQ
```

```
ans =
0.6038    0.0153    0.9318
0.2722    0.7468    0.4660
0.1988    0.4451    0.4186
EDU»
```

If we want to change a specific field, we use a dot and the field name. Here we change the exam scores:

```
EDU» student.Exams = [99 98]
student =
Name: 'John Q Smith'
ID: 296725
Semester: 'Fall 2000'
Exams: [99 98]
HW: [10 9 13 15 4 9 11]
IQ: [3x3 double]
```

We use the fields as we would any other variable. The following command concatenates a new homework score onto the already existing homework scores:

```
EDU» student.HW = [student.HW, 8]
student =
Name: 'John Q Smith'
ID: 296725
Semester: 'Fall 2000'
Exams: [99 98]
HW: [10 9 13 15 4 9 11 8]
IQ: [3x3 double]
```

To add a second student, we index variable *student*:

```
EDU» student(2).Name = 'Marc E. Herniter'
student =
1x2 struct array with fields:
Name
ID
Semester
Exams
HW
IQ
```

Variable *student* is now an array of two elements. Each element has the fields Name, ID, Semester, Exams, HW, and IQ. The fields of the first element are all defined:

```
EDU» student(1)
ans =
Name: 'John Q Smith'
ID: 296725
Semester: 'Fall 2000'
Exams: [99 98]
HW: [10 9 13 15 4 9 11 8]
IQ: [3x3 double]
```

Only the name field of the second element is defined:

```
EDU» student(2)
ans =
Name: 'Marc E. Herniter'
ID: []
Semester: []
Exams: []
HW: []
IQ: []
```

Now let's define the remaining fields in the second element:

```
EDU» student(2).ID = 64832686;
EDU» student(2).Semester = 'Spring 2000';
EDU» student(2).Exams = [66 33 99];
EDU» student(2).HW = [1 2 3 4 1 2 3 4 1 2];
EDU» student(2).IQ = rand(3,3);
```

We add the third student to variable *student*:

```
EDU» student(3).Name ='Paul J. Good';
EDU» student(3).ID = 678432;
EDU» student(3).Semester = 'Fall 1999';
EDU» student(3).Exams = [55 88 88];
EDU» student(3).HW = [10 10 20 20 10 10 20];
EDU» student(3).IQ = rand(3,3);
```

Variable *student* has three elements and holds the same data as variable *data* in Section 4.7. Note that variable *student* is a structure array and variable *data* is a cell array. Now we can easily ask for any data from any element.

```
EDU» student(1).Exams
ans =
   99   98
EDU» student(2).Name
ans =
Marc E. Herniter
```

```
EDU» student(3).Semester
ans =
Fall 1999
EDU» student(1).IQ
ans =
   0.6038      0.0153      0.9318
   0.2722      0.7468      0.4660
   0.1988      0.4451      0.4186
```

4.9 Problems

Problem 4-1 Write a MATLAB function called **isoddm(a)** that checks to see if the elements of a are odd integers; a may be an array of numbers or a scalar. The definition of the function is

$$\text{function } x = \text{isoddm}(a)$$

where $x = \begin{cases} 1 & \text{if the corresponding element in } a \text{ is an odd integer} \\ 0 & \text{if the corresponding element in } a \text{ is an even integer} \\ -1 & \text{if the corresponding element in } a \text{ is not an integer} \end{cases}$

The dimensions of x should be the same as the dimensions of a. The elements of x should be set to 1, 0, −1 according to the above equation and the individual elements of a. An individual element of x should be set to −1 if the corresponding element in a is a real number, a complex number, a *NaN*, or a string. Several examples are shown here:

$a = 5 \qquad x = 1$
$a = 3.7 \qquad x = -1$
$a = 44 \qquad x = 0$

$a = [1\ \ 2\ \ 0\ \ 4\ \ 5\ \ 88.6\ \ 97\ \ 0\ \ 3.14] \qquad x = [1\ \ 0\ \ 0\ \ 0\ \ 1\ \ -1\ \ 1\ \ 0\ \ -1]$

$$a = \begin{bmatrix} 1 & 3 & 4.4 \\ NaN & 4 & -6 \\ 27 & 33.1 & 5 \end{bmatrix} \qquad x = \begin{bmatrix} 1 & 1 & -1 \\ -1 & 0 & 0 \\ 1 & -1 & 1 \end{bmatrix}$$

Problem 4-2 Write a function called hist2???.m (where ??? are your initials) to read a data file (file name specified by the user) and generate a histogram that shows how many times the numbers 0–9 occur. Read the file as a text file and ignore spaces. The file may contain any character, but you should only count the characters 0, 1, 2, 3, 4, 5, 6, 7, 8, 9. Use the MATLAB **HIST** function to generate a histogram similar to the ones shown in the notes. The histogram should contain only 10 bins.

Problem 4-3 Write a function called hist1???.m, where ??? are your initials, to read a data file (file name specified by the user) and generate a histogram that shows how many times the numbers 0–9 occur. Read the file as a text file and ignore spaces. The file may contain any character, but you should only count the characters 0, 1, 2, 3, 4, 5, 6, 7, 8, 9. Use Function 4-8 as a model. This function should print out an asterisk (*) for each occurrence. The function should also display how many times each number appears. The following sample output is for a file that contains the text:

```
43237892 64 4732 382 2382 282
h67dl49dj*k35%^jreiusdu^&#${}htythg'1g376r3jkd7342hkfsd832kjfd763
Program output shown below
```

```
0: ***** (5)
1: ********** (11)
2: **** (4)
3: ********* (9)
4: * (1)
5: (0)
6: **************** (17)
7: ** (2)
8: ***(3)
9: ****(4)
```

Problem 4-4 The circuit below uses 5% resistors with a normal probability distribution. Write a MATLAB script file that calculates the nominal value of V_o/V_{in} and then displays a histogram of the gain spread with resistor tolerances. Call the script file HW44_???.m where ??? are your initials.

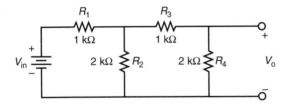

Problem 4-5 Repeat Problem 4-4 using a uniform tolerance distribution rather than a normal distribution. Note that MATLAB function **RANDN** gives a normal distribution and function **RAND** gives a uniform distribution.

Problem 4-6 Repeat Problem 4-4 but instead of a histogram, display a table of the following results:

■ Minimum gain obtained from random results
■ Maximum gain obtained from random results
■ Nominal gain calculated from equation
■ Nominal gain calculated from random results (the mean)
■ Percent of trials within ±1% of the nominal gain
■ Percent of trials within ±2% of the nominal gain
■ Percent of trials within ±3% of the nominal gain
■ Percent of trials within ±4% of the nominal gain
■ Percent of trials within ±5% of the nominal gain
■ The standard deviation of the gain distribution

Use the MATLAB **STD** function.

Problem 4-7 The following voltage and current measurements have been made on a two-terminal device:

Voltage (V)	Current (A)	Voltage (V)	Current (A)	Voltage (V)	Current (A)
1	0.005	1	0.006	1	0.0055
2	0.01	2	0.015	2	0.02
3	0.015	3	0.018	3	0.013
4	0.02	4	0.022	4	0.018
5	0.025	5	0.023	5	0.026
6	0.03	6	0.031	6	0.033
7	0.034	7	0.037	7	0.036

The equation for voltage and current of a resistor is $V_R = R \times I_R$, where V_R is the voltage, I_R is the current, and R is the resistance value of the resistor. Use the MATLAB polynomial curve-fitting function to find the resistance of the device.

Problem 4-8 In Section 4.4.3, we generated a histogram for a voltage divider with a ±5% normal distribution. Redo the plot for a ±5% uniform distribution. Your final plot should graph the number of occurrences versus the gain.

Problem 4-9 A particle starts at coordinates $x = 0$, $y = 0$ and performs a random walk. At steps 1, 2, 3, . . . , the particle is equally likely to take a random step in one of four directions. The four directions are: $x = x + 1$, or $x = x - 1$, or $y = y + 1$, or $y = y - 1$. Use the **RAND** function to calculate random numbers with a uniform distribution. Your program should ask the user for an integer that is the number of steps the particle will take, calculate the random walk, and then plot the random walk with the **COMET** plot function.

Problem 4-10 Continuing with Problem 4-9, write a program that runs a random walk 100 times. Your program should determine the average number of steps required for a particle to travel a distance of z steps from the origin, where $z = \sqrt{x^2 + y^2}$. Your program should request the number z from the user.

Problem 4-11 Write a MATLAB function that sorts the elements of an array in ascending or descending order. The syntax of the function is

$$\text{function } x = \text{sort\_arr}(A, m)$$

Variable A contains the data to be sorted and can be a scalar, or a one- or two-dimensional array. Variable m specifies if the data sorts in ascending or descending order.

- If m is not specified (omitted) or if $m = 0$, then the data sorts in ascending order.
- If $m = 1$, then the data sorts in descending order.
- The dimensions of x are the same as A. If any elements of A are complex, the array sorts according to the magnitude of the numbers (the **ABS** function). If any elements of A are the value *NaN*, the *NaN* values should be placed at the end of the array. If the elements of A are characters, the elements sort in alphabetical order.

- You may use the MATLAB **SORT** function, but this will not completely solve the problem.
- Examples of how the function should work are given here:

$$
\text{Let } A = \begin{bmatrix} 3 & 7 & 9 \\ 7 & 8 & 13 \\ 1 & 2 & 16 \end{bmatrix}, \quad B = [17 \ 16 \ 14 \ 19 \ 23], \quad C = 10, \quad D = \begin{bmatrix} 8 & 9 \\ NaN & 2 \\ 12 & -25 \\ 27 & 56 \\ 33 & 17 \end{bmatrix}
$$

Examples:

$$
\text{sort\_arr}(A, 0) = \begin{bmatrix} 1 & 2 & 3 \\ 7 & 7 & 8 \\ 9 & 13 & 16 \end{bmatrix} \qquad \text{sort\_arr}(A) = \begin{bmatrix} 1 & 2 & 3 \\ 7 & 7 & 8 \\ 9 & 13 & 16 \end{bmatrix}
$$

$$
\text{sort\_arr}(A, 1) = \begin{bmatrix} 16 & 13 & 9 \\ 8 & 7 & 7 \\ 3 & 2 & 1 \end{bmatrix} \qquad \text{sort\_arr}(B, 1) = [23 \ 19 \ 17 \ 16 \ 14]
$$

$$
\text{sort\_arr}(B, 0) = [14 \ 16 \ 17 \ 19 \ 23] \qquad \text{sort}(C, 1) = 10
$$

$$
\text{sort\_arr}(D, 1) = \begin{bmatrix} 56 & 33 \\ 27 & 17 \\ 12 & 9 \\ 8 & 2 \\ -25 & NaN \end{bmatrix} \qquad \text{sort\_arr}('ababab',0) = 'aaabbb'
$$

Problem 4-12   Rewrite the d_sort program to use an insertion sort technique. With this technique, a sorted array starts with no numbers and then numbers are added in their appropriate position. Thus, if we have the numbers 78 14 923 6 19 90 1 22 in an unsorted array, we would step through this array, and add each number in the appropriate place in the sorted array. The sorted array would look like this:

[]	Empty at the start.
[**78**]	First number added to the sorted array.
[**14** 78]	The number 14 added in the correct position.
[14 78 **923**]	The number 923 added in the correct position.
[**6** 14 78 923]	The number 6 added in the correct position.
[6 14 **19** 78 923]	The number 19 added in the correct position.
[6 14 19 78 **90** 923]	The number 90 added in the correct position.
[**1** 6 14 19 78 90 923]	The number 1 added in the correct position.
[1 6 14 19 **22** 78 90 923]	The number 22 added in the correct position.

Problem 4-13 Write the MATLAB functions and script files described below. In the name of the functions, the letters 'xxx' refer to your initials.

A) **Function xxx_cnt1.** Write a function called xxx_cnt1 using this syntax:

$$\text{function } count = \text{xxx\_cnt1}(numbers)$$

Input argument *numbers* is an $n \times 3$ array containing real numbers. Return variable *count* is a 1×3 array that contains integers. Function xxx_cnt1 counts the following:

1. How many times all numbers in a row are different. This count is stored in variable *count*(1, 1).
2. How many times two numbers in a row are the same. This count is stored in variable *count*(1, 2).
3. How many times all numbers in a row are the same. This count is stored in variable *count*(1, 3).

An example of the function is shown:

```
EDU» data = [1 2 3; 7 1 2; 0 7 0; 0 0 7; 7 7 0; 7 0 7; 0 7 7; ➥
7 7 7; 7 7 7]
data =
     1     2     3
     7     1     2
     0     7     0
     0     0     7
     7     7     0
     7     0     7
     0     7     7
     7     7     7
     7     7     7
EDU» x=meh_cnt1(data)
x =
     2     5     2
EDU» data = [ 1 2 3; 1 1 3; 1 1 1; 2 1 2; 3 1 3; 0 1 1; 9 9 9; ➥
1 8 5; 8 3 3; 2 1 5; 7 7 7]
data =
     1     2     3
     1     1     3
     1     1     1
     2     1     2
     3     1     3
     0     1     1
     9     9     9
     1     8     5
     8     3     3
     2     1     5
     7     7     7
```

```
EDU» x=meh_cnt1(data)
x =
      3      5      3
EDU»
```

B) Function xxx_cnt2. Write a function called xxx_cnt2 using this syntax:

$$\text{function } count = \text{xxx\_cnt2}(numbers)$$

Input argument *numbers* is an $n \times 3$ array containing real numbers. Return variable *count* is a 1×4 array that contains integers. Function xxx_cnt2 counts the following:

1. How many times the number 7 does not occur in a row. Store this count in variable *count*(1, 1).
2. How many times the number 7 occurs in a row once. Store this count in variable *count*(1, 2).
3. How many times the number 7 occurs in a row twice. Store this count in variable *count*(1, 3).
4. How many times the number 7 occurs in a row three times. Store this count in variable *count*(1, 4).

An example of the function is shown here:

```
EDU» data = [ 1 2 3; 1 1 3; 1 1 1; 2 1 2; 3 1 3; 0 1 1; 9 9 9; ➡
1 8 5; 8 3 3; 2 1 5; 7 7 7]
data =
      1      2      3
      1      1      3
      1      1      1
      2      1      2
      3      1      3
      0      1      1
      9      9      9
      1      8      5
      8      3      3
      2      1      5
      7      7      7
EDU» x=meh_cnt2(data)
x =
     10      0      0      1
EDU» data = [1 2 3; 7 1 2; 0 7 0; 0 0 7; 7 7 0; 7 0 7; 0 7 7; ➡
7 7 7; 7 7 7]
data =
      1      2      3
      7      1      2
      0      7      0
      0      0      7
      7      7      0
      7      0      7
```

```
        0       7       7
        7       7       7
        7       7       7
EDU» x=meh_cnt2(data)
x =
        1       3       3       2
EDU»
```

C) **Script file xxx_main.** This script file performs the following tasks:

1. Generates a 5461×3 array filled with random integers 0–9.
2. Generates a bar graph that displays how many times all numbers in a row are the same, two numbers in a row are the same, and all the numbers in a row are different. After the plot is displayed, MATLAB should pause.
3. Displays in text form how many times all the numbers in a row are the same, two numbers in a row are the same, and all numbers in a row are different. After the information is displayed, MATLAB should pause.
4. Generates a bar graph that displays how many times the numbers in a row are all 7, two numbers in a row are 7, one number in a row is 7, and how many times no numbers in a row are 7. After the plot is displayed, MATLAB should pause.
5. Displays in text form how many times the numbers in a row are all 7, how many times two numbers in a row are 7, how many times one number in a row is 7, and how many times no numbers in a row are 7. After the information is displayed, MATLAB should pause.
6. Displays the message "End of Homework."

Here is an example:

```
EDU» meh_main
```

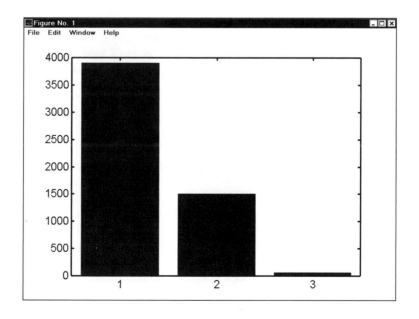

```
Press any key to continue.
The number of times all of the numbers in a row were the same was 61.
The number of times two of the numbers in a row were the same was 1499.
The number of times all of the numbers in a row were  different was 3901.
Press any key to continue.
```

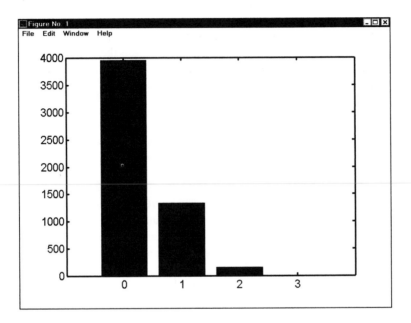

```
Press any key to continue.
The number of times 7 occurred 3 times in a row was 6.
The number of times 7 occurred 2 times in a row was 156.
The number of times 7 occurred 1 time in a row was 1339.
The number of times 7 did not occur in a row was 3960.
Press any key to continue.

End of Homework.
EDU»
```

Problem 4-14 Write a program that reads a text file and generates a histogram of the lengths of the words contained in the file.

Problem 4-15 Write a program that asks the user for an integer and then creates a palindrome. A palindrome is a number whose digits read the same backward and forward. For example, if the user enters 1532, the computer should display 15322351.

Problem 4-16 Write a program that generates a string of n random characters from a to z.

Problem 4-17   Write a program that looks through the string generated in Problem 4-16 and counts how many times the following words appear in the string:

one   pig
two   pony
dog   blue
cat   to be or not to be*

*Words in this phrase need to be in the correct order.

Problem 4-18   Write a program that reads an input string from the user and determines if the string is in the form of a valid web address. For example, the first four characters should be "www." and the last four characters could be .edu, .com, .org, or .gov.

File Input and Output

5

Thus far, we have only looked at reading data from the keyboard and writing data to the screen. Now we will look at reading data from files and writing data to files. When we read data from a file, the file can be one created by MATLAB or by another program. This allows us to exchange data with other programs. For example, if we create a data file with a data acquisition program, we could still use MATLAB to read the file, perform numerical calculations on the data, and graph the data. By writing data to files, MATLAB can export data that can be read by other programs.

Another advantage of reading and writing files is that files stored on disk are nonvolatile. That is, when we shut off the computer, the data is saved in the file. The next time we turn on the computer and run MATLAB, we can read the file and begin where we left off.

5.1 Opening and Closing Files

MATLAB Predefined
fopen function
fclose command
pwd command
uigetfile function
uiputfile function

Whether we are reading or writing a file, the first thing we must do is open the file. This is done in MATLAB with the **FOPEN** command. The syntax is

```
fid=fopen(filename, permissions)
```

Variable *filename* is a text string that contains the file name. If you are opening a file for reading, the file must already exist. If you are opening a file for writing and the file doesn't exist, a new file with the name contained in variable *filename* will be created. Variable *permissions* is a code that specifies how a file is opened. The valid codes are contained in Table 5-1.

You can have more than one file open at a time. In the line **fid=fopen(filename, permissions)**, variable *fid* is a code that identifies the file just opened. Each file you open will be identified by a unique **fid** code. The *fid* codes returned by the **FOPEN** function start at 3 because codes −1 through 2 have specific meanings in MATLAB. Code −1 indicates an error in opening the file. This code is usually returned when you open a file for reading and the file doesn't exist. Instead of generating an error message, MATLAB returns a code of −1 so that your program can check for this error and take corrective action. A code of 0 stands for standard input. This input is usually the keyboard and is always open for writing (permission **'r'**). Code 1 stands for standard output. This output is the MATLAB command window and is always open for output with append permission. Append permission means that data is added to the end of a file, leaving the existing data unchanged. Code 2 stands for standard error and is always open for output with append permission. Standard input, output, and error cannot be opened or closed with the **FOPEN** and **FCLOSE** functions.

When you are done with a file, you should always close the file before terminating MATLAB. This is done with the MATLAB **FCLOSE** function. Its syntax is

status=fclose(fid)

Variable *fid* contains the code of the file you want to close. If you have more than one file open, *fid* specifies which file you want to close. Variable *status* is a code returned by the **FCLOSE** function and indicates whether or not the file was closed properly. A code of −1 indicates that the operation was unsuccessful. A code of 0 means that the file closed properly.

Let's open and close a few files. When opening a file, you can specify the entire pathname or just the file name. If you only specify the file name, the file must reside in the current directory. To see the current directory, use the **PWD** command:

```
EDU» pwd
ans =
C:\MATLAB\TOOLBOX\LOCAL
```

Our current directory is **C:\MATLAB\TOOLBOX\LOCAL**. To see its contents, use the **DIR** function:

```
EDU» dir
    . ..
```

Our directory is currently empty. Let's attempt to open a file for reading:

```
EDU» file_id_1 = fopen('Test1.txt', 'r');
EDU» file_id_1
file_id_1 =
     -1
```

Note that the variable *file_id_1* has a value of −1. This means that there was an error opening the file. This occurred because the permission **'r'** requires the file to exist in order to open it. Since our current directory is empty, an error is indicated by the **FOPEN** function.

TABLE 5-1 Permission codes available in MATLAB for opening a file

Permission Text String	Action
`'r'`	Open the file for reading. The file must exist in order to open it. An error occurs if the file does not exist.
`'r+'`	Open the file for reading and writing. The file must exist in order to open it. An error occurs if the file does not exist.
`'w'`	Open the file for writing. If the file already exists, delete its contents. If the file does not exist, create a new file.
`'w+'`	Open the file for reading and writing. If the file already exists, delete its contents. If the file does not exist, create a new file.
`'a'`	Open the file for writing. If the file already exists, append new data to the end of the file. If the file does not exist create a new file.
`'a+'`	Open the file for reading and writing. If the file already exists, append new data to the end of the file. If the file does not exist, create a new file.

Now let's use permission `'w'`, which creates a new file and opens it for writing:

```
EDU» file_id_1 = fopen('Test1.txt', 'w');
EDU» file_id_1
file_id_1 =
 3
```

The value of variable *file_id_*1 is now 3. This indicates that the file was opened correctly and is identified by code 3. Let's open a second file:

```
EDU» file_id_2 = fopen('Test2.txt', 'w');
EDU» file_id_2
file_id_2 =
 4
```

This file is opened correctly and is identified by *fid* code 4. Since more than one file is open, the *fid* codes specify which file we want to use in the I/O commands that we'll cover later. We can open more files if we wish:

```
EDU» file_id_3 = fopen('Test3.txt', 'w');
EDU» file_id_3
file_id_3 =
 5
EDU» file_id_4 = fopen('Test4.txt', 'w');
EDU» file_id_4
file_id_4 =
 6
```

We can use any of the open files, in this case for output since all of the files were opened with the **'w'** permission code.

Before we exit MATLAB, we must close all open files with the **FCLOSE** function:

```
EDU» status1=fclose(file_id_1);
EDU» status1
status1 =
 0
```

A status of zero indicates that the file was closed successfully.

```
EDU» status2=fclose(file_id_2)
status2 =
 0
EDU» status3=fclose(file_id_3)
status3 =
 0
EDU» status4=fclose(file_id_4)
status4 =
 0
```

The return value of 0 for each use of the **FCLOSE** function indicates that the files were closed successfully. If we look in the current directory, we'll see the files we just created:

```
EDU» dir
.  .. test1.txt test2.txt test3.txt test4.txt
```

The directory now contains the four files we just opened. If we look at the contents of these files, they will be empty because we did not write to any of them, as you can see from the contents of file test4.txt (we're using the Windows Notepad):

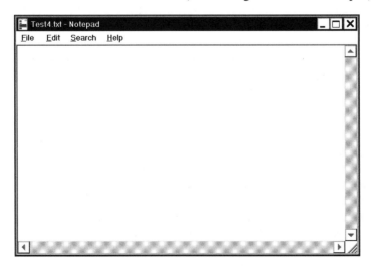

We can specify a complete path when we open files. Let's demonstrate by opening a file in directory c:\transfer. You may not have this directory on your computer, in which case just use a different directory. First let's look at the contents of this directory:

```
EDU» dir c:\transfer
. ..
```

The directory is empty. Now let's open a file in that directory:

```
EDU» fid_a=fopen('c:\transfer\xx.txt', 'w')
fid_a =
 3
```

Notice that the value of variable *fid_a* is now 3. This is because we closed all of the previously open files so the *fid* code starts again at the next available code, 3 in this case. If we look in the transfer directory, this new file is present:

```
EDU» dir c:\transfer
. .. xx.txt
```

Let's close the file before we exit MATLAB:

```
EDU» fclose(fid_a)
ans =
 0
```

The file was successfully closed.

5.1.1 **UIGETFILE** and **UIPUTFILE**

The **UIGETFILE** and **UIPUTFILE** functions provided by MATLAB allow you to use dialog boxes to request file names. Function **UIGETFILE** looks for a file that already exists and finds files to open for reading. Function **UIPUTFILE** finds files that already exist or creates new files. These two functions don't open the file, they are only used to retrieve file names and pathnames. The syntax of the commands is

```
[filename, pathname]=uigetfile('wild card string', 'dialog box title');
[filename, pathname]=uiputfile('wild card string', 'dialog box title');
```

Text string **'wild card string'** reduces the number of files displayed in the dialog box. Examples are **'*.*'** for all files, **'*.txt'** for all files that use extension **'.txt'**, and **'a*.*'** for all files that start with the letter a. Many more wild cards are available. See your Windows or DOS documentation for more information.

The text string **'dialog box title'** is displayed in the title bar of the dialog box. Variable *filename* is a text string returned by the **UIGETFILE** and **UIPUTFILE** functions that contains the name of the file. Variable *pathname* is a text string that contains the path to the file name. Concatenating the two strings specifies the complete location of the file. Both *filename* and *pathname* can be used with the **FOPEN** function to open the file.

Let's examine a few examples.

```
EDU» [fn,pn]=uigetfile('*.*', 'Specify a File Name')
```

When we execute this command, a dialog box appears:

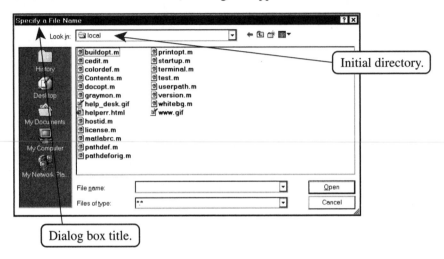

Initial directory.

Dialog box title.

Note that the text **Specify a File Name** appears in the title bar of the dialog box. The initial directory is the present working directory, D:\MATLAB\toolbox\local in our case:

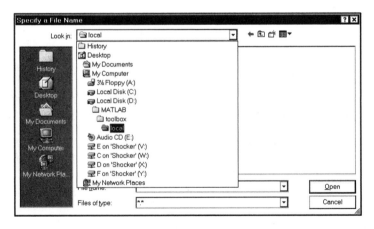

Now let's select a file:

In the above dialog box, the file name test.m is selected and the directory of the file is D:\MATLAB\toolbox\local. When we click the **Open** button, variable *fn* is set to 'test.m' and variable *pn* is set to 'D:\MATLAB\toolbox\local.'

```
EDU» [fn,pn]=uigetfile('*.*', 'Specify a File Name')
fn =
test.m
pn =
D:\MATLAB\toolbox\local\
EDU»
```

Note that if we concatenate variables *pn* and *fn*, we get a string that contains the pathname.

```
EDU» [pn,fn]
ans =
D:\MATLAB\toolbox\local\test.m
EDU»
```

Let's open a file using the **FOPEN** command. To be safe, we'll use both the pathname and the file name. Since the file exists, we can open it for reading:

```
EDU» fid_b=fopen([pn,fn], 'r')
fid_b =
  3
```

We can now use the file for reading.

Note that function **UIGETFILE** does not allow the user to specify a file name that does not already exist.

```
EDU» [fn,pn]=uigetfile('*.*', 'Specify Another File Name')
```

After executing the command, a dialog box appears:

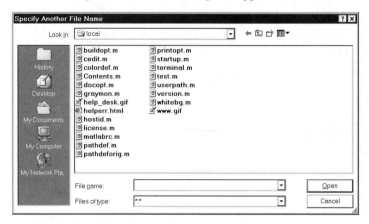

Just to see what happens, let's type the name of a file that doesn't exist in the present directory:

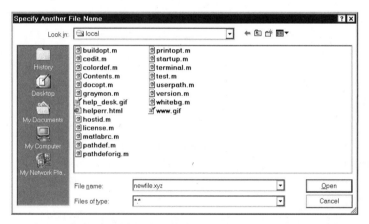

When we click the **OK** button, an error dialog box appears:

The dialog box indicates that the file cannot be found. Thus, the **UIGETFILE** function can't be used to specify a file name of a file that does not exist. Usually, when we create a new file, **UIGETFILE** can't be used to obtain the new file name. Click the **OK** button to acknowledge the mistake:

We won't continue with this example, so click the **Cancel** button:

```
EDU» [fn,pn]=uigetfile('*.*', 'Specify a File Name')
fn =
 0
pn =
 0
EDU»
```

Note that when we click the **Cancel** button, variables *fn* and *pn* have values of 0. This is the code returned by the **UIGETFILE** function that indicates "no file specified."

To create a file name for a new file, we use the **UIPUTFILE** function:

```
EDU» [fn,pn]=uiputfile('*.*', 'Specify a New File Name')
```

When we execute the command, a dialog box is opened:

We can now enter the name of the new file:

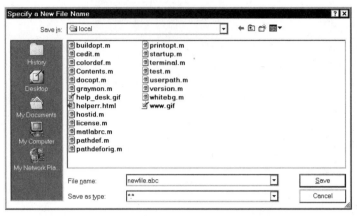

When we click the **Save** button, the function returns successfully with the file name and pathname of the new file:

```
EDU» [fn,pn]=uiputfile('*.*', 'Specify a New File Name')
fn =
newfile.abc
pn =
D:\MATLAB\toolbox\local\
EDU»
```

Note that variable *fn* contains the file name and variable *pn* contains the pathname. We can now open the file with the **FOPEN** function. Always make sure that you use a permission that creates a new file since the file does not exist until you open it:

```
EDU» fid_c=fopen([pn,fn], 'w')
fid_c =
 4
```

We can now write to this file.

We are done with this section, so we'll close all of the open files before exiting MATLAB.

```
EDU» fclose(fid_b)
ans =
 0
EDU» fclose(fid_c)
ans =
 0
```

5.2 **Writing Formatted Output to Files**

MATLAB Predefined
fprintf function
fopen function

The first method we'll use to write to a file is the **FPRINTF** function. We've used this function extensively to write output to the MATLAB screen. When we don't specify a file identifier (*fid* returned by using the **FOPEN** function), the **FPRINTF** function automatically sends the output to the screen. However, if we specify a file identifier, the output is sent to the file specified by the *fid*. We'll show an example, but first we must open the files for writing. Since we are creating new files, we must use either the **'a'** or **'w'** permission.

```
EDU» fid1=fopen('file1.txt', 'w');
EDU» fid2=fopen('file2.txt', 'w');
EDU» fid3=fopen('file3.txt', 'w');
```

We won't specify a file identifier in the **FPRINTF** function at this point. Our output will be written to the command window:

```
EDU» fprintf('This is a test. No fid specified\n');
This is a test. No fid specified
```

Now we write to file1.txt:

```
EDU» fprintf(fid1, ' This is text sent to the file indicated by fid1.\n');
EDU» fprintf(fid1, ' We will generate a few lines.\n');
EDU» fprintf(fid1, 'The value of fid1 is %g\n', fid1);
```

Notice that output is not displayed on the command window; all of the output is sent to the file. Now let's send text to the other two files:

```
EDU» fprintf(fid2, ' Send this stuff to file indicated by fid2.\n');
EDU» fprintf(fid2, ' We will generate a few lines in this file as well.\n');
EDU» fprintf(fid2, 'The value of fid2 is %g\n', fid2);
EDU» fprintf(fid3, ' Lastly, we will send this stuff to file indicated by ➥
fid3.\n');
EDU» fprintf(fid3, ' We will generate a few lines in this file too.\n');
EDU» fprintf(fid3, 'The value of fid3 is %g\n', fid3);
```

We'll close the files here so that we can look at them with a text editor.

```
EDU» fclose(fid1);
EDU» fclose(fid2);
EDU» fclose(fid3);
```

Let's use the Windows Wordpad to view the text files. Each file contains different text:

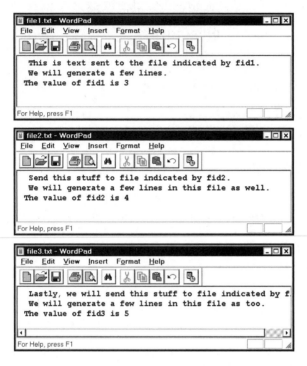

Any text we generated with the **FPRINTF** command in other examples can be used to generate the same text in files. Thus, we can generate formatted output in files as well as on the screen.

5.3   Reading Formatted Data from Files

MATLAB Predefined
fprintf function
while loop
who command

Let's write and read formatted data to and from files. For our example, we'll maintain data for an address book. Data will be stored in the seven arrays listed below. Note: If the data contains less than the specified characters, the unused characters will be filled with spaces (blanks).

- *First_name*—A text string of 20 characters
- *Last_name*—A text string of 20 characters
- *Address*—A text string of 30 characters
- *City*—A text string of 20 characters
- *State*—A text string of 2 characters
- *Zipcode*—A 5-digit integer
- *Zip_p4*—A 4-digit integer

If there are 20 entries in our address book, each array will have 20 rows. Each row contains the information for the same person. For example, *First_name*(5,:), *Last_name*(5,:), *Address*(5,:), *City*(5,:), *State*(5,:), *Zipcode*(5), and *Zip_p4*(5) contain the address information for the fifth person in the address book.

Let's create some data and store it in the variables:

```
EDU» First_name = ['Marc           ';'Beth           ';'David          ']

First_name =

Marc
Beth
David
EDU» Last_name=['Smith           ';'Prudhomme          ';'Zimmerman           ']

Last_name =

Smith
Prudhomme
Zimmerman
EDU» Address=['2070 S. Del Mar Dr.          ';'P.O. Box 12388          ';'112
Thunderbird Av          ']

Address =

2070 S. Del Mar Dr.
P.O. Box 12388
112 Thunderbird Av
EDU» City=['Flagstaff          ';'Los Angeles          ';'Boston          ']

City =

Flagstaff
Los Angeles
Boston

EDU» State=['AZ';'CA';'MA']

State =

AZ
CA
MA

EDU» Zipcode=[86003;96543;02765]

Zipcode =

       86003
       96543
        2765
```

```
EDU» Zip_p4=[5679;5393;1209]

Zip_p4 =

     5679
     5393
     1209
```

EDU»

The information for the second person can be obtained by looking at the second row of each array:

```
EDU» fprintf('%s %s\n%s\n%s %s \n%5.0f-%4.0f\n',First_name(2,:),
          Last_name(2,:),Address(2,:), City(2,:), State(2,:),
          Zipcode(2,:), Zip_p4(2,:));
Beth                Prudhomme
P.O. Box 12388
Los Angeles         CA
96543-5393
EDU»
```

> These three lines would be a single line in a MATLAB script file.

Now let's write the information into a file. Using the **FPRINTF** function will create a human-readable file that we can view with a text editor. We'll separate each field with a comma so that we can read the file easily.

Program 5-1   Write information to the address book

```
% This is file write_text_file.m
%
fid_a=fopen('Address_book.txt','a');
for I = 1:length(Zipcode)
   fprintf(fid_a, '%20s,%20s,%30s,',First_name(I,:),Last_name(I,:), ➥
Address(I,:));
   fprintf(fid_a, [City(I,:), ',', State(I,:), ',']);
   fprintf(fid_a, '%5.0f,%4.0f\n', Zipcode(I), Zip_p4(I));
end
fclose(fid_a);
```

The code segment shows two ways to print strings. The first **FPRINTF** statement uses the **%s** formatting command, **fprintf(fid_a, '%20s,%20s,%30s,', First_name(I,:),Last_name(I,:), Address(I,:)**. The **%20s** formatting string means print the variable as a string using 20 spaces. This **FPRINTF** command prints the first string with 20 characters, then prints a comma, a second string with 20 characters, another comma, a third string with 30 characters, and then another comma. The second **FPRINTF** command, **fprintf(fid_a, [City(I,:), ',', State(I,:), ',']**), uses concatenation to form a single string composed of the *City* string, followed by a comma, the *State* string, and another comma. The last

FPRINTF command, `fprintf(fid_a, '%5.0f,%4.0f\n', Zipcode(I), Zip_p4(I))`, prints the *Zipcode* and *Zip_p4* variables as integers and then starts a new line with the `\n` command.

Let's run the code segment and then display the contents of file Address_book.txt with the Wordpad:

A structured file like this can now be read easily assuming we know the format of the file. Now we'll create a code segment that reads the data. Since we know the formats that were used to write the file, we can read the file easily.

Program 5-2 Read the address book

```
% This is file read_text_file.m
%
fid_x=fopen('Address_book.txt','r');
First_name=[ ];
Last_name=[ ];
Address=[ ];
City=[ ];
State=[ ];
Zipcode=[ ];
Zip_p4=[ ];
while ~feof(fid_x)

    str=fscanf(fid_x, '%20c',1);
    First_name = [First_name;str];
    comma=fscanf(fid_x,'%1c',1);

    str=fscanf(fid_x, '%20c',1);
    Last_name = [Last_name;str];
    comma=fscanf(fid_x,'%1c',1);

    str=fscanf(fid_x, '%30c',1);
    Address = [Address;str];
    comma=fscanf(fid_x,'%1c',1);

    str=fscanf(fid_x, '%20c',1);
    City = [City;str];
    comma=fscanf(fid_x,'%1c',1);

    str=fscanf(fid_x, '%2c',1);
    State = [State;str];
    comma=fscanf(fid_x,'%1c',1);
```

```
num=fscanf(fid_x, '%5d',1);
Zipcode=[Zipcode;num];
comma=fscanf(fid_x,'%1c',1);

num=fscanf(fid_x, '%4d',1);
Zip_p4=[Zip_p4;num];
end_of_line=fscanf(fid_x,'%1c',1);


end
fclose(fid_x);
```

Let's examine a few of these lines. The line **fid_x=fopen('Address_book.txt',
'r');** opens the file for reading. Function **FEOF** checks if we are at the end of a specified file, and if we are, the function returns a value of 1 (true). If we aren't at the end of
the file, the function returns a value of 0 (false). The line **while ~feof(fid_x)** creates a **WHILE** loop that continues looping until we reach the end of the file.

Next, we'll look at the code segment:

```
str=fscanf(fid_x, '%20c',1);
First_name = [First_name;str];
comma=fscanf(fid_x,'%1c',1);
```

The **%c** formatting character means read any character, including spaces. We didn't use
the **%s** formatting character because **%s** reads a string but ignores spaces. Spaces are an
integral part of the structure of the file, so we can't ignore them. The format string
%20c means read a 20-character text string. In the line **str=fscanf(fid_x,
'%20c',1);** the **1** means read one element, or one 20-character text string. Thus, the
line **str=fscanf(fid_x, '%20c',1);** means read one 20-character text string
from file *fid_x* and place it in variable *str*.

The line **comma=fscanf(fid_x,'%1c',1);** means read one 1-character text
string and place it in variable *comma*. The three lines above read a 20-character text
string, add that text string as a new row to variable *First_name*, and then read a comma.
We never do anything with the comma, but we need to read it so that a comma doesn't
appear in the next text string. We use a similar code segment for reading the *Last_name*,
Address, *City*, and *State*.

Lastly, let's look at the code segment:

```
num=fscanf(fid_x, '%5d',1);
Zipcode=[Zipcode;num];
comma=fscanf(fid_x,'%1c',1);
```

The **%d** formatting character means read a decimal number. The formatting string **%5d**
means read a 5-digit decimal number. If any of the leading 5 digits are blank, they are
read as zeros. Thus, "01234" and " 1234" are read as the same number. The space
counts as one digit. In the line **num=fscanf(fid_x, '%5d',1);**, the **1** means

read one element. Thus, the line **num=fscanf(fid_x, '%5d',1);** means read one 5-digit number and place it in variable *num*.

The last line inside the **WHILE** loop is **end_of_line=fscanf(fid_x, '%1c',1);**. The end of a line is marked by a special character. It must be read so that the position indicator for the file sets to point at the first character of the next string. If we don't read the end-of-line character, the first character of the next string will contain the end-of-line character.

To demonstrate that the file works, we'll clear the MATLAB work space.

```
EDU» clear
EDU» who
```

The **WHO** function shows that no variables are currently defined. Let's run the function and then see what variables are defined:

```
EDU» read_text_file
EDU» who
Your variables are:
Address       State        comma         str
City          Zipcode      end_of_line
First_name    Zip_p4        fid_x
Last_name     ans          num
EDU»
```

Let's display a few of the variables to see that they contain the data in the file:

```
EDU» Address
Address =
2070 S. Del Mar Dr.
P.O. Box 12388
112 Thunderbird Av
EDU» First_name
First_name =
Marc
Beth
David
EDU» Zipcode
Zipcode =
     86003
     96543
      2765
EDU»
```

Remember: The structure with which the file is written must be known, and the file must be read using the same formatting with which it was written.

5.4   **Writing and Reading Binary Files**

MATLAB Predefined
fprintf function
fread function
fwrite function
setstr function

The **FPRINTF** function places human-readable text in a file. This file type is also called an ASCII text file. We can easily read this text, but it uses a lot of disk space. If we need to create large data files that will only be read by the computer, binary files are much more space-efficient. The numbers are written in their binary form, and will be unintelligible if we look at them with a text editor. However, the files are smaller than the equivalent ASCII files. The functions for reading and writing binary files are **FREAD** and **FWRITE**.

The syntax of the **FWRITE** command is

```
count=fwrite(FID, A, 'precision')
```

Variable *A* contains the data to be written; it can be an array or a scalar. Variable *count* contains the number of elements successfully written to the file. Variable *fid* is the file identifier; it states to which file the information will be written. Variable *fid* is obtained with the **FOPEN** function. The **'precision'** text string tells how we want to store the information. There are several formats for this, but we'll only discuss a few here:

- 'char'—an 8-bit character; used for text
- 'short'—a 16-bit integer; assuming a 2's complement, this format represents numbers in the range -2^{15} to $2^{15} - 1$
- 'long'—a 32-bit integer; assuming a 2's complement, this format represents numbers in the range -2^{31} to $2^{31} - 1$.
- 'float'—a single-precision real number
- 'double'—a double-precision real number

To see what other precision types are available, use the help facility on the **FREAD** function, but be aware that some types are machine-dependent and may not work on all platforms.

To be safe and to store data with their maximum accuracy, you should always use double precision for real numbers. This results in a larger data file but saves data with its maximum precision. The only reason you may want to use other types of precision is to reduce the size of the file, or if data is in integer or text form.

Now let's write some data. We must first open a file for writing:

```
EDU» fid_bin=fopen('bin_file.abc', 'w');
```

We'll now create some data and write it to the file.

```
EDU» AA=magic(5)
AA =
  17 24 1 8 15
  23 5 7 14 16
  4 6 13 20 22
  10 12 19 21 3
  11 18 25 2 9
```

```
EDU» B=rand(7,4)
B =
 0.2190 0.8310 0.0668 0.0920
 0.0470 0.0346 0.4175 0.6539
 0.6789 0.0535 0.6868 0.4160
 0.6793 0.5297 0.5890 0.7012
 0.9347 0.6711 0.9304 0.9103
 0.3835 0.0077 0.8462 0.7622
 0.5194 0.3834 0.5269 0.2625
EDU» C=27
C =
 27
EDU» k='This is a test'
k =
This is a test
EDU» d=pi
d =
 3.1416
```

We write this data to the file.

```
EDU» fwrite(fid_bin, AA, 'double')
ans =
 25
```

The function returns a value of 25 indicating that the 25 elements of variable *AA* were written successfully. Remember that *AA* is a 5×5 array.

```
EDU» fwrite(fid_bin, B, 'double')
ans =
 28
```

All 28 elements of variable *B* were written successfully. (*B* is a 7×4 array.)

Now we will write variable *C* to the file as a 16-bit integer since its value is 27, and 27 is less than $2^{15} - 1$:

```
EDU» fwrite(fid_bin, C, 'short')
ans =
 1
```

The function indicates that one value was successfully written to the file.

Next, we write variable *k* to the file with *precision* **'char'** because *k* is a text string.

```
EDU» fwrite(fid_bin, k, 'char')
ans =
 14
```

There are 14 characters in the text string and the result shows that 14 elements were successfully written to the file.

Lastly, we write variable *d* to the file as a double-precision real number:

```
EDU» fwrite(fid_bin, d, 'double')
ans =
 1
```

We'll close the file now and attempt to look at it with a text editor.

```
EDU» fclose(fid_bin)
ans =
 0
```

Let's use the Windows Notepad to look at the file:

Since the file contains binary data, it is unreadable. As noted earlier, the purpose of a binary data file is to store data efficiently. Binary files are not meant to be viewed with a text editor.

Next, we'll open the file for reading and see if we can retrieve the data with the **FREAD** function. The syntax of the function is

$$[A, \ count] \ = \ fread(FID, \ size, \ 'precision')$$

To retrieve the data properly, we must read variables in the same order and with the same precision as we used to write them with the **FWRITE** function. We discussed variable *precision* earlier. Variable *A* contains the data read from the file. Variable *count* contains how many elements were read successfully. Variable *fid* is the file identifier that indicates which file we want to read data from. Variable *size* contains the dimensions of the array we want to read. To read a 5×5 array, *size* should be [5, 5]. To read a 7×4 array, *size* should be [7, 4]. To read a scalar, *size* should be [1]. Note that in order to read the data correctly, we need to know exactly how it was written. This includes the order in which variables were written to the file, the precision with which they were written, and the dimensions of arrays.

Let's read the data from the file. First we clear the workspace to show that data is obtained from the file.

```
EDU» clear;
EDU» who
Your variables are:
EDU»
```

The line is blank, indicating that no variables are defined. Now we'll open file bin_file.abc for reading:

```
EDU» fid_q=fopen('bin_file.abc', 'r')
fid_q =
  3
```

First we read variable *A* from the file. This variable is a 5×5 array of double-precision real numbers:

```
EDU» [A, count] = fread(fid_q, [5,5], 'double')
A =
    17    24     1     8    15
    23     5     7    14    16
     4     6    13    20    22
    10    12    19    21     3
    11    18    25     2     9
count =
   25
EDU»
```

The values are the same as when we wrote the file. Next, we read the data for variable *B*. This is a 7×4 array of real numbers:

```
EDU» [B,count]=fread(fid_q, [7,4], 'double')
B =
    0.9501    0.0185    0.1763    0.3529
    0.2311    0.8214    0.4057    0.8132
    0.6068    0.4447    0.9355    0.0099
    0.4860    0.6154    0.9169    0.1389
    0.8913    0.7919    0.4103    0.2028
    0.7621    0.9218    0.8936    0.1987
    0.4565    0.7382    0.0579    0.6038
count =
   28
EDU»
```

Next in the file is variable *C*, which is written as a 16-bit integer ('short').

```
EDU» C=fread(fid_q, [1], 'short')
C =
   27
```

After variable *C*, text string *k* is read from the file; it is a 14-character text string:

```
EDU» [k, count]=fread(fid_q, [14], 'char')
k =
 84
 104
 105
 115
 32
 105
 115
 32
 97
 32
 116
 101
 115
 116
count =
 14
```

Note that the text string is read as ASCII codes into a column array. To change it back to a string, we take the transpose of the column array and use the MATLAB **SETSTR** function:

```
EDU» k=k'
k =
 Columns 1 through 12
 84 104 105 115 32 105 115 32 97 32 116 101
 Columns 13 through 14
 115 116
EDU» k=setstr(k)
k =
This is a test
```

Lastly, we read in variable *d*, which is a double-precision real number.

```
EDU» [d, count]=fread(fid_q, [1], 'double')
d =
 3.1416
count =
 1
```

We are done with the file, so we close it before exiting:

EDU» **fclose(fid_q)**
ans =
  0

As a second example, let's save the address book information of the previous section as a binary file. We'll write the information to the file and then read the file. The code for writing the file is shown in Program 5-3.

Program 5-3 Write a binary file

```
% This is file write_binary_file.m.
%
fid_a=fopen('Address_book.bin','w');
for I = 1:length(Zipcode)
   fwrite(fid_a, First_name(I,:), 'char');
   fwrite(fid_a, Last_name(I,:), 'char');
   fwrite(fid_a, Address(I,:), 'char');
   fwrite(fid_a, City(I,:), 'char');
   fwrite(fid_a, State(I,:), 'char');
   fwrite(fid_a, Zipcode(I,:), 'long');
   fwrite(fid_a, Zip_p4(I,:), 'short');
end
fclose(fid_a);
```

The variables *First_name*, *Last_name*, *Address*, *City*, and *State* are written with *precision* **'char'**. Each row of these variables is a text string. Each element in the text string is written with *precision* **'char'** or as 8-bit codes. For example, the line **fwrite(fid_a, First_name(I,:), 'char');** writes the *I*th row of variable *First_name* with *precision* **'char'**. Since each row contains 20 characters, this line writes the 20 elements with *precision* **'char'**.

The line **fwrite(fid_a, Zipcode(I,:), 'long');** writes the zipcode as a 32-bit integer. The largest number we can write with a short integer is $2^{15} - 1$ or 32,767. The largest zipcode we can write is 99999, which is too large to be written as a short integer. Thus, zipcodes are written to the file with *precision* **'long'**.

The zip+4 code can have values from 0 to 9999. We can use short integers for numbers in this range. Short integers use 16 bits and long integers use 32 bits. Short integers save us 2 bytes of disk space for each address entry.

When we run the code segment above, no output is written to the MATLAB command window. A file named Address_book.bin is created; we cannot view the file with a text editor. However, we can compare the size of the file Address_book.txt with Address_book.bin and see the difference in their size:

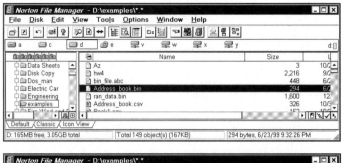

As you can see, the text version, Address_book.txt, uses 324 bytes of disk space, and the binary version, Address_book.bin, uses 294 bytes. To reduce disk space, you should always use binary files when possible.

Next, let's look at a code segment shown in Program 5-4 for reading the binary version of the address book file:

Program 5-4   Read a binary file

```
%This is file read_binary_file.m
%
fid_x=fopen('Address_book.bin','r');
First_name=[ ];
Last_name=[ ];
Address=[ ];
City=[ ];
State=[ ];
Zipcode=[ ];
Zip_p4=[ ];
while ~feof(fid_x)

   str=fread(fid_x, [20],'char');
   First_name = [First_name; setstr(str')]

   str=fread(fid_x, [20],'char');
   Last_name = [Last_name;setstr(str')]

   str=fread(fid_x, [30],'char');
   Address = [Address;setstr(str')]
```

```
    str=fread(fid_x, [20],'char');
    City = [City;setstr(str')]

    str=fread(fid_x, [2],'char');
    State = [State;setstr(str')]

    num=fread(fid_x, 1,'long');
    Zipcode=[Zipcode;num]

    num=fread(fid_x, 1,'short');
    Zip_p4=[Zip_p4;num]
    fprintf('Wait!!!\n')
    pause

end
fclose(fid_x);
```

Files must be read with the same precision and order as they are written. In Program 5-4, the line **str=fread(fid_x, [20],'char');** reads 20 elements with *precision* **'char'**. This is the data from variable *First_name,* which is a 20-element string written with *precision* 'char'. After reading the information, the data is stored in variable *First_name*. All strings are read in a similar manner. Thus, **str=fread(fid_x, [30],'char');** reads 30 characters from the file as a text string. The line **Last_name = [Last_name;setstr(str')]** appends the text string as a new row to variable *Last_name*; **str=fread(fid_x,[2], 'char');** reads two characters from the file as a text string; and **State = [State;setstr(str')]** adds the string as a new row to variable *State*.

The zipcode data and zip+4 data are written as numeric values and must be read as such. The line **num=fread(fid_x, 1,'long');** reads one number as a long integer. This is the same format as the zipcode information is written in. The line **num=fread(fid_x, 1,'short');** reads one number as a short integer; again in the same format as the zip_p4 information.

Note that when we write and read binary data, we don't put in commas and new line characters like we did with the formatted text files. These characters are added to make the file easy for us to read. When writing binary data files, the files contain a continuous stream of data. No commas or new line characters are needed because the files won't be viewed with a text editor. The only way to view the file is to read it with a code segment similar to the one above, and then write more code to display the data in a form we that can easily read.

5.5 Exchanging Data with Other Programs—Comma-Separated Values

MATLAB Predefined **csvwrite** function

In Section 5.3, we created a text file that separated the data fields by commas. The comma was not part of the data. It was only used to separate the various fields. A character used in this manner is called a delimiter because it "delimits" the data. The comma is not a special character. The only requirement of a delimiter is that it is a unique character not used in the data. For example, the data in the address book contains spaces. Therefore, a space can't be used as a delimiter. If the data contained decimal numbers, a period would not be an appropriate delimiter. Some common characters used for delimiters are commas, spaces, semicolons, and tabs.

For our example, we'll use a comma as the delimiter. It works because none of the data in our address book contains a comma. However, the comma is ordinarily a risky choice because people's names and addresses are often set up in ways that use commas.

We'll use the address book data that was written to a file as text. Recall that the fields in the data are separated by commas. The contents of the file are shown here:

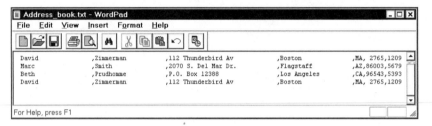

Let's use Microsoft Excel to read the file. Run Excel:

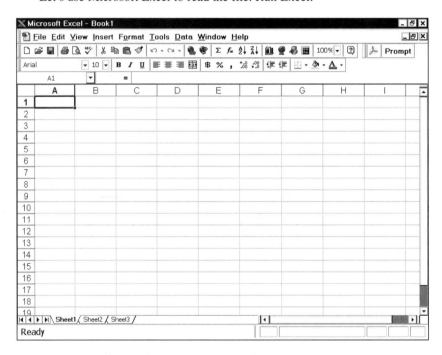

Select **File** and then **Open** from the Excel menus to open a file. Select file Address_book.txt:

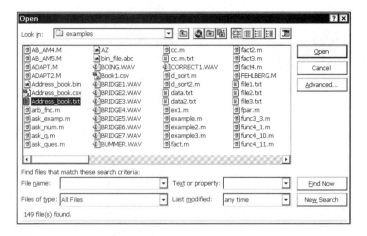

When Excel opens the file and reads the data, it recognizes that the data may be delimited text fields:

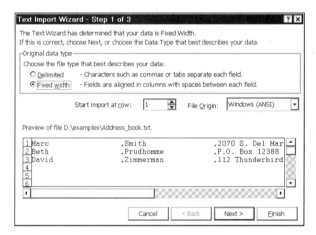

Select **Delimited**:

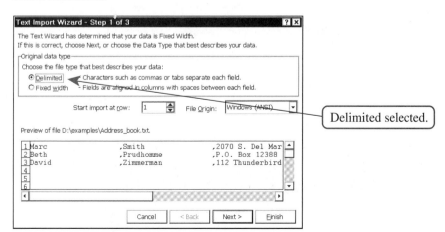

When we click the **Next** button, we are asked to specify a delimiter:

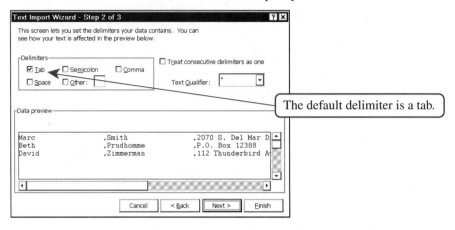

Select the **Comma** and click the **Next** button:

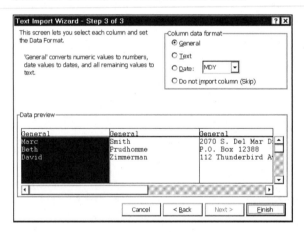

The dialog box shows how the data will be parsed. We are also allowed to select which data we want to import. We select all of the data:

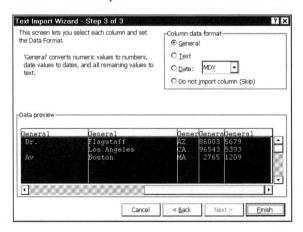

When we click the **Finish** button, Excel imports the data:

	A	B	C	D	E	F	G	H	I
1	Marc	Smith	2070 S. D	Flagstaff	AZ	86003	5679		
2	Beth	Prudhomm	P.O. Box 1	Los Angele	CA	96543	5393		
3	David	Zimmerma	112 Thund	Boston	MA	2765	1209		

We can now use Excel to manipulate the data.

In Section 5.3, we wrote a specific code segment to create a text file delimited by commas. This was necessary because the data written to the file contained text information and numeric information. MATLAB has a function for writing numeric data to comma-separated-value files, and a function for reading data from a comma-separated-value file into an array. The function is only used for writing and reading numeric data.

The function for writing data to a file is called **CSVWRITE**. The syntax is **csvwrite (filename,m)**. Variable *m* is an array of numeric data. Each row of the array is placed on a new line of the file. Elements of the array that are on the same row are separated by commas. Note that if array *m* contains any zeros, the zeros will be replaced in the file by blanks. Suppose $a = [1\ 0\ 2\ 0\ 3]$. When *a* is written to a file using **CSVWRITE**, the file looks like **1,,2,,3**. We'll use the following array as our example:

```
EDU» a = [1 2 3 0 5;6 7 8 9 10; 11 12 13 14 15; 16 17 18 19 20; ➡
21 22 23 24 25]
a =
     1     2     3     0     5
     6     7     8     9    10
    11    12    13    14    15
    16    17    18    19    20
    21    22    23    24    25
EDU»
```

We write this data to a file called xx.csv:

```
EDU» csvwrite('xx.csv', a)
EDU»
```

Now we can look at the contents of the file using a text editor:

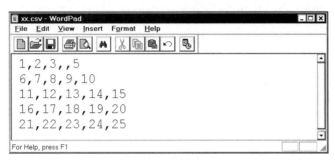

Excel recognizes files with comma-separated values. If you open a file with the extension **.csv**, Excel assumes that it's a file with comma-separated values. It automatically separates the data into columns based on the location of the commas:

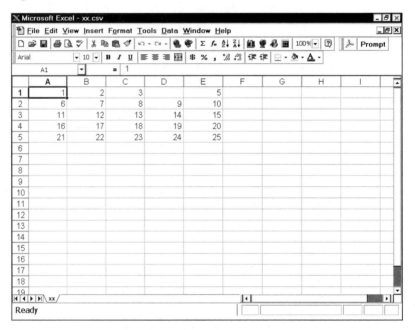

CSV (comma-separated values) files are a standard format that easily allows different programs to exchange data. Excel allows you to save spreadsheets in a CSV format. Let's use the following data in our example:

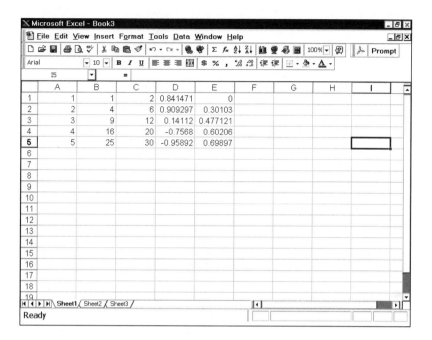

When we select **File** and **Save As** from the Excel menus, we can specify a file type of CSV:

Let's save and close the file, so we can look at it with a text editor:

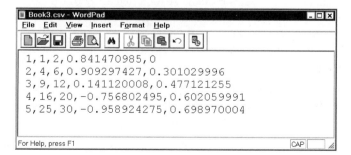

A file of this type can be read by MATLAB using the **CSVREAD** function:

```
EDU» d=csvread('book3.csv')
d =
      1.0000    1.0000    2.0000    0.8415         0
      2.0000    4.0000    6.0000    0.9093    0.3010
      3.0000    9.0000   12.0000    0.1411    0.4771
      4.0000   16.0000   20.0000   -0.7568    0.6021
      5.0000   25.0000   30.0000   -0.9589    0.6990
EDU»
```

5.6 Problems

Problem 5-1 Write a MATLAB function that requests address information for an entry into an address book database. The form of the function is

function enter_data

This function will perform the following tasks. Note that in tasks 1–5, if the entry is longer than the maximum allowed number of characters, it should be truncated to that number of characters. If the entry is less than the maximum number of characters, it should be padded with blanks to make that number.

1. Ask the user for the first name of the person. The first name is a character string of 20 characters.
2. Ask the user for the last name of the person. The last name is a character string of 20 characters.
3. Ask the user for the address of the person. The address is a character string of 30 characters.
4. Ask the user for the city where the person lives. The city name is a character string of 20 characters.
5. Ask the user for the state where the person lives. The state name is a character string of 2 characters.
6. Ask the user for the person's zipcode. The zipcode is a character string of 5 or 10 characters. If the zipcode is neither 5 nor 10 characters, the zipcode string should be set to the string **EntryError**. If the user enters a 5-digit zipcode, add the characters "-0000" to the zipcode to make a total of 10 characters. For example, if the user enters 86011, the string should be modified to "86011-0000".
7. After all of the above have been entered, the function should display the information in a readable fashion.
8. Next, the program should ask if the user wishes to modify the information. If no, go to task 9. If yes,

   - Ask which item the user would like to modify. The user should be able to modify any of the items in the address information.
   - Ask the user to enter the new information.
   - Display the complete information for the entry.
   - Go to task 8 again.

9. Ask the user if they would like to add the entry to the database file. If no, go to task 10. If yes,

- Open the file address_book.txt with append permission.
- Write the information on a single line. Place a \n at the end of the line.
- Close the file.

10. Ask the user if they would like to make another address entry.

- If yes, go to task 1.
- If no, terminate.

You should take advantage of the **ask_q** function discussed earlier in the text. Carefully examine the following transcript, which is one solution to the problem. It should illustrate more clearly how your function should work.

```
EDU» enter_data

Specify information for the address book entry.

Specify the first name, 0 to 20 characters: Marc
Specify the last name, 0 to 20 characters: Herniter
Specify the address, 0 to 30 characters: 7840 Candi Lane
Specify the City name, 0 to 20 characters: Flagstaff
Specify the State, 2 characters: AZ
Specify the Zipcode, 5 or 10 characters: 86004




The information you entered was:
    First Name: Marc
    Last Name: Herniter
    Address: 7840 Candi Lane
    City: Flagstaff
    State: AZ
    Zipcode: 86004-0000


Do you wish to modify the entry? (y/n): n

Do you wish to add this entry to the data file?
y
Do you wish to make another entry? (y/n): y
Specify information for the address book entry.

Specify the first name, 0 to 20 characters: Joe
Specify the last name, 0 to 20 characters: Smith
Specify the address, 0 to 30 characters: 2750 Aspen Glenn
Specify the City name, 0 to 20 characters: Wilcox
```

```
Specify the State, 2 characters: AZ
Specify the Zipcode, 5 or 10 characters: 84320-9645



The information you entered was:
    First Name: Joe
    Last Name: Smith
    Address: 2750 Aspen Glenn
    City: Wilcox
    State: AZ
    Zipcode: 84320-9645


Do you wish to modify the entry? (y/n): y
```

```
Which Item would you like to modify?
    1 = First Name
    2 = Last Name
    3 = Address
    4 = City
    5 = State
    6 = Zipcode

Specify a number: 6
Specify the Zipcode, 5 or 10 characters: 84320-9963



The information you entered was:
    First Name: Joe
    Last Name: Smith
    Address: 2750 Aspen Glenn
    City: Wilcox
    State: AZ
    Zipcode: 84320-9963



Do you wish to modify the entry? (y/n): n

Do you wish to add this entry to the data file?
y
Do you wish to make another entry? (y/n): Anne
Error - Input one of the following: y,n.
```

```
Do you wish to make another entry? (y/n): y
Specify information for the address book entry.

Specify the first name, 0 to 20 characters: Becky
Specify the last name, 0 to 20 characters: Mazze
Specify the address, 0 to 30 characters: NAU Box 893
Specify the City name, 0 to 20 characters: Tucson
Specify the State, 2 characters: AZ
Specify the Zipcode, 5 or 10 characters: 789




The information you entered was:
    First Name: Becky
    Last Name: Mazze
    Address: NAU Box 893
    City: Tucson
    State: AZ
    Zipcode: EntryError


Do you wish to modify the entry? (y/n): y


Which Item would you like to modify?
    1 = First Name
    2 = Last Name
    3 = Address
    4 = City
    5 = State
    6 = Zipcode

Specify a number: 6
Specify the Zipcode, 5 or 10 characters: 86011-89478



The information you entered was:
    First Name: Becky
    Last Name: Mazze
    Address: NAU Box 893
    City: Tucson
    State: AZ
    Zipcode: EntryError
```

```
Do you wish to modify the entry? (y/n): 6
Error - Input one of the following: y,n.

Do you wish to modify the entry? (y/n): 86011-1678
Error - Input one of the following: y,n.

Do you wish to modify the entry? (y/n): n

Do you wish to add this entry to the data file?
y
Do you wish to make another entry? (y/n): y
Specify information for the address book entry.

Specify the first name, 0 to 20 characters: ghjds
Specify the last name, 0 to 20 characters: hfkds
Specify the address, 0 to 30 characters: fhkjds
Specify the City name, 0 to 20 characters: hfkjds
Specify the State, 2 characters: fhkjd
Specify the Zipcode, 5 or 10 characters: hkfd




The information you entered was:
    First Name: ghjds
    Last Name: hfkds
    Address: fhkjds
    City: hfkjds
    State: fh
    Zipcode: EntryError


Do you wish to modify the entry? (y/n): y



Which Item would you like to modify?
    1 = First Name
    2 = Last Name
    3 = Address
    4 = City
    5 = State
    6 = Zipcode

Specify a number: 1
Specify the first name, 0 to 20 characters: Levar
```

```
The information you entered was:
    First Name: Levar
    Last Name: hfkds
    Address: fhkjds
    City: hfkjds
    State: fh
    Zipcode: EntryError


Do you wish to modify the entry? (y/n): y


Which Item would you like to modify?
    1 = First Name
    2 = Last Name
    3 = Address
    4 = City
    5 = State
    6 = Zipcode

Specify a number: 2
Specify the last name, 0 to 20 characters: Lamberpizza



The information you entered was:
    First Name: Levar
    Last Name: Lamberpizza
    Address: fhkjds
    City: hfkjds
    State: fh
    Zipcode: EntryError



Do you wish to modify the entry? (y/n): y


Which Item would you like to modify?
    1 = First Name
    2 = Last Name
    3 = Address
    4 = City
    5 = State
    6 = Zipcode
```

```
Specify a number: 3
Specify the address, 0 to 30 characters: 1786 North Hallow Rd



The information you entered was:
    First Name: Levar
    Last Name: Lamberpizza
    Address: 1786 North Hallow Rd
    City: hfkjds
    State: fh
    Zipcode: EntryError



Do you wish to modify the entry? (y/n): y



Which Item would you like to modify?
    1 = First Name
    2 = Last Name
    3 = Address
    4 = City
    5 = State
    6 = Zipcode

Specify a number: 4
Specify the City name, 0 to 20 characters: Williams



The information you entered was:
    First Name: Levar
    Last Name: Lamberpizza
    Address: 1786 North Hallow Rd
    City: Williams
    State: fh
    Zipcode: EntryError



Do you wish to modify the entry? (y/n): y



Which Item would you like to modify?
    1 = First Name
    2 = Last Name
```

```
    3 = Address
    4 = City
    5 = State
    6 = Zipcode

Specify a number: 5
Specify the State, 2 characters: MA




The information you entered was:
    First Name: Levar
    Last Name: Lamberpizza
    Address: 1786 North Hallow Rd
    City: Williams
    State: MA
    Zipcode: EntryError




Do you wish to modify the entry? (y/n): y




Which Item would you like to modify?
    1 = First Name
    2 = Last Name
    3 = Address
    4 = City
    5 = State
    6 = Zipcode

Specify a number: 6
Specify the Zipcode, 5 or 10 characters: 02861-1786




The information you entered was:
    First Name: Levar
    Last Name: Lamberpizza
    Address: 1786 North Hallow Rd
    City: Williams
    State: MA
    Zipcode: 02861-1786
```

```
Do you wish to modify the entry? (y/n): n

Do you wish to add this entry to the data file?
y
Do you wish to make another entry? (y/n): n
```

type address_book.txt

Marc	,Herniter	,7840 Candi Lane	,Flagstaff	,AZ,86004-0000
Joe	,Smith	,2750 Aspen Glenn	,Wilcox	,AZ,84320-9963
Becky	,Mazze	,NAU Box 893	,Tucson	,AZ,EntryError
Levar	,Lamberpizza	,1786 North Hallow Rd	,Williams	,MA,02861-1786

Problem 5-2   Write a MATLAB function that reads the file address_book.txt created in Problem 5-1, sorts the data by the last name, and then writes the sorted data to a file named addr_sort.txt. The form of the function is

function sort_data

Here is an example.

File address_book.txt:

Marc	,Herniter	,7840 Candi Lane	,Flagstaff	,AZ,86004-0000
Joe	,Smith	,2750 Aspen Glenn	,Wilcox	,AZ,84320-9963
Becky	,Mazze	,NAU Box 893	,Tucson	,AZ,EntryError
Levar	,Lamberpizza	,1786 North Hallow Rd	,Williams	,MA,02861-1786

File addr_sort.txt after function sort_data was run:

Marc	,Herniter	,7840 Candi Lane	,Flagstaff	,AZ,86004-0000
Levar	,Lamberpizza	,1786 North Hallow Rd	,Williams	,MA,02861-1786
Becky	,Mazze	,NAU Box 893	,Tucson	,AZ,EntryError
Joe	,Smith	,2750 Aspen Glenn	,Wilcox	,AZ,84320-9963

Miscellaneous MATLAB Functions and Variables

6

OBJECTIVES

☐ Use a number of specialized MATLAB functions, including
  - time/date—**CLOCK** and **DATE**
  - serial date numbers—**DATESTR**, **DATENUM**, and **NOW**
  - execution time—**FLOPS**, **CPUTIME**, **ETIME**, **TIC**, and **TOC**
  - evaluation—**EVAL** and **FEVAL**
  - multimedia—**WAVERAD** and **SOUND**
  - miscellaneous—**SPRINTF** and **PAUSE**

We have used a number of MATLAB functions in many examples and also in other functions. If you look in many of the m-files provided with MATLAB, or visit The MathWorks website and download some m-files, you'll see these functions contained in the m-files. The functions we cover here are not an exhaustive list, rather, we're examining them because they are useful to introductory programming students, aid in solving problems, or are just fun to use.

6.1 Time, Date, and Execution Time

MATLAB provides a number of functions that help you determine the present date and time, or how long a program or function takes to execute.

6.1.1 CPUTIME

MATLAB Predefined
cputime function
clock function
etime function
for loop
tic and toc functions
datestr function
datenum function
now function
flops function

The **CPUTIME** function is a running counter of how many seconds you've been using MATLAB in the current session. The first time you use the **CPUTIME** function, the value returned by the function is zero:

```
<Student Edition> MATLAB Command Window
File   Edit   Window   Help

  To get started, type one of these commands: helpwin, helpdesk, or demo

  Setting defaults for plots in startup.m
EDU» cputime
ans =
     0
EDU»
```

Let's keep the window open without doing anything, in this case, about 2 minutes. We then run the **CPUTIME** function again:

On this screen, the function returns the value of 123.78 seconds. Since we didn't do anything that required large amounts of time (in fact, we did nothing), we can conclude that the time displayed by the **CPUTIME** function is the elapsed time in seconds since we first used the **CPUTIME** function.

We can use the **CPUTIME** function to time how long a function takes to execute in real time. For example, let's see how long it takes Matlab to find the inverse of a matrix 1000 times:

```
t=cputime;
for i=1:1000
   a=inv(rand(32));
end
cputime-t
ans =
     5.8700
```

This code segment calculates the inverse of a 32×32 matrix 1000 times. The line **t=cputime** records the time when the code segment starts. The line **cputime-t** calculates the difference between the time when the code segment starts and when it finishes. Thus, **cputime-t** is how much real time the code segment used.

The result of 5.87 shows that the preceding code segment took 5.87 seconds to execute on the computer used to write this book. Your execution time will be different. The **CPUTIME** function is only accurate to 1/100 of a second. Smaller amounts of time are not measurable:

```
a=rand(32);
t=cputime;
b=a*a;
dif=cputime-t;
dif
dif =
     0
```

Now let's run the **CPUTIME** function again to see what it displays:

This screen shows a result of 233.83 seconds. This is almost 4 minutes. The simple calculations we've been running certainly don't require 4 minutes of CPU time. Thus, the value returned by the **CPUTIME** function is the number of seconds since we first ran the **CPUTIME** command. Even though the **CPUTIME** function tracks how long MATLAB has been running, by using the function before and after a code segment, we can also time how long the segment takes to execute.

6.1.2 CLOCK

The **CLOCK** function is similar to the **CPUTIME** function except that it returns a row array of values for the current date and time:

```
EDU» clock
ans =
  1.0e+003 *
    1.9970    0.0030    0.0170    0.0210    0.0290    0.0429
```

The **CLOCK** function prints a row array with six elements. Year appears first, then month, then day of the month. In our example, the date is 1997/3/17 or March 3, 1997. The last three elements show time in a 24-hour format. Hours is the fourth element, minutes is next, and seconds is last. In this example, the time is 21:29:04.29 or 9:29 P.M. plus 4.29 seconds.

6.1.3 ETIME

The **CLOCK** function can also be used to time functions. However, because of the format in which the **CLOCK** function return results, it's difficult to calculate elapsed times. Let's see how long the following code segment takes to execute:

```
t=clock;
for i=1:1000
   a=inv(rand(32));
end
clock-t
ans =
        0        0        0        0        0    7.4700
```

The **CLOCK** function returns a row array of 6 elements. When we take the difference between two row arrays returned by the **CLOCK** function, we get a row array as a result. The first three elements are the differences in the date, and elements 4–6 are the time. Since our code segment is small and fast, the elements are all zero, except for the last one, which is seconds. The result shows that the code segment took 7.47 seconds to run. Note that the clock is only accurate to 1/100 of a second. If you display the clock frequently, you'll note that it displays only two decimal places for seconds.

Taking the difference between two results returned by the **CLOCK** function can be used to calculate small time differences easily. However, if the elapsed time crosses a date boundary, or an hour or minute boundary, the function is harder to use. Thus if we started a program at 11:59 P.M. on December 31 and let it run for a few minutes, the year, month, day, hour, and minute will all change.* Calculating the elapsed time in this case is more difficult.

To avoid this problem, MATLAB provides the **ETIME** function. This function calculates the time difference in seconds between two row arrays returned by the **CLOCK** function:

```
t=clock;
for i=1:1000
   a=inv(rand(32));
end
etime(clock,t)
ans =
    7.5200
```

The result shows that the code segment required 7.52 seconds to run. Let's see what happens when a code segment takes over a minute to run. Note that the **FOR** loop executes 15,000 times rather than 1000 times as in the last example.

```
t1=clock;
for i=1:15000
   a=inv(rand(32));
end
t2=clock;
t2-t1
etime(t2,t1)
ans =
         0        0        0        0   1.0000   26.7800
ans =
   86.7800
```

The result of **t2-t1** tells us that the code segment took 1 minute 26.78 seconds to complete, while the **ETIME** function tells us that the code segment took 86.78 seconds. Using the difference between values returned by the **CLOCK** function is not difficult, but it can take some thought to determine the time difference. Here's another example:

---

*This could be serious entertainment at an engineering New Year's Eve party!

```
t1=clock;
for i=1:10000
   a=inv(rand(32));
end
t2=clock;
t2-t1
etime(t2,t1)
ans =
          0          0          0          0     1.0000     -2.2200
ans =
    57.7800
```

This result tells us that the code segment took 1 minute minus 2.22 seconds to run. The **ETIME** function tells us that the code segment took 57.78 seconds to run.

We can easily convert the row array result to a difference in seconds. We simply ignore the year and month, and calculate the time difference for the days, hours, and minutes. Note that a day has 86,4000 seconds, an hour has 3600 seconds, and a minute has 60 seconds. We can easily calculate the time difference in seconds as

$$\text{time difference} = \left(86,400 \times \frac{\text{difference}}{\text{in days}}\right) + \left(3600 \times \frac{\text{difference}}{\text{in minutes}}\right) + \left(60 \times \frac{\text{difference}}{\text{in minutes}}\right) + \text{seconds}$$

If we let $delta = t_2 - t_1$, then the difference in seconds can be expressed as $seconds = 86400 * delta(3) + 3600 * delta(4) + 60 * delta(5) + delta(6)$. In MATLAB, this product is easily expressed as the product of a row vector times a column vector (the dot product between a row and column vector):

```
seconds = [delta(1) delta(2) delta(3) delta(4) delta(5) delta(6)]*[0; 0;
86400; 3600; 60; 1]
```

Note that *delta* is a row vector with six values.

We modify the code segment accordingly:

```
t1=clock;
for i=1:10000
   a=inv(rand(32));
end
t2=clock;
(t2-t1)*[0; 0; 86400; 3600; 60; 1]
etime(t2,t1)
ans =
    55.6900
ans =
    55.6900
```

The two methods return the same result. This multiplication is exactly what the **ETIME** function does. Before we look at the **ETIME** function, note that

`[0;0;86400;3600;60;1]` is equivalent to `[0,0,86400,3600,60,1]'`. Note that the `'` character means transpose, and that the transpose of a row vector is a column vector:

```
EDU» [0,0,86400,3600,60,1]'
ans =
            0
            0
        86400
         3600
           60
            1
```

To use the transpose operator, we modify the code segment only slightly:

```
t1=clock;
for i=1:10000
   a=inv(rand(32));
end
t2=clock;
(t2-t1)*[0,0, 86400, 3600, 60, 1]'
ans =
   55.7500
etime(t2,t1)

ans =
   55.7500
```

Thus, using `(t2-t1)*[0,0, 86400, 3600, 60, 1]'` or `(t2-t1)*[0; 0; 86400; 3600; 60; 1]` is really a matter of personal preference. The **ETIME** function uses `(t2-t1)*[0,0, 86400, 3600, 60, 1]'`. Let's look at the **ETIME** function:

```
EDU» type etime.m
function t = etime(t1,t0)
%ETIME    Elapsed time.
% ETIME(T1,T0) returns the time in seconds that has elapsed ➠
between
% vectors T1 and T0. The two vectors must be six elements long, in
% the format returned by CLOCK:
%
%      T = [Year Month Day Hour Minute Second]
%
% Here's an example of using ETIME to time an operation:
%
%    t0 = clock;
```

```
%    operation
%    etime(clock,t0)
%
% Caution: As currently implemented, it won't work across month ➥
or year
% boundaries. It is an M-file, so it can be fixed if you ➥
require this.
%
% See also TIC, TOC, CLOCK, CPUTIME.
% Copyright (c) 1984-94 by The MathWorks, Inc.
t = (t1 - t0) * [0 0 86400 3600 60 1]';
```

The last line performs the same function as the one in our code segment.

6.1.4 TIC and TOC

The **TIC** and **TOC** functions use the **CLOCK** and **ETIME** functions to calculate the elapsed time between using the **TIC** function and using the **TOC** function. Here's an example:

```
tic;
for i=1:10000
    a=inv(rand(32));
end
t2=clock;
toc
elapsed_time =
    57.7800
```

The result is saved in a variable named *elapsed time*. As you can see, the code segment required 57.78 seconds to run. If we want to hide the result of the **TOC** function and store it in a variable, we use the function in the form **x=toc**:

```
tic;
for i=1:10000
    a=inv(rand(32));
end
t2=clock;
x=toc;
```

The code segment runs and displays no results. However, if we ask for the value of *x*, it contains the time difference:

```
x
x =
    55.8000
```

In this case, the time difference was 55.8 seconds. This result differs from the previous example because Windows is a multitasking environment and different programs may have been running in the background during our run. The displayed time is the real time required to execute the code segment, not the processor time.

Now let's look at the **TIC** and **TOC** functions. The **TIC** function is

```
EDU» type tic.m
function tic
%TIC     Start a stopwatch timer.
% The sequence of commands
%     TIC
%     any stuff
%     TOC
% prints the time required for the stuff.
%
% See also TOC, CLOCK, ETIME, CPUTIME.
% Copyright (c) 1984-94 by The MathWorks, Inc.
% TIC simply stores CLOCK in a global variable.
global TICTOC
TICTOC = clock;
```

Function **TIC** declares variable *TICTOC* as global and then sets its value to the current value of the clock. Function **TOC** is

```
EDU» type toc.m
function t = toc
%TOC     Read the stopwatch timer.
% TOC, by itself, prints the elapsed time since TIC was used.
% t = TOC; saves the elapsed time in t, instead of printing it ➡
out.
%
% See also TIC, ETIME, CLOCK, CPUTIME.
% Copyright (c) 1984-94 by The MathWorks, Inc.
% TOC uses ETIME and the value of CLOCK saved by TIC.
global TICTOC
if nargout < 1
    elapsed_time = etime(clock,TICTOC)
else
    t = etime(clock,TICTOC);
end
```

Note that function **TOC** also declares variable *TICTOC* as global. Thus, functions **TIC** and **TOC** know the value of variable *TICTOC* and can use and modify its value. Next, function **TOC** calculates the elapsed time using the **ETIME** function. The **IF** statement determines how the function returns its value. If function **TOC** is used as **x=toc**, then the line **t = etime(clock,TICTOC);** executes. If function **TOC** is used by itself, as in the following example,

```
tic;
for i=1:10000
   a=inv(rand(32));
end
t2=clock;
toc        ◄──────────────[ toc used here ]
```

then the line **elapsed_time=etime(clock,TICTOC)** executes. The **IF** statement determines which form to use from a special variable named *nargout*. This variable is discussed in detail in Section 3.6. Briefly, variable *nargout* contains the number of output variables used when the function was called. For example, if we call the **TOC** function using the line **x=toc**, *nargout* = 1. If we use the **toc** function by itself, as in the code segment above, then *nargout* = 0.

6.1.5 DATE

The **DATE** function returns a text string that contains the date in dd-mmm-yy format.

```
EDU» str=date
str =
2-Apr-97
```

We can modify the string using any of the earlier techniques that apply to strings.

```
EDU» str(3:7)
ans =
Apr-9
```

6.1.6 Serial Date Numbers (DATESTR, DATENUM, and NOW)

A serial date number is a real number that represents the present date and time. Thus, 0 represents the date December 31, 1 BCE at 00:00 hours (12:00 A.M.); 0.5 represents December 31, 1 BCE at 12:00 hours (12 P.M.), and 1 represents January 1, in the year 0 at 00:00 hours (12:00 A.M.). A serial date number of 1.25 is six hours later or January 1, in the year 0 at 06:00 hours (6:00 A.M.). January 1 in the year 1 is 366 days later (the year 0 was a leap year) than the code for January 1, in the year 0:

```
EDU» datenum('1-Jan-0000')
ans =
     1
EDU» datenum('1-Jan-0001')
ans =
   367
```

Thus, a serial date number is the number of days since the date December 31, 1 BCE at 00:00 hours. The fractional part of the number represents a fraction of a day. For example, a date number xxx.5 means a half day, or the time 12:00 hours (12:00 P.M.). A date number of xxx.75 means three-quarters of a day or 18:00 hours (6:00 P.M.).

DATESTR

The **DATESTR** function converts a serial date number into one of the common formats for dates and time, and stores the result in a text string. The syntax of the function is

$$x = \text{datestr}(d, \textit{dateform})$$

where x is the resultant string and d is the date number. Variable *dateform* can be a string or a code number that indicates how the date will be displayed. The possible codes are listed in Table 6-1.

TABLE 6-1   Formatting codes for the **DATESTR** function

Dateform *Number*	Dateform *String*	*Example*
0	'dd-mmm-yyyy HH:MM:SS'	01-Mar-1995 15:45:17
1	'dd-mmm-yyyy'	01-Mar-1995
2	'mm/dd/yy'	03/01/95
3	'mmm'	Mar
4	'm'	M
5	'mm'	3
6	'mm/dd'	03/01
7	'dd'	1
8	'ddd'	Wed
9	'd'	W
10	'yyyy'	1995
11	'yy'	95
12	'mmmyy'	Mar95
13	'HH:MM:SS'	15:45:17
14	'HH:MM:SS PM'	3:45:17 PM
15	'HH:MM'	15:45
16	'HH:MM PM'	3:45 PM
17	'QQ-YY'	Q1-96
18	'QQ'	Q1

We can use any of the predefined *dateform* numbers, or we can use the equivalent *dateform* string. Let's display the date 0 with a format of 0:

```
EDU» datestr(0,0)
ans =
31-Dec--001 00:00:00
EDU»
```

As you can see, a serial date number of 0 represents the date 31 December in the year 1 BCE at 12:00 A.M. or 00:00 hours. A serial date number of 0.5 should be the same day at 12:00 hours:

```
EDU» datestr(0.5,0)
ans =
31-Dec--001 12:00:00
EDU»
```

A serial date number of 1 is thus January 1, in the year 0 at 00:00 hours (12:00 A.M.):

```
EDU» datestr(1,0)
ans =
01-Jan-0000 00:00:00
EDU»
```

A serial date number of 1.25 is six hours later or January 1, in the year 0 at 06:00 hours.

```
EDU» datestr(1.25,0)
ans =
01-Jan-0000 06:00:00
EDU»
```

DATENUM

The **DATENUM** function converts a date into a serial date number. It can be used in three forms:

$$N = \text{datenum}(S) \quad \text{or} \quad N = \text{datenum}(Y, M, D) \quad \text{or} \quad N = \text{datenum}(Y, M, D, H, MI, S)$$

In the first form, S is a string that is in one of the date formats 0, 1, 2, 6, 13, 14, 15, 16 as defined in Table 6-1. In the second and third forms, Y, M, D, H, MI, S represent the year, month, day, hours, minutes, and seconds as numbers. For an example, we will find the date code for January 1 in the year 0 at 12:00 midnight:

```
EDU» datenum('01-Jan-0000')
ans =
     1
EDU» datenum(0,1,1,0,0,0)
ans =
     1
EDU»
```

Next, let's find the date code for a year later:

```
EDU» datenum('01-Jan-0001')
ans =
   367
EDU»
```

You can easily see that the year 0 is a leap year by subtracting serial date numbers:

```
EDU» datenum('01-Jan-0001')-datenum('01-Jan-0000')
ans =
   366
EDU»
```

That is, the year has 366 days and therefore must be a leap year.

NOW

The MATLAB **NOW** function returns the serial date number for the present time and date:

```
EDU» format long
EDU» now
ans =
     7.296969152519676e+005
EDU»
```

The result is a large number because it's the number of days between the present date and the date Dec 31 in the year 1 BCE.

We can use the **DATESTR** and **NOW** functions together to display the current date and/or time easily:

```
EDU» datestr(now,0)
ans =
02-Nov-1997 22:00:29
EDU» datestr(now,1)
ans =
02-Nov-1997
EDU» datestr(now,13)
ans =
22:01:15
EDU»
```

6.1.7 FLOPS

In the world of computing, flops is an acronym that stands for floating-point operations per second, and is sometimes used as a benchmark for comparing computer processing speed. A flop is any operation that involves a mathematical operation between floating-point numbers such as addition, subtraction, multiplication, and division. A faster computer has a higher flops rating.

In MATLAB, the **FLOPS** function returns the number of floating-point operations performed in the current MATLAB session. When you start MATLAB, the number of operations is zero:

```
EDU» flops
ans =
     0
```

Each time you perform a numerical calculation that involves floating-point numbers, the count is incremented.

```
EDU» y=1.3+5;
EDU» flops
ans =
    1
```

In MATLAB, addition is counted as one flop, as are multiplication, division, and powers:

```
EDU» y=1*2/3^4;
EDU» flops
ans =
    4
```

Working with complex numbers requires more operations to perform the basic operations. Note that using the **FLOPS** function as **flops(0)** resets the flops counter to zero.

```
EDU» flops(0)
EDU» flops
ans =
    0
EDU» a=1+2i;
EDU» flops
ans =
    2
EDU» b=3+4i;
EDU» flops
ans =
    4
EDU» a+b;
EDU» flops
ans =
    6
EDU» a/b;
EDU» flops
ans =
    12
EDU» a*b;
EDU» flops
ans =
    18
```

From these results, you can see that defining a complex number (**a=1+2i**) requires two flops; addition and subtraction of complex numbers (**a+b**) require two flops, and multiplication and division of complex numbers (**a*b** and **a/b**) require six flops.

Since the number of flops used determines the execution time, knowing how many flops are required thus gives you an idea of how long your run will take.

To see how many flops are required, we ask for the flops before and after running a code segment:

```
x=flops;
for i=1:10000
   a=inv(rand(32));
end
ops=flops-x;
ops
ops =
   698143280
```

Note that integer operations don't require flops:

```
EDU» flops
ans =
   698143281
EDU» fix(3.9);
EDU» ceil(4.8);
EDU» flops
ans =
   698143281
```

Incrementing a counter in a **FOR** loop is also an integer operation:

```
x=flops;
for i=1:10000
end
ops=flops-x;
ops
ops =
     0
```

6.2 Evaluation

6.2.1 **EVAL**

The **EVAL** function evaluates a text string that contains a MATLAB command. Here's an example:

```
EDU» eval('sin(pi/2)')
ans =
   1
```

We can evaluate a string that contains any legal MATLAB command:

```
EDU» str='5+4*3/2';
EDU» eval(str)
ans =
     11
```

As an example of the **EVAL** function, we will rewrite the parallel resistor function, on page 187. If you review Function 3-7, you will see that it takes a fair amount of code to implement the function. The implementation in Function 6-1 is much more compact.

Function 6-1 Calculating the equivalent parallel resistance of resistors using the **EVAL** function

```
function Req=parallel2(R1,R2,R3,R4,R5,R6,R7,R8,R9,R10)
% This is file Parallel2.m
% Function Parallel2 - Calculates the equivalent
% resistance of up to 10 parallel resistors.
% Req = parallel resistance if 1 through 10 resistors are specified.
%     = -1 if no resistors were specified.
%     = NaN - if a resistor has a value of zero.

if nargin == 0
    Req = -1;
else
    str='Geq=(1/R';
    for i = 1:(nargin-1)
        a=num2str(i);
        str=[str,a,') + (1/R'];
    end
    a=num2str(nargin);
    str=[str,a,');'];

    eval(str);
    Req=1/Geq;
end
```

The code segment

```
str='Geq=(1/R';
for i = 1:(nargin-1)
    a=num2str(i);
    str=[str,a,') + (1/R'];
end
a=num2str(nargin);
str=[str,a,');'];
```

creates a text string based on the value of variable *nargin*. Examples of the text string (*str*) are shown below. Note that value of *str* depends on how many input arguments are passed to the function:

`parallel2(1)`	`str =` `Geq=(1/R1);`
`parallel2(2,2)`	`str =` `Geq=(1/R1) + (1/R2);`
`parallel2(5,3,7,2,5)`	`str =` `Geq=(1/R1) + (1/R2) + (1/R3) + (1/R4) + (1/R5);`

`parallel2(5,3,7,2,5,9,1,3,4)`	`str =`
	`Geq=(1/R1) + (1/R2) + (1/R3) + (1/R4) +`
	`(1/R5) + (1/R6) + (1/R7) + (1/R8) + (1/R9);`

As you can see, the text string we're creating is a valid MATLAB command. When we pass this text string to the **EVAL** command, MATLAB executes the command and generates a value for variable *Geq*. The **EVAL** function allows the equation for *Geq* to change depending on how we call function parallel2.

6.2.2 FEVAL

The name **FEVAL** stands for "function evaluate" and evaluates MATLAB functions whose names are contained in strings. An example would be **feval('sin',pi)**, which evaluates the function sin.m with an input of π:

```
EDU» feval('sin',pi)
ans =
  1.2246e-016
```

The answer is approximately 0. Note that the statements **feval('sin',pi)** and **sin(pi)** are equivalent:

```
EDU» feval('sin',pi)
ans =
  1.2246e-016
EDU» sin(pi)
ans =
  1.2246e-016
EDU»
```

The **FEVAL** function allows you to write a function that works with any arbitrary MATLAB function. You then specify the name of the function specific to your needs in a text string. A good example of using the **FEVAL** function is the MATLAB **FZERO** function. The **FZERO** function solves any equation in one variable.

Suppose we want to find x such that $x^3 - 2x^2 + x + 18 \ln(|x|) + 27 = 0$. The first thing we do is write a MATLAB function (m-file) that calculates the equation (Function 6-2).

Function 6-2 Arbitrary MATLAB function

```
function y=arb_fnc(x)
% This is file arb_fnc.m
y=x.^3 - 2.*x.*x + x + 18.*log(abs(x))+27;
```

This function is saved as arb_fnc.m. We then plot the function to see what it looks like:

```
EDU» x=linspace(-3,3,200);
EDU» y=arb_fnc(x);
EDU» plot(x,y);
EDU» grid
```

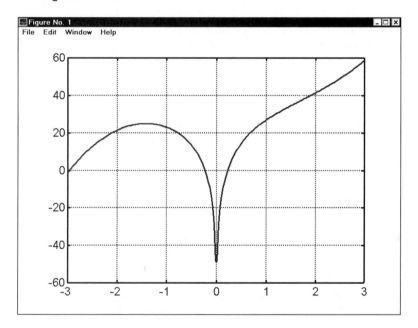

We can now use the **FZERO** function to solve for x such that the function equals zero. The **FZERO** function requires that we give it a seed value where it will start searching for the root. Let's find the root closest to $x = 2$. Note that the name of the function we want to evaluate is passed to the **FZERO** function as a text string.

```
EDU» format long
EDU» root=fzero('arb_fnc', 2)
Zero found in the interval: [0.18981, 3.28]
root =
 0.22147234924274
```

For the result to be correct, it should be true that arb_fnc(0.22147234924274) equals zero:

```
EDU» arb_fnc(0.22147234924274)
ans =
    -7.105427357601002e-015
```

The result is close to zero.

How is it that the **FZERO** function uses any arbitrarily named function and finds where it is equal to zero? The answer is that it uses the **FEVAL** function. Note that when we called the **FZERO** function, we passed the name of our function as a text string: **root=fzero('arb_fnc', 2)**. In Function 6-3, we edit the **FZERO** function with a text editor and look at its contents.

Function 6-3 MATLAB **FZERO** function

> Variable *FunFcn* is a text string that contains the name of the function to be evaluated.

```
function b = fzero(FunFcn,x,tol,trace)
%FZERO Find a zero of a function of one variable.
%  FZERO(F,X) finds a zero of f(x).  F is a string containing the
%  name of a real-valued function of a single real variable.   X is
%  a starting guess.  The value returned is near a point where F
%  changes sign.  For example, FZERO('sin',3) is pi.  Note the
%  quotes around sin.  Ordinarily, functions are defined in M-files.
%
%  An optional third argument sets the relative tolerance for the
%  convergence test.   The presence of an nonzero optional fourth
%  argument triggers a printing trace of the steps.

%  C.B. Moler 1-19-86
%  Revised CBM 3-25-87, LS 12-01-88.
%  Copyright (c) 1984-94 by The MathWorks, Inc.

%  This algorithm was originated by T. Dekker.  An Algol 60 version,
%  with some improvements, is given by Richard Brent in "Algorithms for
%  Minimization Without Derivatives", Prentice-Hall, 1973.  A Fortran
%  version is in Forsythe, Malcolm and Moler, "Computer Methods
%  for Mathematical Computations", Prentice-Hall, 1976.

% Initialization

if nargin < 3, trace = 0; tol = eps; end
if nargin == 3, trace = 0; end
if trace, clc, end
if (length(x) > 1) | (~finite(x))
   error('Second argument must be a finite scalar.')
end
if x ~= 0, dx = x/20;
else, dx = 1/20;
end
a = x - dx;  fa = feval(FunFcn,a);
if trace, home, init = [a fa], end
```

> Function **FEVAL** evaluates the function whose name is contained in variable *FunFcn* at point *a*.

```
b = x + dx;   fb = feval(FunFcn,b);
if trace, home, init = [b fb], end

% Find change of sign.

while (fa > 0) == (fb > 0)
   dx = 2*dx;
   a = x - dx;   fa = feval(FunFcn,a);
   if trace, home, sign = [a fa], end
   if (fa > 0) ~= (fb > 0), break, end
   b = x + dx;   fb = feval(FunFcn,b);
   if trace, home, sign = [b fb], end
      .
      .(this is not a complete listing)
      .
```

> **FEVAL** function used

The first line of the function (**function b = fzero(FunFcn,x,tol, trace)**) defines variable *FunFnc*. This string variable contains the name of the function we want to evaluate. Note that several lines use the **FEVAL** function:

```
a = x - dx;   fa = feval(FunFcn,a);
b = x + dx;   fb = feval(FunFcn,b);
a = x - dx;   fa = feval(FunFcn,a);
b = x + dx;   fb = feval(FunFcn,b);
```

Thus, **FZERO** can evaluate any function because it uses **FEVAL** to evaluate the function specified by the user in string variable *FunFnc*.

6.3 PAUSE Function

The **PAUSE** function causes MATLAB to stop and wait for a specified amount of time. **PAUSE** by itself causes MATLAB to wait indefinitely, and **PAUSE(n)** causes MATLAB to wait *n* seconds. The different uses of the **PAUSE** function are

- **PAUSE**—Wait indefinitely for a user response. Resume when the user strikes any key.
- **PAUSE(n)**—Pause for *n* seconds before continuing.
- **PAUSE OFF**—Specifies that any future uses of **PAUSE** or **PAUSE(n)** should be ignored.
- **PAUSE ON**—Specifies that MATLAB should pause for any future uses of **PAUSE** or **PAUSE(n)**.

6.4 Sound

6.4.1 WAVREAD Function

The **WAVREAD** function allows you to read a Windows wav sound file. The syntax is **[y,Fs]=wavread(fname)**. Variable *fname* is a string that contains the name of the file you want to read. Variable *y* is a row array that contains the sound data. Variable *Fs* is a scalar that contains the sampling frequency at which the wav file was recorded. In earlier student versions of MATLAB, variable *y* is limited to 16,384 elements. This places a severe limitation on the length of sounds that we can read and play. Most of the sounds that come with Windows are too large to be read using the student version. The later student versions of MATLAB don't have this limitation and can play longer sound files.

For an example, let's read a sound file that is small enough to read with the older student versions:

```
EDU» [y,fs]=wavread('c:\windows\wave\pop.wav');
```
The sampling frequency at which this sound was recorded is:
```
EDU» fs
fs =
      22050
```

This sound was recorded at a sampling frequency of 22.05 kHz. The number of elements in *y* is

```
EDU» length(y)
ans =
      4442
```

To see what the data looks like, we'll look at 20 elements in variable *y*:

```
EDU» y(4000:4020)
ans =
   -0.3438
   -0.3438
   -0.3438
   -0.3438
   -0.3438
   -0.3438
   -0.1719
   -0.1719
   -0.1719
   -0.1719
        0
        0
        0
        0
    0.1719
    0.1719
    0.1719
    0.1719
    0.1719
```

```
            0.1719
            0.1719
```

Further examination of *y* shows that all of its elements are between –1 and 1.

6.4.2 SOUND

The **SOUND** function plays a wav file with a specified sampling frequency. To play the sound we loaded with the **WAVREAD** function, all we do is pass the variables to the function:

EDU» **sound(y,fs);**

After issuing the command, the sound plays through your sound card.

Now that you know how to read wav files, you can use the data in many of the signal processing functions provided by MATLAB. You can also use the sound functions anywhere in a MATLAB function. For example, you could play a sound when the user responds improperly to a question. In Program 6-1, we modify Program 2-27 to play a sound if the user types an improper response.

Program 6-1 Modification to Program 2-27

```
%
% Program VI_1
%
% This program asks for a number between 1 and 5 inclusive.
% The program then performs the tasks assigned to the number.
% The tasks are not specified for this program.

exit=0;
[err_sound, fs]=wavread('c:\windows\wave\pop.wav');
while ~exit
   command=-99;
   str_cmd=input('Specify your command [1-5,E]: ', 's');

   if strcmp(str_cmd,'E')
       exit=1;
   else
       command=str2num(str_cmd);
   end

   if isempty(command)
       command = -99;
   end

   if command==1
       fprintf('You chose option 1.\n');
       % Command for option 1 entered here.
   elseif command==2
       fprintf('You chose option 2.\n');
       % Command for option 2 entered here.
```

```
    elseif command==3
        fprintf('You chose option 3.\n');
        % Command for option 3 entered here.
    elseif command==4
        fprintf('You chose option 4.\n');
        % Command for option 4 entered here.
    elseif command==5
        fprintf('You chose option 5.\n');
        % Command for option 5 entered here.
    else
        if exit ~= 1
            fprintf('Error - You must enter an integer from 1 to 5, or the ➡
    letter E.\n');
            sound(err_sound,fs);
        end
    end
end
fprintf('Good Bye.\n');
```

The second line of the program **[err_sound, fs]=wavread('c:\windows \wave\pop.wav')** reads a short sound into variable *err_sound*. Near the end of the program, the line **sound(err_sound,fs)** was added to play the sound if the user enters an invalid response.

Another annoying example would be playing the same sound 10 times:

```
[z, fs]=wavread('c:\windows\wave\pop.wav');
for I = 1:10
    sound(z,fs);
    pause(1);
end
```

6.5 SPRINTF Function

The **SPRINTF** function performs the same functions as the **FPRINTF** function except the output is written to a string rather than to the command window or an output file. Many MATLAB functions display text strings created by the user. Using the **SPRINTF** function makes all of the functionality of the **FPRINTF** function available to create those strings. The **SPRINTF** function is especially useful when we want to create strings that contain the numerical values of variables.

We'll show a short example using the **PLOT** function.

```
EDU» x=linspace(-10*pi,10*pi,200);
EDU» y=sin(x)./x;
EDU» plot(x,y);
EDU» xx=3*pi/4;
EDU» yy=sin(xx)/xx;
EDU» str=sprintf('The value of sinc(x)\nat x = %3.2f is %3.2f',xx, yy);
EDU» text(-38, 0.4, str);
```

This code segment produces the following plot:

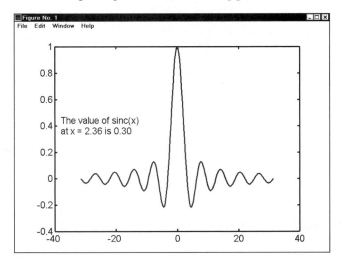

The **TEXT** function places text on a plot window. The first two arguments specify the location where the text will be placed. The location is specified in units of the *x*- and *y*-axes. The third argument is the text string we would like to place. First, we use the **SPRINTF** function to create a text string that contains numerical values, `str=sprintf('The value of sinc(x)\nat x = %3.2f is %3.2f',xx, yy)`. Note that variable *str* contains numerical values and a new line character (`\n`). We then use the command `text(-38, 0.4, str)` to place this string on the plot window at coordinates *x* = –38 and *y* = 0.4.

Here's another example of the **TEXT** function:

```
EDU» text(0,-0.3, 'This is a SINC function.');
```

Executing this statement places the text string `'This is a SINC function.'` at axes coordinates *x* = 0, and *y* = –0.3:

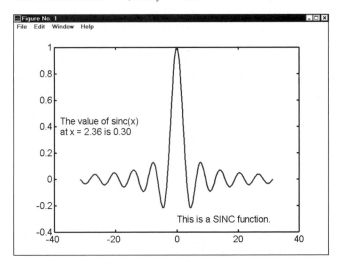

As another example of the **SPRINTF** function, we'll use three different methods to generate a text string:

```
EDU» str2='X ranges from -10pi to 10pi';
EDU» str3= sprintf('\n              OR\n');
EDU» str4=sprintf('X ranges from %4.1f to %4.1f', -10*pi,10*pi);
EDU» str5=[str2,str3,str4];
EDU» text(10, 0.6, str5);
```

In this example we use three different ways of creating a string. *str2* is created using quotes. To create *str3,* a text string with new line characters (**\n**), the **SPRINTF** function is used, and *str4,* a text string with formatted numbers, uses the **SPRINTF** function. Finally, *str5* is created by concatenating the first three strings. The final string is placed at coordinates $x = 10$ and $y = 0.6$:

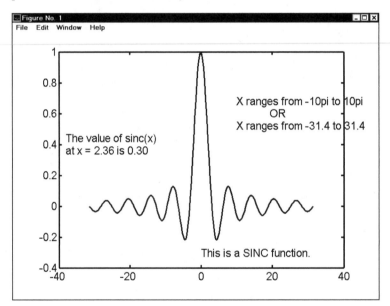

The **SPRINTF** function allows us to use the formatting power of the **FPRINTF** function with any MATLAB function that uses text strings.

6.6 Problems

Problem 6-1   Find all leap years for the next 100 years. Use serial date numbers to find all years that have 366 days.

Problem 6-2   Write the MATLAB functions described in parts A and B. In the function name, xxx refers to your initials.

A) **Function xxx_d1**   This function creates diary entries that can be saved on the computer. The diary files are text files and the name of the files is the date and time at which the entries were created. The function should do the following:

1. Use the **DATESTR** and **NOW** functions to create a text string composed of the date and time at which the entry is created. Use the **STRREP** function to replace colons (`:`) with a dash (`-`). Use concatenation to add `'.txt'` to the text string. An example text string is `06-Nov-1998 21-05-39.txt`.
2. Open a file using the text string created in item 1 as the file name. The file should be placed in a directory named c:\diary.
3. Ask the user for text input and print the text to the file.
4. Keep asking for text until the user types a line that contains a single character, which is a period (`.`). Do not print this line to the file.
5. After the user ends the entry with a period (`.`), close the file. If the file is closed correctly, display the message `Entry saved`. If the file is not closed properly, display the error message `Problem closing the file. Entry not saved`.

A session of using an example function is shown here:

```
EDU» meh_d1
Compose your diary entry. To end the entry, enter a line
that contains only a period.

:Today I wrote a homework assignment for my EGR 222 students.
:It was a very happy assignment.
:
:I also walked the dog today.
:It was cold and she barked.
:.
Entry saved.
EDU»
```

If you display directory c:\diary, you should see the file for the entry:

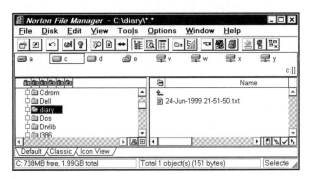

Since the diary file is a text file, you can also use the Wordpad to view the file:

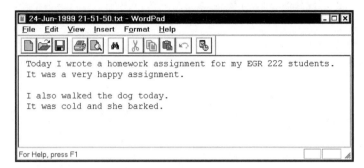

B) Function xxx_d2. Create a function that asks for a diary file and then displays the entry in the MATLAB command window. The function should do the following:

1. Use the **UIGETFILE** function to select a diary file. The default directory for searching for files should be set to C:\diary.
2. Open the file.
3. From the file name, determine the date at which the entry was created. Display the date and time the entry was created. An example would be **Date of Entry: 24-Jun-1999 21-51-50**.
4. Read the file and display the entry on the MATLAB command window.

A session of an example solution is shown here:

EDU» **meh_d2**

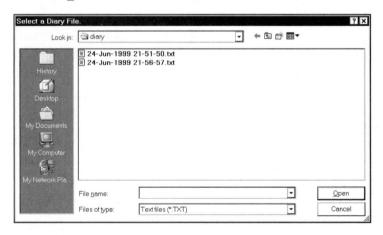

The user then selects a file.

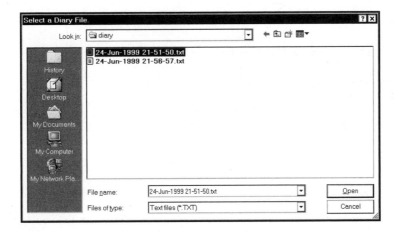

When the user clicks the **Open** button, the diary entry is displayed:

```
Date of Entry: 24-Jun-1999 21-51-50
Today I wrote a homework assignment for my EGR 222 students.
It was a very happy assignment.

I also walked the dog today.
It was cold and she barked.
EDU»
```

Problem 6-3 Modify the ask_q function discussed in Function 3-3 so that an error sound is played if the user enters an invalid response.

Problem 6-4 Locate the directory on your system where wav files are stored (Windows comes with some built-in sounds that you can use). Write a script file to perform the following operations:

a. Read the wave file using the **WAVREAD** function.
b. Use the **TIC** and **TOC** functions to time how long it takes to play the sound. Play the sound five times and display the statistics: max play time, min play time, and average play time.
c. Plot the sound signal versus time using the MATLAB **PLOT** function.

7

Plotting in MATLAB

OBJECTIVES

☐ Examine plotting, including
- Basic two-dimensional plots
- Modifying line styles, markers, and colors
- Grids and modifying the axes
- Placing text on a plot
- Obtaining numerical values from a plot

☐ Use **SEMILOGX**, **SEMILOGY**, **LOGLOG**, **POLAR**, **COMET**, and **SUBPLOT**.

☐ Handle graphics for completely controlling plot windows.

In this chapter we cover the plotting capabilities of MATLAB in more detail. We will introduce different plot types such as polar, semi-log, and log-log, as well as use MATLAB to generate presentation-quality plots. We'll also plot different types of functions and choose plots that best display those functions. This chapter only covers a fraction of the plotting and graphics capabilities of MATLAB, but it should give you a good grasp of MATLAB plotting and graphics so that you can use the help facilities to discover and explore the many other capabilities available.

7.1 Basic Two-Dimensional Plots

MATLAB Predefined
plot function
title function
xlabel function
ylabel function
hold on command
hold off command
legend function

The basic plotting function in MATLAB is **plot**(x, y) where x contains a one-dimensional array of data and y contains a one-dimensional array of data with the same number of elements. The **PLOT** function opens a new window and plots the data:

```
EDU» x=linspace(0,4*pi,200);
EDU» y=sin(x);
EDU» plot(x,y);
```

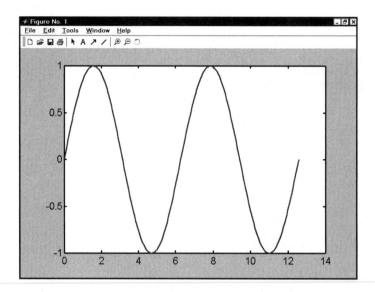

We can also give the plot a title and label the *x*- and *y*-axes using the **TITLE**, **XLABEL**, and **YLABEL** functions:

```
EDU» title('A simple plot using MATLAB')
EDU» ylabel('SIN(x)');
EDU» xlabel('x (Radians)');
```

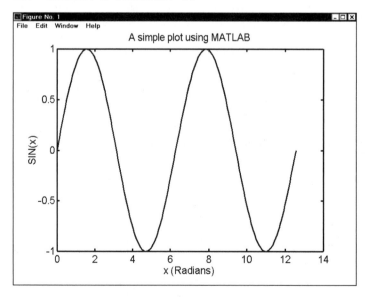

Suppose we want to plot the cosine function on the same plot and we want the plot of sin(*x*) to still show. If we use the **PLOT** function again, the original plot is erased and a new plot is generated:

EDU» `plot(x,cos(x))`

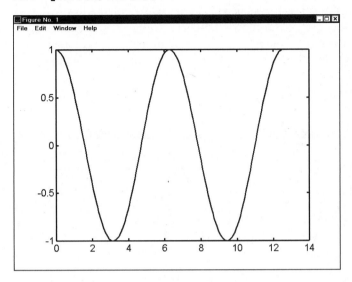

As you can see, the original plot is erased, including the title and the *x*- and *y*-axis labels. So remember that if you use the **PLOT** function without taking precautions to retain the previous plot, the **PLOT** function will always erase the previous plot and generate a completely new one.

We can generate a plot with several graphs in several ways. First, let's get our original plot back.

```
EDU» x=linspace(0,4*pi,200);
EDU» y=sin(x);
EDU» plot(x,y);
EDU» title('A simple plot using MATLAB')
EDU» ylabel('SIN(x)');
EDU» xlabel('x (Radians)');
```

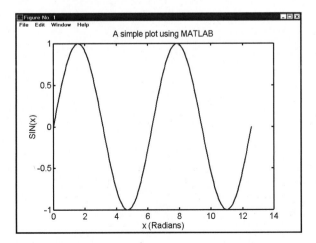

To keep this plot intact when the **PLOT** command is used again, we type in the command **HOLD ON**. This command holds the current plot so that future commands add to the existing plot. **HOLD OFF** returns the setting to its default state where subsequent uses of the **PLOT** command erase the entire plot and start a new one. We'll use the **HOLD ON** command and then plot the cos(*x*) function on the same plot:

```
EDU» hold on
EDU» plot(x,cos(x))
```

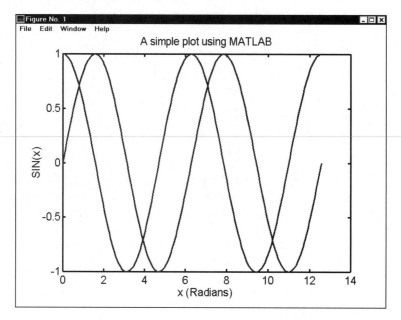

The **HOLD** setting remains in the "on" state until we switch it to the "off" state. Any uses of the **PLOT** command will add plots to the current plot:

```
EDU» z=abs(sin(x));
EDU» plot(x,z);
EDU» zz=sin(x).*sin(x);
EDU» plot(x,zz);
```

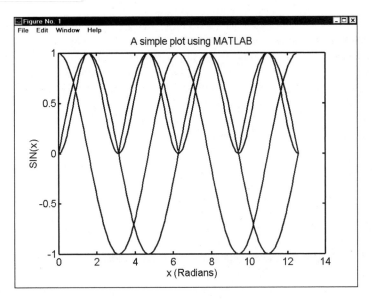

Now let's look at other ways to generate a plot with several traces. The **PLOT** function can be specified with several *x-y* pairs of data. First, we type in **HOLD OFF** so that the next **PLOT** command creates a completely new plot. We'll generate a plot with two traces:

```
EDU» hold off
EDU» x=linspace(0,4*pi,200);
EDU» y=sin(x);
EDU» plot(x,y,x,cos(x));
EDU» xlabel('x - (Radians)');
EDU» title('Sine and Cosine Functions');
```

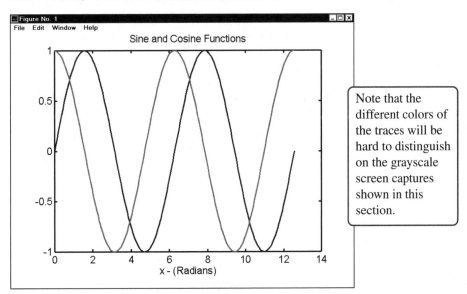

Note that the different colors of the traces will be hard to distinguish on the grayscale screen captures shown in this section.

As you can see, both functions are plotted on the same graph. The one difference between this plot and the previous one is that they are displayed in different colors. When used in this manner, each trace in the plot is automatically displayed in a different color. We can plot several sets of data in one plot function. Since **HOLD OFF** was specified in the last code segment, it is still in effect and each use of the **PLOT** function will create a new plot:

```
EDU» x=linspace(0,4*pi,200);
EDU» y=sin(x);
EDU» z=abs(sin(x));
EDU» zz=sin(x).*sin(x);
EDU» plot(x,y, x,cos(x), x,z, x,zz);
EDU» xlabel('x - (Radians)');
EDU» title('Various Functions');
```

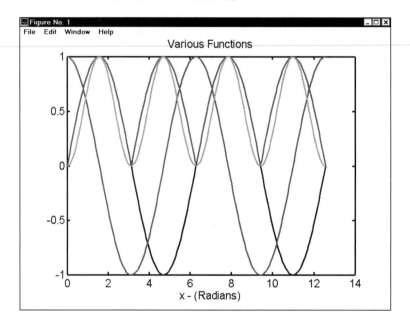

To specify which trace is which, we use the **LEGEND** function:

```
EDU» legend('sin(x)', 'cos(x)', 'abs(sin(x))', '(sin(x)*sin(x)')
```

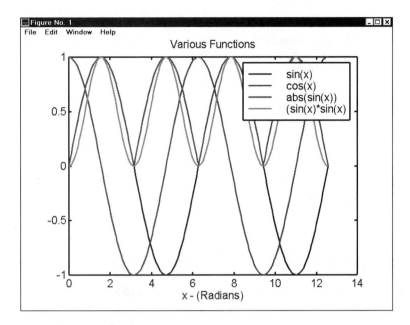

In the previous example, the *x*-coordinates for all four traces were the same. We can generate the same plot by making *x* a column array, and then making *y* an array with four columns and the same number of rows as *x*. Let's reduce the size of *x* so that we can view the structures of *x* and *y* more easily:

```
EDU» x=linspace(0,pi,10)'
x =
          0
     0.3491
     0.6981
     1.0472
     1.3963
     1.7453
     2.0944
     2.4435
     2.7925
     3.1416
```

As you can see, *x* is a column array with 10 rows. Now let's create *y*:

```
EDU» y=[sin(x), cos(x), abs(sin(x)), sin(x).*sin(x)]
y =
          0     1.0000          0          0
     0.3420     0.9397     0.3420     0.1170
     0.6428     0.7660     0.6428     0.4132
     0.8660     0.5000     0.8660     0.7500
     0.9848     0.1736     0.9848     0.9698
```

```
0.9848    -0.1736     0.9848     0.9698
0.8660    -0.5000     0.8660     0.7500
0.6428    -0.7660     0.6428     0.4132
0.3420    -0.9397     0.3420     0.1170
0.0000    -1.0000     0.0000     0.0000
```

So *y* has four columns and 10 rows, and we can now plot these data. Each column of *y* will be plotted against the one column of data in *x*. Thus, the same *x*-coordinate values are used in all four traces. We'll add more points to *x* to make the traces a bit smoother:

```
EDU» x=linspace(0,4*pi,200)';
EDU» y=[sin(x), cos(x), abs(sin(x)), sin(x).*sin(x)];
EDU» plot(x,y)
EDU» legend('sin(x)', 'cos(x)', 'abs(sin(x))', '(sin(x)*sin(x)')
EDU» xlabel('x - Radians');
EDU» title('Various Trigonometric Functions');
```

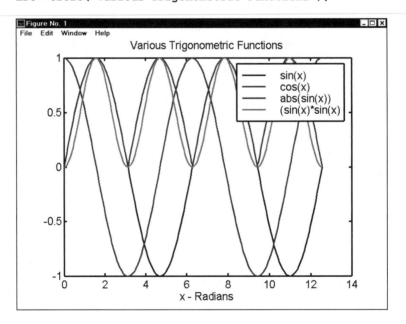

In the method just presented, each trace must have the same number of points, and in the previous example, the same values for the *x*-coordinate were used for all four traces. A slight variation of that example allows each trace to have different *x*- and *y*-coordinates. In the previous example, *y* had four columns and *x* had one column. The values of each column of *y* were plotted against the one column in *x*. However, if *x* has more than one column and *y* has the same dimensions as *x*, the first column in *y* is plotted against the first column in *x*, the second column in *y* against the second column in *x*, the third column in *y* against the third column in *x*, and so on. Thus, it must be true that *x* is the same size as *y* (same number of rows and columns). Let's create a small data set so that we can display the structure of *x* and *y*.

```
EDU» x1=linspace(0,pi,10)';
EDU» x2=linspace(pi,2*pi,10)';
EDU» x3=linspace(2*pi,3*pi,10)';
EDU» x4=linspace(3*pi,4*pi,10)';
EDU» x=[x1,x2,x3,x4];
EDU» y1=sin(x1);
EDU» y2=cos(x2);
EDU» y3=abs(sin(x3));
EDU» y4=sin(x4).*sin(x4);
EDU» y=[y1,y2,y3,y4];
```

This code segment shows *x* and *y* as follows:

```
EDU» x
x =
         0     3.1416     6.2832     9.4248
    0.3491     3.4907     6.6323     9.7738
    0.6981     3.8397     6.9813    10.1229
    1.0472     4.1888     7.3304    10.4720
    1.3963     4.5379     7.6794    10.8210
    1.7453     4.8869     8.0285    11.1701
    2.0944     5.2360     8.3776    11.5192
    2.4435     5.5851     8.7266    11.8682
    2.7925     5.9341     9.0757    12.2173
    3.1416     6.2832     9.4248    12.5664
EDU» y
y =
         0    -1.0000     0.0000     0.0000
    0.3420    -0.9397     0.3420     0.1170
    0.6428    -0.7660     0.6428     0.4132
    0.8660    -0.5000     0.8660     0.7500
    0.9848    -0.1736     0.9848     0.9698
    0.9848     0.1736     0.9848     0.9698
    0.8660     0.5000     0.8660     0.7500
    0.6428     0.7660     0.6428     0.4132
    0.3420     0.9397     0.3420     0.1170
    0.0000     1.0000     0.0000     0.0000
```

We can now plot these data. The **4** in the second line of the following code segment instructs the **LEGEND** function to place the legend in the lower-right corner:

```
EDU» plot(x,y)
EDU» legend('sin(x)', 'cos(x)', 'abs(sin(x))', '(sin(x)*sin(x)',4)
EDU» xlabel('x - Radians');
EDU» title('Various Trigonometric Functions');
```

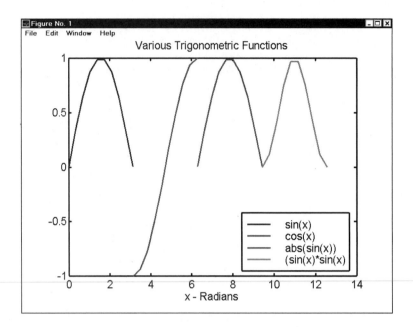

In all of our previous examples, the traces had the same number of points. If we want to plot several traces on the same plot, with each trace having a different number of points, we must specify the *x* and *y* pairs separately:

```
EDU» x1=linspace(0,2*pi,10);
EDU» x2=linspace(0,2*pi,50);
EDU» x3=linspace(pi,2*pi,100);
EDU» x4=linspace(pi/3,2*pi,150);
EDU» y1=sin(x1);
EDU» y2=cos(x2);
EDU» y3=abs(sin(x3));
EDU» y4=sin(x4).*sin(x4);
EDU» plot(x1,y1, x2,y2, x3,y3, x4,y4);
EDU» legend('sin(x)', 'cos(x)', 'abs(sin(x))', '(sin(x)*sin(x)',4)
EDU» xlabel('x - Radians');
EDU» title('Various Trigonometric Functions');
```

In this example, the data in y_1 is plotted against x_1, the data in y_2 against x_2, the data in y_3 against x_3, and the data in y_4 against x_4. Each set of data pairs has unique *x*-coordinates and each set has different numbers of points. The traces will all be placed on the same plot window and use the same axes:

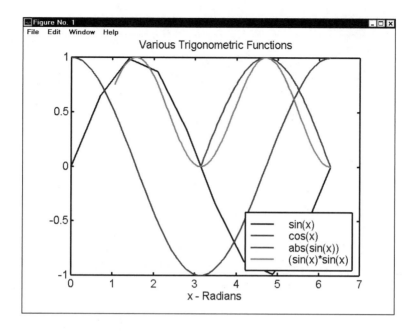

7.2 Line Styles, Markers, and Colors

In the previous plots, MATLAB chose the color and style of the traces for us. With the **PLOT** command, we can also specify line styles. The general form of the **PLOT** command is

$$\texttt{plot}(x_1,\ y_1,\ s_1,\ x_2,\ y_2,\ s_2,\ x_3,\ y_3,\ s_3,\ \ldots)$$

where x_i and y_i contain the *x*- and *y*-coordinates of the data as shown previously. Variables s_i are text strings that specify the color, markers, and line style for a particular trace. A group of variables for a specific trace (x_i, y_i, s_i) is called a triple in MATLAB. The text strings, s_i, can be a text string of one to four characters. If the text string is not specified, MATLAB chooses the color and line style for the trace. s_i is a text string of the form "cmls" where c is a single character that specifies the trace color, m is a single character that specifies the marker type, and ls is one or two characters that specify the line style. Since the codes for specifying c, m, and ls are unique, you don't need to specify all three parameters. That is, if you only want to specify a marker type and a line style, you don't need to specify a color. If you only want to specify a line style, you don't need to specify a color or a marker. The codes are specified in Table 7.1.

TABLE 7-1 Line styles, markers, and colors for the **PLOT** command

Color Code	Color	Marker Code	Marker	Line Style Code	Line Style
y	yellow	.	point	-	solid
m	magenta	o	circle	:	dotted
c	cyan	x	x-mark	-.	dashdot
r	red	+	plus	--	dashed
g	green	*	star		
b	blue	s	square		
w	white	d	diamond		
k	black	v	triangle (down)		
		^	triangle (up)		
		<	triangle (left)		
		>	triangle (right)		
		p	pentagram		
		h	hexagram		

Let's look at an example using the last plot.

```
EDU» x1=linspace(0,2*pi,10);
EDU» x2=linspace(0,2*pi,50);
EDU» x3=linspace(pi,2*pi,100);
EDU» x4=linspace(pi/3,2*pi,150);
EDU» y1=sin(x1);
EDU» y2=cos(x2);
EDU» y3=abs(sin(x3));
EDU» y4=sin(x4).*sin(x4);
EDU» plot(x1,y1,'gs', x2,y2,'kp:', x3,y3, x4,y4,'r--');
EDU» legend('sin(x)', 'cos(x)', 'abs(sin(x))', '(sin(x)*sin(x)',3)
EDU» xlabel('x - Radians');
EDU» title('Various Trigonometric Functions');
```

The changes in the **LEGEND** command instruct MATLAB to place the legend in the lower-left corner of the plot. The line **plot(x1,y1,'gs', x2,y2,'kp:', x3,y3, x4,y4,'r:');** specifies the following:

1. Plot the first trace with green squares, no line drawn between points.
2. Plot the second trace in black, mark points with pentagons (stars), and draw a dotted line to connect the points.
3. Nothing is specified for the third trace. MATLAB will choose the color of the trace and use a solid line without markers.
4. Plot the fourth trace with a red dashed line.

The code segment generates this plot:

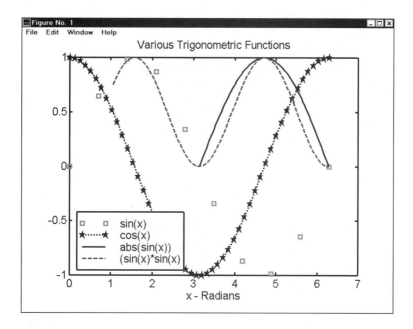

7.3   Plot Color

MATLAB Predefined **whitebg** function   To change the color of the plot, we use the **WHITEBG** command. This command typically toggles between a white background and a black one. For example, the previous plot has a white background. If we use the **WHITEBG** command, the plot changes to this one:

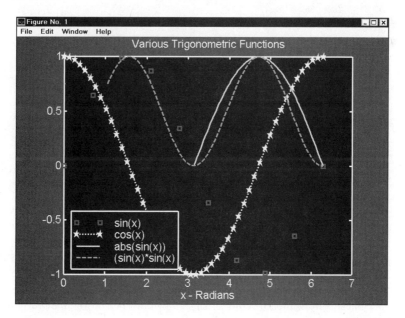

The colors of the elements change to complement the new background. If we use the **WHITEBG** command again, the plot changes back to the white background:

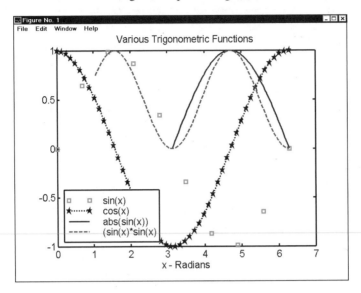

We can also specify a background color with the **WHITEBG** command. The color codes are the same as shown in Table 7-1. For example, **WHITEBG('y')** produces a plot with a yellow background (shown on the left below), and **WHITEBG('m')** produces a plot with a magenta background (on the right):

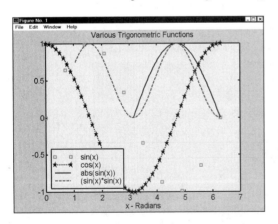 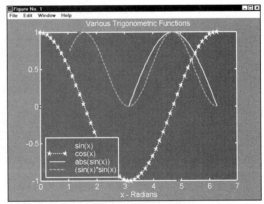

One problem with using the **WHITEBG** command to toggle between black and white is that you may not know the current state of a plot. If you don't know the current state, you also won't know what will happen when you toggle the color. To avoid this problem, you can always specify white (**'w'**) or black (**'k'**) with the **WHITEBG** function. Instead of toggling the color, the color then changes to the specified color. If the plot is already white and you specify white, nothing happens, but if the plot is black, and you specify white, the color changes to white. Thus, from the preceding plot, using **WHITEBG('w')** produces a white plot:

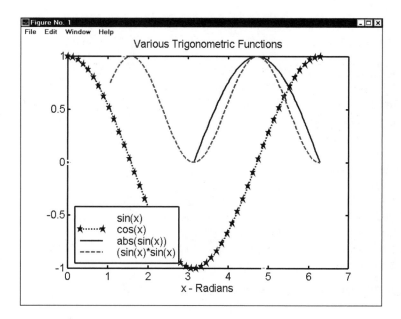

Through all of our color gyrations, we lost the markers for the first trace. Usually this problem doesn't occur because you won't be changing the background color as much as we have here.

EXERCISE 7-1 The following code segment is an interesting application of the **FOR** loop and the **WHITEBG** command. What does this segment do?

```
x=linspace(0,pi,200);
y=sin(x);
plot(x,y);
ylabel('Sin(x)');
xlabel('x');
title('Plot of Sine Function.')
for i = ['y','m', 'c', 'r', 'g', 'b', 'k', 'w']
   pause(1);
   whitebg(i);
end
```

7.4 Plotting **Grid**

The MATLAB **GRID** command places a grid on the current plot. The command can be used in three forms:

- **GRID ON**—Places a grid on the current plot.
- **GRID OFF**—Removes the grid from the current plot.
- **GRID**—Toggles the grid on or off.

Here's an example—we've used a previous plot and added the **GRID** command at the end:

```
EDU» x1=linspace(0,2*pi,10);
EDU» x2=linspace(0,2*pi,50);
EDU» x3=linspace(pi,2*pi,100);
EDU» x4=linspace(pi/3,2*pi,150);
EDU» y1=sin(x1);
EDU» y2=cos(x2);
EDU» y3=abs(sin(x3));
EDU» y4=sin(x4).*sin(x4);
EDU» plot(x1,y1,'gs', x2,y2,'kp:', x3,y3, x4,y4,'r--');
EDU» legend('sin(x)', 'cos(x)', 'abs(sin(x))', '(sin(x)*sin(x)',3)
EDU» xlabel('x - Radians');
EDU» title('Various Trigonometric Functions');
EDU» grid on
```

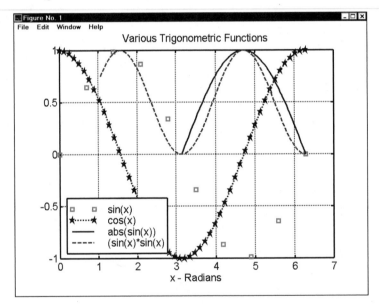

7.5 The AXIS Command

MATLAB Predefined
axis function

The **AXIS** command allows us to specify properties for a plot's axes. This command has several features that we won't cover. To view these options, see the help information for this command. We'll only use this command to set the axes limits. Here are two forms of the command:

- **AXIS([Xmin Xmax Ymin Ymax])**—This form specifies min and max limits for the *x*- and *y*-axes.
- **AXIS AUTO**—This form returns the axis control back to MATLAB and allows MATLAB to choose the axes limits.

As an example, let's specify limits for the previous plot:

```
EDU» x1=linspace(0,2*pi,10);
EDU» x2=linspace(0,2*pi,50);
EDU» x3=linspace(pi,2*pi,100);
EDU» x4=linspace(pi/3,2*pi,150);
EDU» y1=sin(x1);
EDU» edu y2=cos(x2);
EDU» y3=abs(sin(x3));
EDU» y4=sin(x4).*sin(x4);
EDU» plot(x1,y1,'gs', x2,y2,'kp:', x3,y3, x4,y4,'r--');
EDU» legend('sin(x)', 'cos(x)', 'abs(sin(x))', '(sin(x)*sin(x)',3)
EDU» xlabel('x - Radians');
EDU» title('Various Trigonometric Functions');
EDU» grid on
EDU» axis([3 5 0 1])
```

The command **AXIS([3 5 0 1])** specifies the x-axis to be between 3 and 5, and the y-axis to be between 0 and 1. Points outside of this range won't be displayed. The code segment produces the following plot:

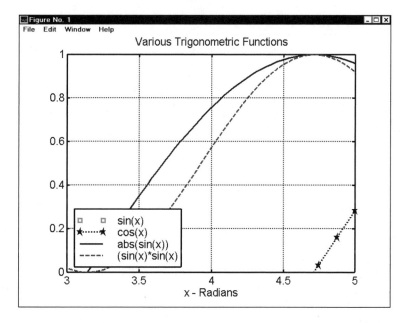

The **AXIS** command can also be used to zoom in on a specific portion of the plot. To allow MATLAB to set the axes limits, we use the **AXIS AUTO** command:

```
EDU» axis auto
```

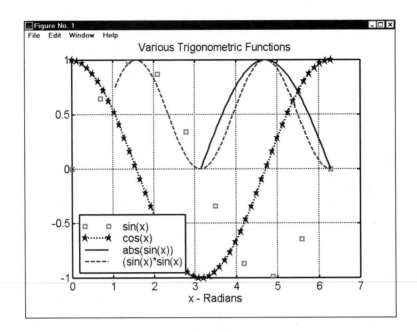

7.6 Placing Text on a Plot

MATLAB Predefined
gtext function
text function

The MATLAB functions **GTEXT** and **TEXT** place text on a plot. The format of the **TEXT** function is

$$text(x, \ y, \ str)$$

where *x* and *y* specify where the text is to be placed. The location is specified in units of the current *x*- and *y*-axes. Variable *str* is the text string to be displayed. A few examples of placing text on the screen using the **TEXT** function were shown on pages 354–356, but we'll give another example here. First, let's generate the previous plot:

```
EDU» x1=linspace(0,2*pi,10);
EDU» x2=linspace(0,2*pi,50);
EDU» x3=linspace(pi,2*pi,100);
EDU» x4=linspace(pi/3,2*pi,150);
EDU» y1=sin(x1);
EDU» y2=cos(x2);
EDU» y3=abs(sin(x3));
EDU» y4=sin(x4).*sin(x4);
EDU» plot(x1,y1,'gs', x2,y2,'kp:', x3,y3, x4,y4,'r-');
EDU» legend('sin(x)', 'cos(x)', 'abs(sin(x))', '(sin(x)*sin(x)',3)
EDU» xlabel('x - Radians');
EDU» title('Various Trigonometric Functions');
```

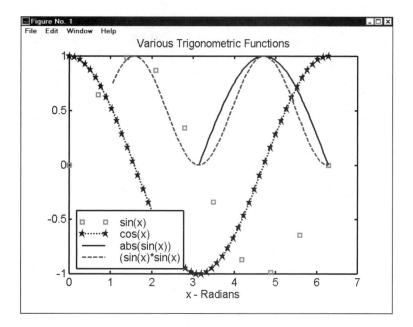

As before, the *x*-axis ranges from 0 to 7, so when we specify the text's *x*-coordinate, it should be a number between 0 and 7. The *y*-axis ranges from –1 to 1 and when we specify the text's *y*-coordinate, it should be a number between –1 and 1. We'll place a text string at location $x = 0.5$ and $y = -0.25$:

EDU» **text(0.5,-0.25,'LEGEND')**

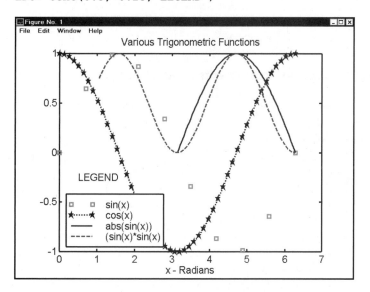

Note that if you follow this example, your text may be placed differently or may overlap the legend due to different monitor resolutions.

The **GTEXT** function also places text on the screen; however, the command uses the mouse to specify the location of the text. When the **GTEXT** command is executed, the current figure window is brought to the top and a crosshair is displayed. You can position the crosshair with the mouse. If you press the mouse button, the text is placed at the location of the crosshair. If you press a key on the keyboard, the text will not be placed, and control will be returned to MATLAB. The syntax of the **GTEXT** command is gtext(*str*) where *str* is a text string.

Let's use the command **gtext('This is a test.');**. When we execute the command, the figure window pops to the top and a crosshair appears and moves with the mouse:

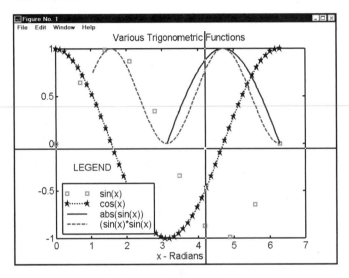

Place the cursor in a convenient location for the text string and press the mouse button. The text will be placed at the location of the cursor:

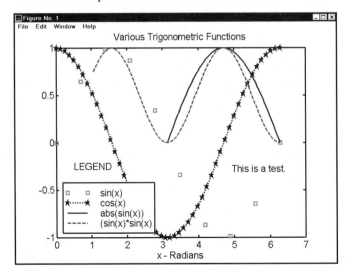

7.7 Modifying Text with T$_E$X Commands

MATLAB *Predefined*
xlabel function
ylabel function
title function
text function
gtext function
set function

T$_E$X is a powerful typesetting system for technical manuscripts. If you are familiar with T$_E$X, you can use some of the math commands in the MATLAB **XLABEL**, **YLABEL**, **TITLE**, **TEXT**, and **GTEXT** commands to include mathematical formulas. We'll show a few examples with Greek letters, subscripts, and superscripts. Many other T$_E$X commands are possible. To see more of the available math commands, see *The T$_E$X Book.*[*]

First let's generate a plot on which to place text.

```
EDU» x=linspace(-4,4,200);
EDU» y=exp(-x.*x);
EDU» plot(x,y);
```

To get Greek letters, use a backslash followed by the name of the letter. If the first letter of the name is uppercase, the Greek letter will be uppercase. If the name is all lowercase letters, the Greek letter will be lowercase. Here are some examples:

```
EDU» text(-3, 0.7, '\alpha\beta\gamma\omega\zeta\epsilon')
EDU» text(-3, 0.6, '\Gamma\Omega\Sigma\Lambda\Theta')
```

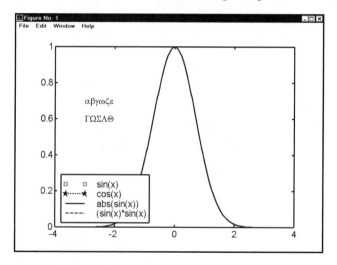

Special mathematical symbols are also available. You'll need to refer to *The T$_E$X Book* to find out the commands for these symbols:

```
EDU» text(-3, 0.5, '\Uparrow\downarrow\ldots\rightarrow\approx\equiv\subset')
```

---

[*]*The T$_E$X Book,* Donald E. Knuth, Addison-Wesley, 1986.

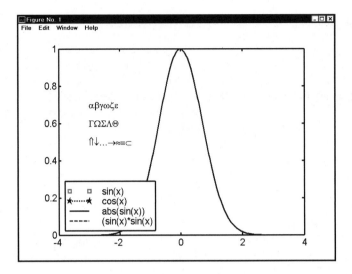

Subscripts and superscripts are generated with the _ and ^ characters. The character immediately following the _ or ^ displays as a subscript or superscript. If you need more than one character sub- or superscripted, enclose the character string in curly brackets. As an example, let's generate the equation

$$y = x^{3x} + x^2 + x^{-100}$$

and place it in the title. The **TITLE** command below creates this equation. The **SET** command shrinks the size of the plot area and allows more space for the superscripts. Without the **SET** command, the superscripts would be clipped off at the top of the plot.

```
EDU» set(gca,'PlotBoxAspectRatio',[1,0.7,1])
EDU» title('Plot of y=x^{3x} + x^2 + x^{-100}')
```

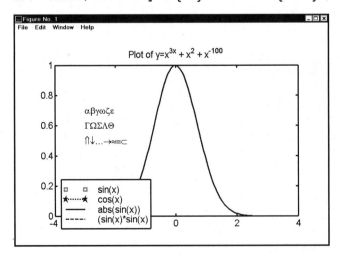

Let's also generate some labels for the x- and y-axes:

```
EDU» xlabel('X_{axis} – x_1+x_2+x_{333}')
EDU» ylabel('Y_{axis} – A Plot of an Equation')
```

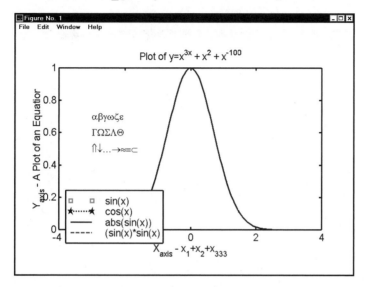

With the available T$_E$X commands, you can easily annotate your plots with mathematical equations.

7.8 Obtaining Numerical Values from a Plot

MATLAB Predefined
ginput function
msgbox function
waitfor function

Numerical values can be obtained from a plot using the **GINPUT** function. The syntax of the function is **[x, y] = ginput(N)** where *N* is the number of points requested and *x* and *y* contain the *x*- and *y*-coordinates of the recorded points. If *N* is omitted, points are collected until you press the ENTER key.

When you run the **GINPUT** function, the current figure window pops to the top and a set of crosshairs appears on the figure window. The crosshairs are moved with the mouse (or the arrow keys on some systems). When you click the left mouse button or press any key other than the ENTER key, the coordinates of the cursor are recorded. MATLAB will keep recording points until the specified number of points have been recorded or until you press the ENTER key. The values returned by the **GINPUT** function are the coordinates of the cursor in the current *x*- and *y*-axes scales.

Let's look at an example. First we generate a plot:

```
x1=linspace(0,2*pi,10);
x2=linspace(0,2*pi,50);
x3=linspace(pi,2*pi,100);
x4=linspace(pi/3,2*pi,150);
y1=sin(x1);
y2=cos(x2);
y3=abs(sin(x3));
```

```
y4=sin(x4).*sin(x4);
plot(x1,y1,'gs', x2,y2,'kp:', x3,y3, x4,y4,'r--');
legend('sin(x)', 'cos(x)', 'abs(sin(x))', '(sin(x)*sin(x)',3)
xlabel('x - Radians');
title('Various Trigonometric Functions');
```

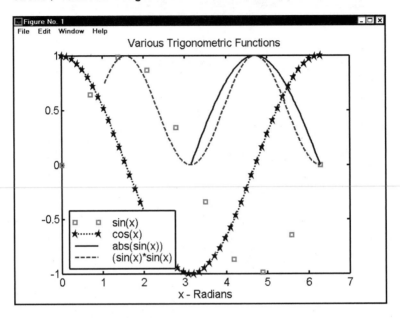

We'll run this code segment:

```
figure(1)
msg1=sprintf('Place the cursor at the location of a point you wish to ');
msg1=[msg1,sprintf('see the numerical values and then press ')];
msg1=[msg1,sprintf('the LEFT mouse button.')];
h=msgbox(msg1);
waitfor(h)
[x1,y1]=ginput(1);
msg2=sprintf('Place the cursor at the location of where you would ');
msg2=[msg2,sprintf('like to display the values on the plot and then ')];
msg2=[msg1,sprintf('press the LEFT mouse button.')];
h=msgbox(msg2);
waitfor(h)
[x2,y2]=ginput(1);
str=sprintf('x=%5.3f, y=%5.3f',x1,y1);
line([x1,x2],[y1,y2]);
text(x2,y2,str);
```

As you examine portions of this code segment, note that since only one figure window is open, its reference number is 1. The line **figure(1)** makes figure 1 the topmost window. If this figure window is not yet visible, it will be after you execute this statement:

```
msg1=sprintf('Place the cursor at the location of a point you wish to');
msg1=[msg1,sprintf(' see the numerical values and then press ')];
msg1=[msg1,sprintf('the LEFT mouse button.')];
```

These lines create a single text string that contains instructions for the user. Note that it takes three lines of code to generate a rather long text string. Concatenation joins smaller text strings into a single long string.

```
h=msgbox(msg1);
```

This line opens a small message box and displays the text string msg_1. Return variable *h* is called a handle and refers to the message box. When we want information about the message box, we use the handle. The message box has a single button labeled **OK**. The message box remains open until the user clicks the **OK** button.

```
waitfor(h)
```

When we use the **MSGBOX** command, MATLAB displays the box and then continues on with the next statement. For this application, we want to display the message box, then pause the execution of the code segment until the user reads the message box and clicks the **OK** button. We do this with the **waitfor(h)** statement, which blocks execution of MATLAB commands until the window specified by *h* is deleted. While the message box is open, *h* is a valid handle and MATLAB continues to block execution. When the user clicks the **OK** button, the message box is deleted and the **WAITFOR** command no longer blocks execution. When we combine the **MSGBOX** command with the **WAITFOR** command, MATLAB displays the message box and waits until the user clicks the **OK** button. When the **OK** button is clicked, the message box is deleted and MATLAB resumes execution of the program.

After the user reads the message box and clicks the **OK** button, the line

```
[x1,y1]=ginput(1);
```

is executed. This line places the crosshair on the current figure and waits for the user to press the mouse button. When the button is pressed, the coordinates of the cursor are stored variables x_1 and y_1. The coordinates are in terms of the current *x*- and *y*-axis scales.

The next few lines create a text string that instructs the user to select a second point.

```
msg2=sprintf('Place the cursor at the location of where you would ');
msg2=[msg2,sprintf('like to display the values on the plot and then ')];
msg2=[msg2,sprintf('press the LEFT mouse button.')];
h=msgbox(msg2);
waitfor(h);
```

The text string is displayed in a message box, and the **waitfor(h);** command causes MATLAB to wait until the user reads the message box and clicks the **OK** button.

After clicking the **OK** button, the line **[x2,y2]=ginput(1);** executes. Crosshairs are placed on the figure and users select where they would like to place the text string. The coordinates are stored in variables x_2 and y_2.

The line **str=sprintf(' x=%5.3f, y=%5.3f',x1,y1);** creates a text string that displays the values of coordinates x_1 and y_1. These were the coordinates of the first data point selected.

The statement **line([x1,x2], [y1,y2]);** draws a line from coordinates x_1, y_1 to coordinates x_2, y_2. This line is used as a pointer to the first data point.

The final statement of the code segment, **text(x2,y2, str);**, places text string **str** at the second point selected by the mouse: x_2, y_2.

A sample run of the code segment follows; we'll start with the plot already created:

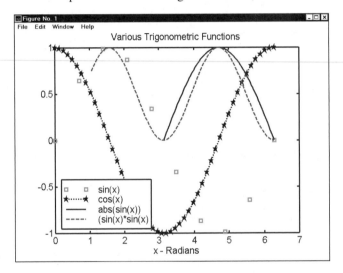

Now we run the entire code segment just discussed:

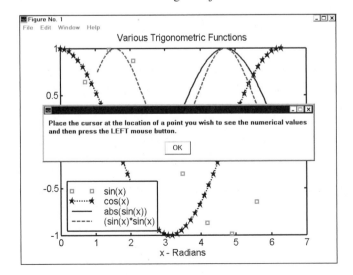

The message box remains until we click the **OK** button. When we click the **OK** button, crosshairs appear on the figure window:

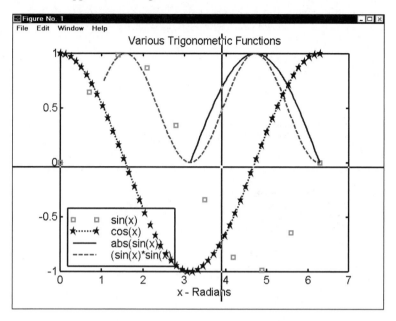

Let's place the cursor as shown in the next figure:

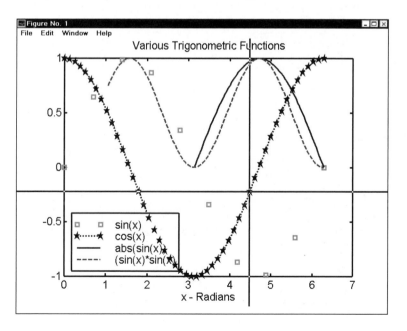

When we click the left mouse button, the second dialog box appears:

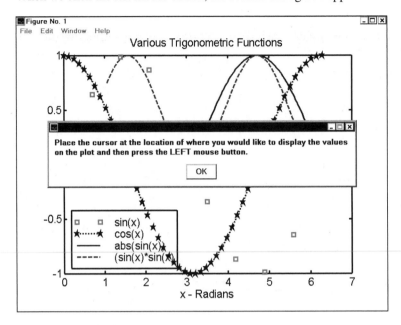

When we click the **OK** button, crosshairs again appear on the figure window:

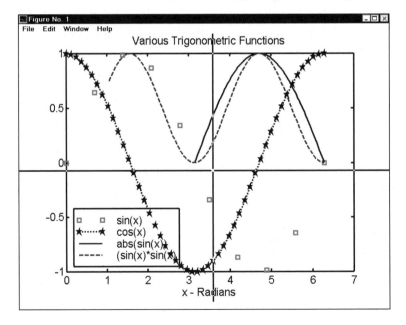

The second point we select is the text location. We'll place the cursor as shown in the next figure:

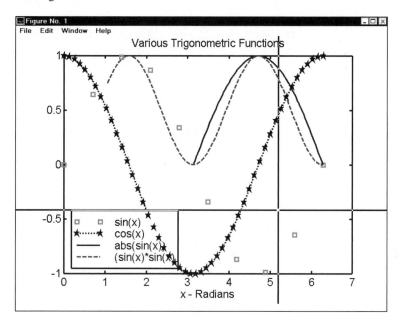

When we click the mouse button, the program draws the line and displays the text string:

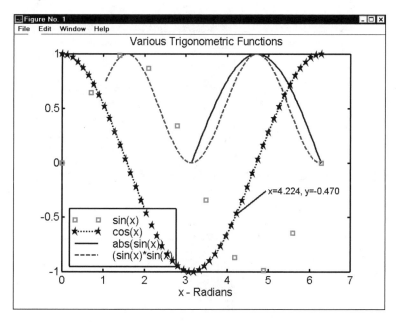

As a second example, let's create a code segment that asks for two points and then has the figure zoom in on the box defined by those two points:

```
EDU» figure(1)
EDU» msg1=sprintf('Place the cursor at the location of a point you wish to');
EDU» msg1=[msg1,sprintf(' use as one corner for the zoom box.')];
EDU» h=msgbox(msg1);
EDU» waitfor(h)
EDU» [x1,y1]=ginput(1);
EDU» msg2=sprintf('Place the cursor at the location of a point you wish to');
EDU» msg2=[msg2,sprintf(' use as the second corner for the zoom box.')];
EDU» h=msgbox(msg2);
EDU» waitfor(h)
EDU» [x2,y2]=ginput(1);
EDU» x=[x1,x2];
EDU» y=[y1,y2];
EDU» axis([min(x), max(x), min(y), max(y)]);
```

Let's run the code segment:

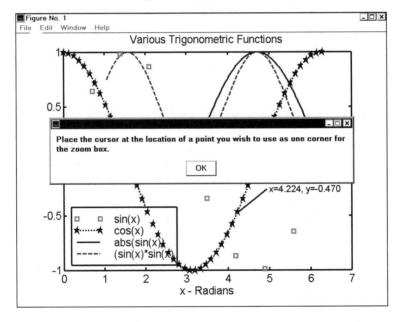

After clicking the **OK** button, cursors appear on the figure window:

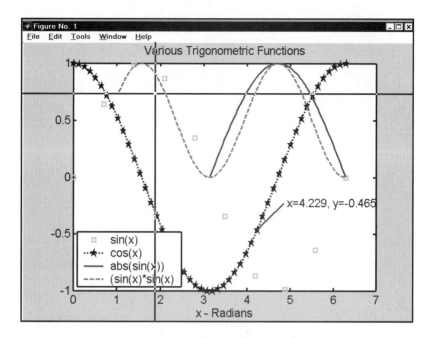

Place the cursor as shown above. When we click the left mouse button, the coordinates of the cursor are recorded and the second message box opens:

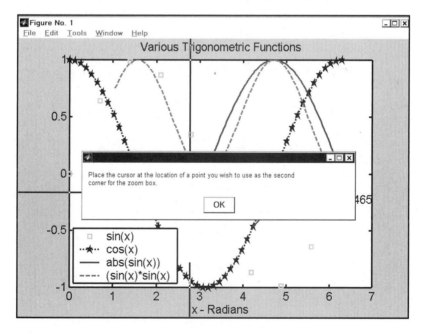

After clicking the **OK** button, cursors appear on the figure window:

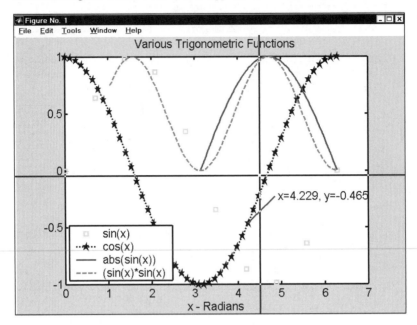

Place the cursor as shown above and click the mouse button. After clicking the left button, the coordinates of the mouse location are recorded. The statement `axis([min(x), max(x), min(y), max(y)]);` then changes the axes scales so that only the area enclosed by the rectangle defined by the two points is displayed.

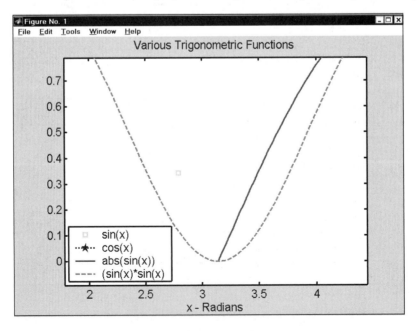

7.9 Various MATLAB 2-D Plot Types

7.9.1 SEMILOGX

MATLAB Predefined
semilogx
function
plot function
logspace
function
linspace
function
subplot function
semilogy
function
loglog function
polar function
comet function
figure function
shg function
celf function
close function

The **SEMILOGX** command works just like the **PLOT** command except that the *x*-axis is a log scale. As an example, let's generate Bode magnitude and phase plots for the gain expression $\dfrac{V_o}{V_{in}} = \dfrac{1000}{j\omega + 1000}$. The Bode magnitude plot is a plot of $20 \log_{10}(|V_0/V_{in}|)$ versus ω. A Bode phase plot plots the angle V_0/V_{in} versus ω. We first generate the magnitude plot:

```
EDU» omega=logspace(1,4,200);
EDU» mag=20*log10(abs(1000./(i*omega+1000)));
EDU» semilogx(omega, mag);
EDU» axis([10,10000,-20 5]);
EDU» grid on;
EDU» ylabel('Gain(dB)');
EDU» xlabel('Frequency (rad/sec)');
EDU» title('Bode Magnitude plot of Vo/Vin');
```

The **LOGSPACE** command is similar to the **LINSPACE** command except that it generates logarithmically spaced points. The syntax is logspace(d_1, d_2, n) where d_1 and d_2 specify the starting and ending decades for the points. The function generates n points from 10^{d_1} to 10^{d_2}. To show the difference, let's generate points from 10 to 100 using the **LOGSPACE** and **LINSPACE** functions:

```
EDU» x1=linspace(10,100,10)';
EDU» x2=logspace(1,2,10)';
EDU» x=[x1,x2]
x =
    10.0000    10.0000
    20.0000    12.9155
    30.0000    16.6810
    40.0000    21.5443
    50.0000    27.8256
    60.0000    35.9381
    70.0000    46.4159
    80.0000    59.9484
    90.0000    77.4264
   100.0000   100.0000
```

linspace results

logspace results

The points generated by **LINSPACE** are separated by a constant amount, 10 in this case. However, the points generated by **LOGSPACE** are not separated by constant amounts.

It turns out that if we take $\log_{10}$ of the points generated by **LOGSPACE**, we get points separated by a constant amount:

```
x=logspace(1,2,10)';
y=log10(x)
for i = 1:9
  delta=y(i+1)-y(i);
  fprintf('y(%1d)-y(%1d)=%g\n', i+1, i, delta);
end
y =
    1.0000
    1.1111
    1.2222
    1.3333
    1.4444
    1.5556
    1.6667
    1.7778
    1.8889
    2.0000
y(2)-y(1)=0.111111
y(3)-y(2)=0.111111
y(4)-y(3)=0.111111
y(5)-y(4)=0.111111
y(6)-y(5)=0.111111
y(7)-y(6)=0.111111
y(8)-y(7)=0.111111
y(9)-y(8)=0.111111
y(10)-y(9)=0.111111
```

As you can see, values of *y* range from 1 to 2 and are separated by a constant amount, 0.111111 in this case. Thus, the **LOGSPACE** function generates a set of points such that the log of the values is separated by a constant amount. When these points are plotted logarithmically, they are separated by equal distances.

The values generated by the **LOGSPACE** function can easily be generated using the **LINSPACE** function. We'll generate the same points using the **LINSPACE** function.

```
EDU» x=linspace(1,2,10)';
EDU» x=10.^x;
EDU» y=logspace(1,2,10)';
EDU» [x,y]
ans =
   10.0000   10.0000
   12.9155   12.9155
   16.6810   16.6810
   21.5443   21.5443
   27.8256   27.8256
   35.9381   35.9381
   46.4159   46.4159
```

```
 59.9484   59.9484
 77.4264   77.4264
100.0000  100.0000
```

As you can see, the two lines **x=linspace(1,2,10)';** and **x=10.^x;** generate the same points as **logspace(1,2,10)**.

Now let's plot the Bode magnitude plot:

```
EDU» omega=logspace(1,4,200);
EDU» mag=20*log10(abs(1000./(i*omega+1000)));
EDU» semilogx(omega, mag);
EDU» axis([10,10000,-20 5]);
EDU» grid on;
EDU» ylabel('Gain(dB)');
EDU» xlabel('Frequency (rad/sec)');
EDU» title('Bode Magnitude plot of Vo/Vin');
```

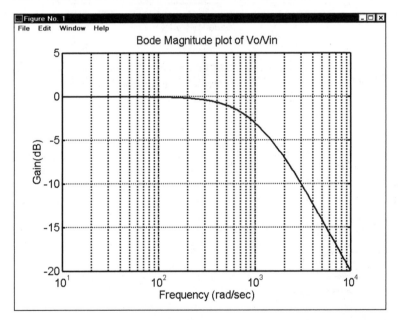

The phase plot is easily generated by modifying a few lines of the magnitude plot:

```
EDU» omega=logspace(0,5,200);
EDU» rad_to_deg = 360/(2*pi);
EDU» phase = rad_to_deg*angle(1000./(i*omega+1000));
EDU» semilogx(omega, phase);
EDU» grid on;
EDU» ylabel('Angle (Degrees)');
EDU» xlabel('Frequency (rad/sec)');
EDU» title('Bode Phase plot of Vo/Vin');
```

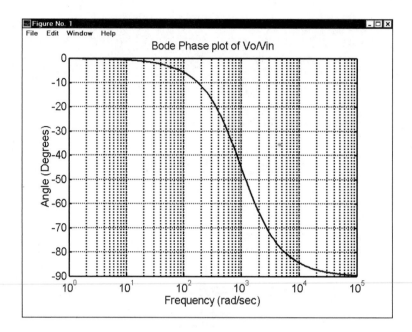

We can use the MATLAB **SUBPLOT** function to place both plots on the same figure. We'll discuss this function in more detail in Section 7.9.5.

```
EDU» subplot(2,1,1)
EDU» omega=logspace(0,5,200);
EDU» mag=20*log10(abs(1000./(i*omega+1000)));
EDU» semilogx(omega, mag);
EDU» grid on;
EDU» ylabel('Gain(dB)');
EDU» xlabel('Frequency (rad/sec)');
EDU» title('Bode Magnitude plot of Vo/Vin');
EDU» subplot(2,1,2);
EDU» rad_to_deg = 360/(2*pi);
EDU» phase = rad_to_deg*angle(1000./(i*omega+1000));
EDU» semilogx(omega, phase);
EDU» grid on;
EDU» ylabel('Angle (Degrees)');
EDU» xlabel('Frequency (rad/sec)');
EDU» title('Bode Phase plot of Vo/Vin');
```

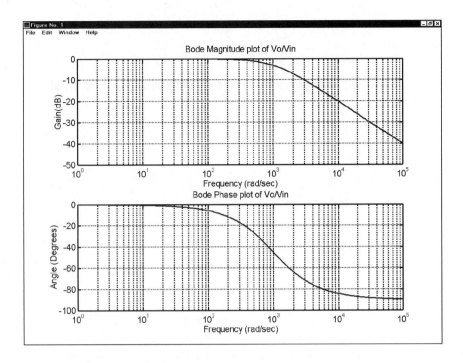

7.9.2 **SEMILOGY**

The **SEMILOGY** function is similar to the **SEMILOGX** and **PLOT** functions except that the y-axis is a log scale and the x-axis is a linear scale. To demonstrate the function, we'll plot the equation $y = 5 \times 10^{3x}$ for values of x from 0 to 100. We first plot the function using the **PLOT** and **SEMILOGX** functions:

```
EDU» x=linspace(0,100,200);
EDU» y=5.*10.^(3.*x);
EDU» plot(x,y)
```

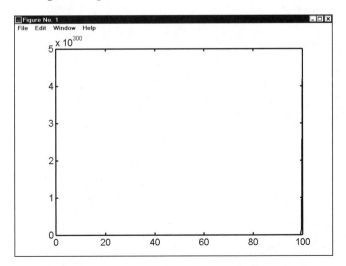

The plot doesn't appear to be to useful. Next, let's use the **SEMILOGX** function:

EDU» **semilogx(x,y)**

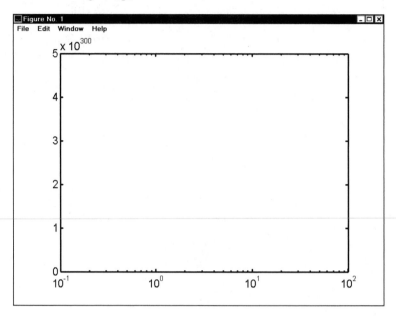

This plot isn't useful either.

The reason the previous two plots don't display the data well is because the values of y range from very small values to extremely large ones: for $x = 0$, $y = 5$, and for $x = 100$, $y = 5 \times 10^{300}$. The y-axis must accommodate all values of y, so it must range from 0 to 5×10^{300}. When the smaller numbers are plotted on this scale, they appear very close to zero because most of them are small compared to 5×10^{300}, and they are displayed as zero.

The **SEMILOGY** function plots $\log_{10}(y)$ versus x, both on linear scales. Even though y ranges from 5 to 5×10^{300}, $\log_{10}(y)$ ranges from 0.6990 to 300.6990. We can easily plot these values versus x. We first generate a plot of $\log_{10}(y)$ versus x using the **PLOT** command:

EDU» **x=linspace(0,100,200);**
EDU» **y=5.*10.^(3.*x);**
EDU» **k=log10(y);**
EDU» **plot(x,k)**
EDU» **ylabel('Log10(y)');**
EDU» **xlabel('x')**

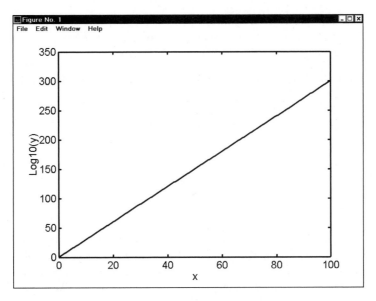

When we plot y versus x using the **SEMILOGX** function [not $\log_{10}(y)$ versus x], the same plot is generated except that the y-axis displays values of y rather than k:

```
EDU» x=linspace(0,100,200);
EDU» y=5.*10.^(3.*x);
EDU» semilogy(x,y)
EDU» ylabel('y');
EDU» xlabel('x');
```

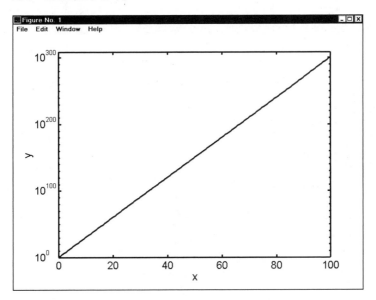

Note that the y-axis is now a log scale; that is, factors of 10 are equally spaced on the axis.

With an equation of the form $y = 5 \times 10^{3x}$, plotting the $\log_{10}(y)$ gives us very useful information. First, let's do some mathematical manipulation on the equation:

$$
\begin{aligned}
k = \log_{10}(y) &= \log_{10}(5 \times 10^{3x}) \\
&= \log_{10}(5) + \log_{10}(10^{3x}) \\
&= 0.69897 + 3x
\end{aligned}
$$

Thus, the equation for k is $k = 0.69897 + 3x$. This equation is a straight line with a y-intercept of 0.69897 and a slope of 3. In general, if we have an equation of the form $y = A \times 10^{Bx}$ and we plot $\log_{10}(y)$ versus x on a linear-linear plot, the slope of the plot will be B and the y-intercept will be $\log_{10}(A)$.

7.9.3   LOGLOG

For an example of a **LOGLOG** plot, we'll continue with the example of $y = 5 \times 10^{3x}$. As in the preceding section, if we plot this function with a **SEMILOGY** plot, it displays the exponential plot as a straight line. This is useful in determining information like constants A and B as discussed in the previous section. However, displaying the equation on a semilog plot changes the characteristic shape from an exponential to a straight line.

For an exponential equation, if the range in the x-variable is small, we can still use a linear-linear plot:

```
EDU» x=linspace(0,1,200);
EDU» y=5.*10.^(3.*x);
EDU» plot(x,y)
EDU» ylabel('y');
EDU» xlabel('x');
```

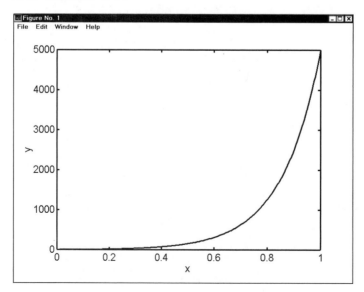

This plot has the characteristic shape we expect from an exponential.

If we plot the function with **SEMILOGY**, we get a straight line:

```
EDU» x=linspace(0,1,200);
EDU» y=5.*10.^(3.*x);
EDU» semilogy(x,y)
EDU» ylabel('y');
EDU» xlabel('x');
```

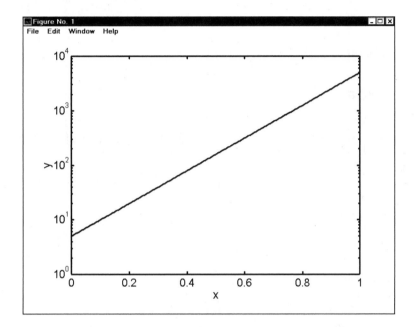

Since the y-axis is a log scale and the x-axis a linear scale, the exponential converts to a straight line.

Next, suppose both x and y have values that vary greatly. We'll let x vary from 0.001 to 100 and plot this on a linear-linear plot:

```
EDU» x=linspace(0.001,100,1000);
EDU» y=5.*10.^(3.*x);
EDU» plot(x,y)
EDU» ylabel('y');
EDU» xlabel('x');
```

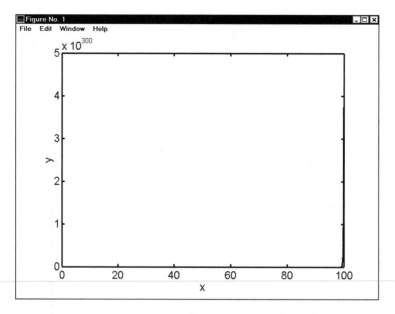

This plot isn't useful because y ranges from 5 up to 5×10^{300}. Most values of y are small compared to 5×10^{300} and are displayed as 0.

Now let's display the same equation and range of values with a **LOGLOG** plot.

```
EDU» x=logspace(-3,2,1000);
EDU» y=5.*10.^(3.*x);
EDU» loglog(x,y)
EDU» ylabel('y');
EDU» xlabel('x');
```

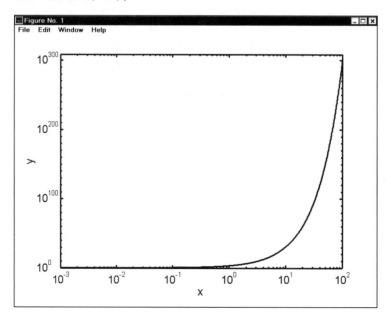

The shape is what we expect for an exponential. Thus, both linear-linear plots (the **PLOT** function) and log-log plots (the **LOGLOG** function) display data with their natural shape. Linear-linear plots are often used when numerical values for data have a small range and domain. Log-log plots are used when the x- and y-coordinates of the data vary by several orders of magnitude. In our example, x ranges from 10^{-3} to 10^{2} (5 decades) and y ranges from 10^{0} to 10^{300} (300 decades).

7.9.4 POLAR

A **POLAR** plot is used for data with polar coordinates. Let's plot a linear spiral, $r = 2\theta$. The MATLAB **POLAR** function has the syntax **polar(*theta*, *r*, *S*)**, where S is a text string that specifies the line style and has the same options as noted in the **PLOT** function (S is optional as discussed in the PLOT function).

```
EDU» theta=linspace(0,8*pi,200);
EDU» r=2*theta;
EDU» polar(theta,r)
```

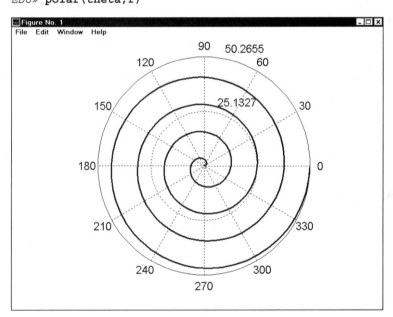

As a second example, we'll plot the logarithmic spiral $r = 5\log_{10}(\theta)$:

```
EDU» theta=linspace(1,20*pi,1000);
EDU» r=5*log10(theta);
EDU» polar(theta,r)
```

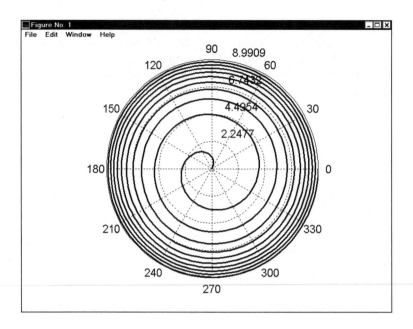

7.9.5 COMET

The **COMET** function displays a plot and shows its trajectory as an animation. As an example, we'll use our logarithmic spiral:

```
EDU» theta=linspace(1,20*pi,1000);
EDU» r=5*log10(theta);
EDU» x=r.*cos(theta);
EDU» y=r.*sin(theta);
EDU» comet(x,y);
EDU» title('Logarithmic Spiral');
```

You must run the code segment to view the real-time animation. Screen captures of the plot at three instances in time are shown here:

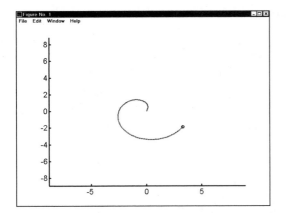

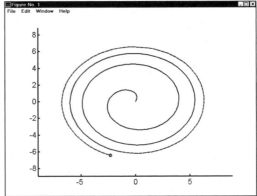

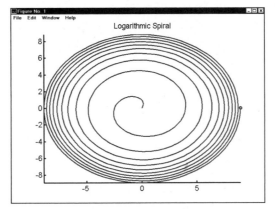

7.9.6 SUBPLOT

The **SUBPLOT** function splits a figure window into several small plots. Each plot is created separately, and can use any of the plotting commands used previously. The syntax of the function is **subplot(m, n, p)**. The function splits the figure into $m \times n$ plots with m rows and n columns. Variable p selects which subplot is active. The subplots are numbered from 1 to $m \times n$ with the upper-left subplot number 1. The numbers that identify plots increase from left to right and first row to last row. Thus, the lower-right plot is numbered $m \times n$.

Let's create a subplot with six plots, organized into two rows and three columns. In each subplot we'll place a different plot. We start with plot 1 active:

```
EDU» subplot(2,3,1)
EDU» theta=linspace(1,20*pi,1000);
EDU» r=5*log10(theta);
EDU» polar(theta,r)
EDU» title('Logarithmic Spiral')
```

The line **subplot(2,3,1)** creates a figure window and makes subplot 1 active. All of the following plot commands now apply to subplot 1. The code segment produces the following figure:

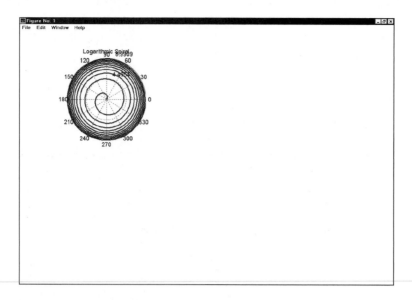

Next, we create a linear spiral on plot 2:

```
EDU» subplot(2,3,2)
EDU» theta=linspace(0,8*pi,200);
EDU» r=2*theta;
EDU» polar(theta,r)
EDU» title('Linear Spiral')
```

The line **subplot(2,3,2)** makes subplot 2 active. The plot commands following this line all apply to subplot number 2. The code segment produces this figure:

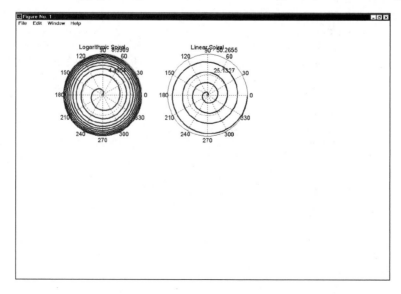

Next, we add the **LOGLOG** plot shown previously in subplot 3.

```
EDU» subplot(2,3,3)
EDU» x=logspace(-3,2,1000);
EDU» y=5.*10.^(3.*x);
EDU» loglog(x,y)
EDU» ylabel('y');
EDU» xlabel('x');
```

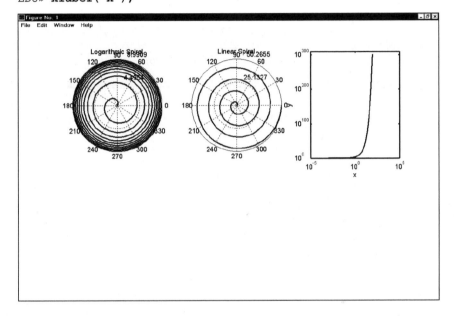

We fill in the remaining three subplots:

```
EDU» subplot(2,3,4)
EDU» x=linspace(0,1,200);
EDU» y=5.*10.^(3.*x);
EDU» semilogy(x,y)
EDU» ylabel('y');
EDU» xlabel('x');
EDU» subplot(2,3,5)
EDU» omega=logspace(1,4,200);
EDU» mag=20*log10(abs(1000./(i*omega+1000)));
EDU» semilogx(omega, mag);
EDU» axis([10,10000,-20 5]);
EDU» grid on;
EDU» ylabel('Gain(dB)');
EDU» xlabel('Frequency (rad/sec)');
EDU» title('Bode Magnitude plot of Vo/Vin');
EDU» subplot(2,3,6)
```

```
EDU» x1=linspace(0,2*pi,10)';
EDU» x2=linspace(0,2*pi,50)';
EDU» x3=linspace(pi,2*pi,100)';
EDU» x4=linspace(pi/3,2*pi,150)';
EDU» y1=sin(x1);
EDU» y2=cos(x2);
EDU» y3=abs(sin(x3));
EDU» y4=sin(x4).*sin(x4);
EDU» plot(x1,y1,'gs', x2,y2,'kp:', x3,y3, x4,y4,'r--');
EDU» legend('sin(x)', 'cos(x)', 'abs(sin(x))', '(sin(x)*sin(x)',3)
EDU» xlabel('x - Radians');
EDU» title('Various Trigonometric Functions');
```

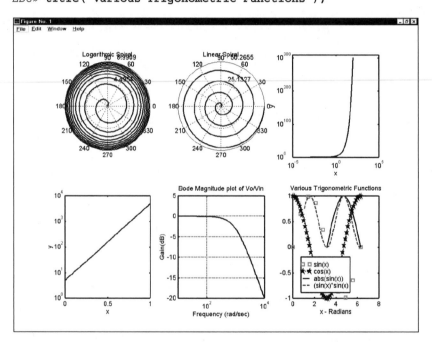

7.9.7   Working with Multiple Figures

Instead of placing multiple plots on a single figure using the **SUBPLOT** function, we can also open multiple-figure windows and create different plots on each window. The **FIGURE** command creates and selects windows. It has two forms, **h=figure**, and **figure(h)**.

When used in the form **h=figure**, a new blank figure window is created and *h* is a handle to "figure." A handle is a variable that is used to identify a figure, and *h* will have integer values starting at 1. Each time we use the **FIGURE** command, a new figure window is created, and the handle for that figure will be a unique integer.

If *h* is a valid handle, using **figure(h)** makes the figure identified by *h* the current figure and forces it to become visible by raising it above all other windows on the screen. If figure *h* does not exist (*h* is not a handle to a figure), and *h* is an integer, then **figure(h)** creates a new figure with handle *h*.

Let's open a few figure windows and create separate plots on the windows. When the following command was executed, no figure windows were open:

```
EDU» f1=figure
f1 =
     1
EDU»
```

As you can see, the handle for the first figure is the integer value 1. When we execute the command, a blank figure window opens:

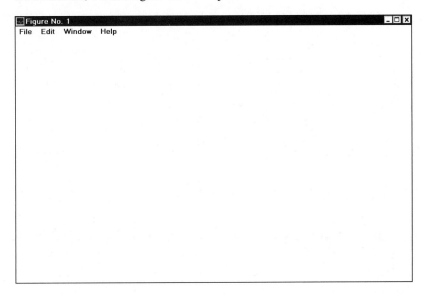

Let's open a second figure:

```
EDU» f2=figure
f2 =
     2
EDU»
```

Here the handle is the integer 2. All figure window handles are integers. When MATLAB creates the handles, the handles are integers that start at 1 and increase for each figure opened. When we executed the preceding command, the following window was created:

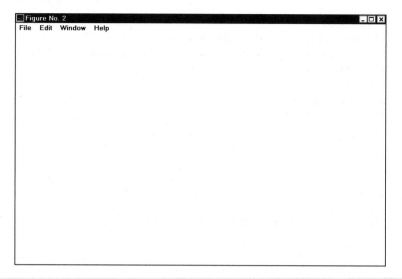

To create the next figure, we'll use the form **figure(h)**:

```
EDU» f3=figure(25)
f3 =
     25
EDU»
```

The handle is set to the number 25, and a new figure is opened and labeled Figure No. 25:

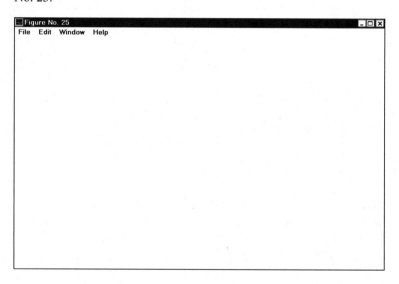

We now have three figures opened, and we'll add plots to all of them. The last figure command we used was **figure(25)**, so this figure window is the active figure. All plot commands will be added to the active figure window. On this figure window, we'll create two Bode plots. The code segment and plot are as follows:

```
EDU» subplot(2,1,1)
EDU» omega=logspace(0,5,200);
EDU» mag=20*log10(abs(1000./(i*omega+1000)));
EDU» semilogx(omega, mag);
EDU» axis([10,10000,-20 5]);
EDU» grid on;
EDU» ylabel('Gain(dB)');
EDU» xlabel('Frequency (rad/sec)');
EDU» title('Bode Magnitude plot of Vo/Vin');
EDU» subplot(2,1,2);
EDU» rad_to_deg = 360/(2*pi);
EDU» phase = rad_to_deg*angle(1000./(i*omega+1000));
EDU» semilogx(omega, phase);
EDU» grid on;
EDU» ylabel('Angle (Degrees)');
EDU» xlabel('Frequency (rad/sec)');
EDU» title('Bode Phase plot of Vo/Vin');
```

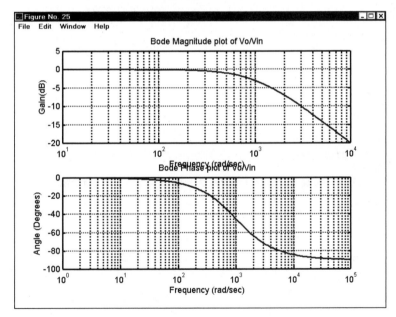

Next we make figure 1 the active figure:

```
EDU» figure(f1)
```

The window is brought to the top and is now the active figure window. All plotting commands will apply to this window. We'll create a spiral plot on this window:

```
EDU» theta=linspace(1,20*pi,1000);
EDU» r=5*log10(theta);
EDU» polar(theta,r)
```

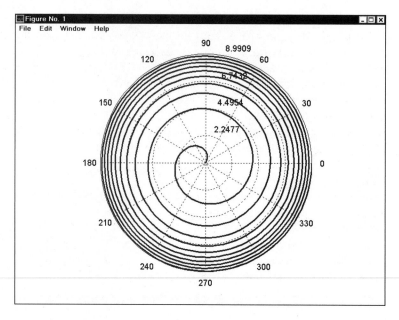

Lastly, let's create a semilog plot on the second window:

```
EDU» figure(f2)
EDU» x=linspace(0,100,200);
EDU» y=5.*10.^(3.*x);
EDU» semilogy(x,y)
EDU» ylabel('y');
EDU» xlabel('x');
```

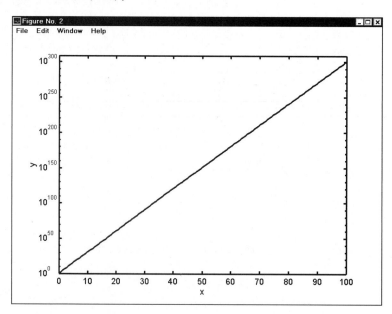

A few other commands can be used with multiple windows. The **SHG** command makes the active figure window the topmost window. The function **CLF** clears the active window. Neither of these functions makes a particular window active. They apply only to the active window. The **FIGURE** command also accomplishes the same thing as the **SHG** command. Using the function as **figure(h)** makes figure *h* the active figure and also makes the window the topmost window. Here is an example:

```
EDU» figure(25)
EDU» clf
```

The MATLAB **CLOSE** function closes windows. When used in the form **CLOSE(h)**, MATLAB closes the window whose handle is *h*. This method applies to any window in general, as well as figure windows. For example, the **CLOSE** function can close a message box. When used in the form **CLOSE**, it closes the active figure window. When used in the form **CLOSE ALL**, it closes all figure windows.

7.10 Handle Graphics for Manipulating Plots

MATLAB *Predefined*
figure function
plot function
semilog function
loglog function
polar function
xlabel function
ylabel function
title function
text function
gca function
set function
uisetfont function
uisetcolor function

We've already used handles in previous examples such as message boxes and figure windows. As noted earlier, a handle is a variable that contains a value that refers to a window or an item in a window. When you look at the MATLAB help information, many commands state that they return a handle to the object just created. If we save this handle, we can use it to refer to the object when we want to modify it.

7.10.1   Obtaining a Handle for an Object

Figure Handles

We have already used figure handles in Section 7.9.7 when we discussed the **FIGURE** command. When we use the command **h=figure**, a new empty window is opened and the handle to the figure is stored in variable h;

```
EDU» hf=figure
hf =
     1
EDU»
```

The value of this handle is an integer. Recall from Section 7.9.7 that all figure handles are integers starting at 1. The preceding command creates this empty window:

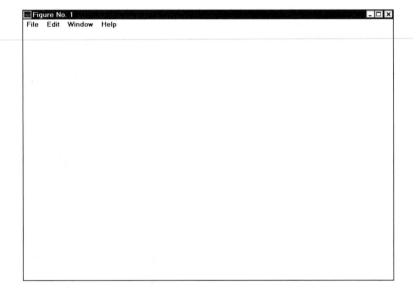

Handles from Plot Commands

All plot commands such as **PLOT**, **SEMILOG**, **LOGLOG**, and **POLAR** will return handles when used in the proper form, **h=plot (...)**. As an example, let's create a semilog plot in the window just opened and obtain the plot's handle:

```
EDU» omega=logspace(1,4,200);
EDU» mag=20*log10(abs(1000./(i*omega+1000)));
EDU» hp=semilogx(omega, mag)
hp =
    15.0007
```

```
EDU» axis([10,10000,-20 5]);
EDU» grid on;
```

The value of this handle is 15.0007, but the value of your handle may be different. We don't need to know the value of the handle, we just need to save the value in a variable so that we can use the handle later. The preceding code segment creates the following plot:

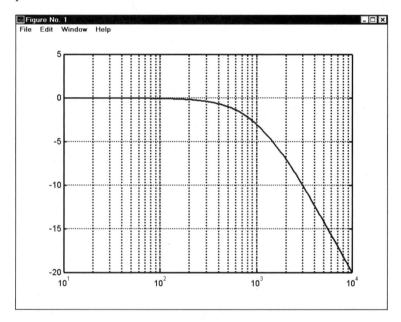

If we create a plot with multiple traces, the **PLOT** commands return a column array of handles:

```
EDU» x=linspace(1,2*pi,100);
EDU» y=sin(x);
EDU» z=cos(x);
EDU» zz=cos(x).*sin(x);
EDU» hp=plot(x,y,x,z,x,zz)
hp =
    17.0007
    18.0007
    19.0007
```

The code segment creates this plot:

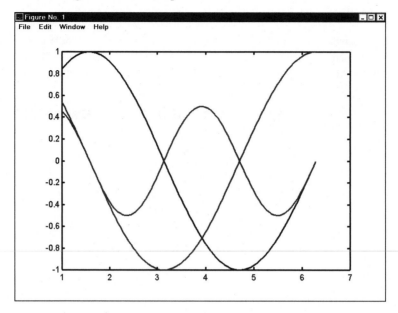

Each handle referes to a different trace on the plot.

Text Objects

The commands **XLABEL**, **YLABEL**, **TITLE**, and **TEXT** return handles to the objects they create. Now we'll create a title, *x*- and *y*- labels, and a text object on the preceding plot and save the handles to those objects. The numerical values of your handles may differ from those shown here:

```
EDU» hy=ylabel('Y-Axis -')
hy =
   20.0007
EDU» hx=xlabel('Time(Seconds)')
hx =
   21.0007
EDU» ht=title('Various Trigonometric Functions')
ht =
   22.0007
EDU» htext=text(4, 0.8, 'This is a Test.')
htext =
   23.0007
```

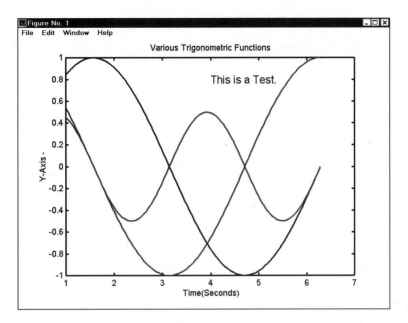

We don't really need to know the numeric values of these handles. However, if we wish to change any property of the objects, such as text, color, or size, we need to save the value of the handle in a variable.

Axes Handle (**GCA**)

The active figure window (and the only figure window we have open thus far in this example) is Figure No. 1. To get the handle to the axes of the active figure window, we use the **GCA** command in the form **h=gca**. In the command below, we save the handle in variable *h_axes:*

```
EDU» h_axes=gca;
```

7.10.2   Modifying an Object with the **SET** Command

Now that we've saved handles for some objects, we can modify those objects using the **SET** command. We can do this in two ways. **SET** can be used to find out what properties of an object can be changed and also to change the object's properties.

Obtaining the Properties of an Object

When used in the form set(*h*), where *h* is a valid handle, the **SET** command displays the available properties of the object. First, let's find the properties of a text object such as the title.

```
EDU» set(ht)
      Color
      EraseMode: [ {normal} | background | xor | none ]
      Editing: [ on | off ]
      FontAngle: [ {normal} | italic | oblique ]
      FontName
      FontSize
      FontUnits: [ inches | centimeters | normalized | {points} | pixels ]
      FontWeight: [ light | {normal} | demi | bold ]
      HorizontalAlignment: [ {left} | center | right ]
      Position
      Rotation
      String
      Units: [ inches | centimeters | normalized | points | pixels | {data} ]
      Interpreter: [ {tex} | none ]
      VerticalAlignment: [ top | cap | {middle} | baseline | bottom ]

      ButtonDownFcn
      Children
      Clipping: [ {on} | off ]
      CreateFcn
      DeleteFcn
      BusyAction: [ {queue} | cancel ]
      HandleVisibility: [ {on} | callback | off ]
      Interruptible: [ {on} | off ]
      Parent
      Selected: [ on | off ]
      SelectionHighlight: [ {on} | off ]
      Tag
      UserData
      Visible: [ {on} | off ]

EDU»
```

Some properties like **FontWeight** list their available options. Thus, the **FontWeight** property can have the options **light**, **normal**, **demi**, or **bold**. Properties like **Rotation** or **FontName** don't list choices because the number of choices for these properties isn't fixed. For **Rotation**, the property can be anywhere between 0 and 360 degrees. For **FontName**, the choices depend on what fonts are installed on your system.

Next, we'll look at the properties of a plot. Our last plot had three traces on it. We can only look at the properties of a single trace at a time. Let's display the properties of the first trace on the plot:

```
EDU» set(hp(1))
      Color
      EraseMode: [ {normal} | background | xor | none ]
```

```
LineStyle: [ {-} | — | : | -. | none ]
LineWidth
Marker: [ + | o | * | . | x | square | diamond | v | ^ | > ➥
| < | pentagram | hexagram | {none} ]
MarkerSize
MarkerEdgeColor: [ none | {auto} ] -or- a ColorSpec.
MarkerFaceColor: [ {none} | auto ] -or- a ColorSpec.
XData
YData
ZData

ButtonDownFcn
Children
Clipping: [ {on} | off ]
CreateFcn
DeleteFcn
BusyAction: [ {queue} | cancel ]
HandleVisibility: [ {on} | callback | off ]
Interruptible: [ {on} | off ]
Parent
Selected: [ on | off ]
SelectionHighlight: [ {on} | off ]
Tag
UserData
Visible: [ {on} | off ]
```

EDU»

As you can see, we can change properties such as **LineStyle**, **LineWidth**, **Marker** type, and **MarkerSize**.

Since each trace has its own handle, we can change the properties of individual traces. For instance, we can make the first trace **LineWidth 2**, the second trace **LineWidth 4**, and the third trace **LineWidth 5**, if we so desire.

Finally, let's look at the properties of the axes:

```
EDU» set(h_axes)
AmbientLightColor
Box: [ on | {off} ]
CameraPosition
CameraPositionMode: [ {auto} | manual ]
CameraTarget
CameraTargetMode: [ {auto} | manual ]
CameraUpVector
CameraUpVectorMode: [ {auto} | manual ]
CameraViewAngle
CameraViewAngleMode: [ {auto} | manual ]
```

```
CLim
CLimMode: [ {auto} | manual ]
Color
ColorOrder
DataAspectRatio
DataAspectRatioMode: [ {auto} | manual ]
DrawMode: [ {normal} | fast ]
FontAngle: [ {normal} | italic | oblique ]
FontName
FontSize
FontUnits: [ inches | centimeters | normalized | {points} | pixels ]
FontWeight: [ light | {normal} | demi | bold ]
GridLineStyle: [ - | — | {:} | -. | none ]
Layer: [ top | {bottom} ]
LineStyleOrder
LineWidth
MinorGridLineStyle: [ - | — | {:} | -. | none ]
NextPlot: [ add | {replace} | replacechildren ]
PlotBoxAspectRatio
PlotBoxAspectRatioMode: [ {auto} | manual ]
Projection: [ {orthographic} | perspective ]
Position
TickLength
TickDir: [ {in} | out ]
TickDirMode: [ {auto} | manual ]
Title
Units: [ inches | centimeters | {normalized} | points | pixels ]
View
XColor
XDir: [ {normal} | reverse ]
XGrid: [ on | {off} ]
XLabel
XAxisLocation: [ top | {bottom} ]
XLim
XLimMode: [ {auto} | manual ]
XScale: [ {linear} | log ]
XTick
XTickLabel
XTickLabelMode: [ {auto} | manual ]
XTickMode: [ {auto} | manual ]
YColor
YDir: [ {normal} | reverse ]
YGrid: [ on | {off} ]
YLabel
YAxisLocation: [ {left} | right ]
YLim
YLimMode: [ {auto} | manual ]
```

```
YScale: [ {linear} | log ]
YTick
YTickLabel
YTickLabelMode: [ {auto} | manual ]
YTickMode: [ {auto} | manual ]
ZColor
ZDir: [ {normal} | reverse ]
ZGrid: [ on | {off} ]
ZLabel
ZLim
ZLimMode: [ {auto} | manual ]
ZScale: [ {linear} | log ]
ZTick
ZTickLabel
ZTickLabelMode: [ {auto} | manual ]
ZTickMode: [ {auto} | manual ]

ButtonDownFcn
Children
Clipping: [ {on} | off ]
CreateFcn
DeleteFcn
BusyAction: [ {queue} | cancel ]
HandleVisibility: [ {on} | callback | off ]
Interruptible: [ {on} | off ]
Parent
Selected: [ on | off ]
SelectionHighlight: [ {on} | off ]
Tag
UserData
Visible: [ {on} | off ]

EDU»
```

In the next section we will show how to change these properties. After you complete the next section, you are encouraged to revisit this section to experiment with changing some of the properties.

Modifying Text Objects

Now that we know the names of some object properties, we can change their values. As an example, we'll change the properties of the title. Let's change the font to Courier New, the size to 24 points, and then make the text bold and rotate it by 90 degrees:

```
EDU» set(ht, 'FontName', 'Courier New');
EDU» set(ht,'FontSize', 24)
EDU» set(ht, 'Rotation', 90, 'FontWeight', 'bold')
```

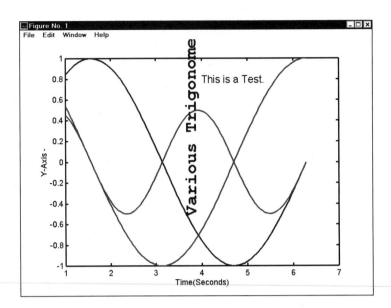

We can change the properties for any object for which we have a handle. If you use the **SET** command with the handles returned for the **XLABEL**, **YLABEL**, **TEXT**, and **GCA** commands, you'll see that they do have different properties, but all of them have the **FontName**, **FontSize**, and **Color** properties. Let's change those properties here:

```
EDU» set(ht,'Rotation',0, 'FontSize',18, 'Color', 'c')
EDU» set(hy,'FontName','Arial','FontSize',18, 'Color','r')
EDU» set(hx,'FontName','Arial','FontSize',18, 'Color','g')
EDU» set(htext,'FontName','Arial','FontSize',18, 'Color','m')
EDU» set(h_axes,'FontName','Arial','FontSize',14, 'Color','y')
```

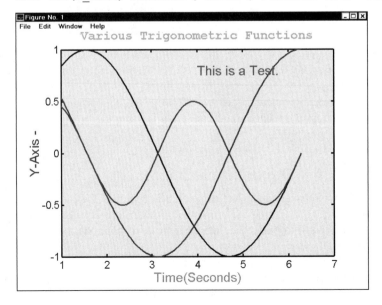

Note that changing the axes color actually changes the plot color. To change the color of the axes numbers, we use the **Ycolor** and **Xcolor** properties:

```
EDU» set(h_axes,'Ycolor','r', 'Xcolor','g')
```

Changing the Default Properties of Figures

You may have noticed that most figures in this text appear differently from what you obtain using the same MATLAB commands on your computer. All figures shown use the Arial font for text, the text size is fairly large, the trace line width is thick, and the axes numbers are large. This is because we have placed some **SET** commands in the startup.m file that set the default values for this object's properties.

File startup.m is a MATLAB file you place in directory C:\MATLAB\toolbox\local (assuming that MATLAB is installed on drive C:). If this file exists, and is located the local directory, it will run every time MATLAB is started. We've placed the following commands in this file:

```
set(0,'defaultlinelinewidth',2);
set(0,'defaultaxesfontname','Arial');
set(0,'defaultaxesfontsize',14);
set(0,'defaultaxeslinewidth',2)
set(0,'defaulttextfontsize',14);
set(0,'defaulttextfontname','Arial');
```

A handle of 0 is specified when we want to set a property's default value. The line **set(0,'defaultlinelinewidth',2);** sets the line width of all traces to 2 (normally it's 1). The commands

```
set(0,'defaultaxesfontname','Arial');
                    and
set(0,'defaultaxesfontsize',14);
```

set the font name and size of the axes numbers to 14-point Arial. The command **set(0,'defaultaxeslinewidth',2)** sets the line width of the axes to 2. The commands

```
set(0,'defaulttextfontsize',14);
                    and
set(0,'defaulttextfontname','Arial');
```

set the defaults for text placed on the figure using the **TEXT** command.

To obtain the list of default properties that you can change, use the command **set(0,'default')**:

```
EDU» set(0,'default')
  textFontName :
  textFontSize :
  axesLineWidth :
```

```
axesFontSize :
axesFontName :
lineLineWidth :
figureColormap :
axesColorOrder :
axesColor :
figureInvertHardcopy : [ {on} | off ]
figureColor :
lineColor :
surfaceEdgeColor : [ none | flat | interp ] -or- {a ColorSpec}.
patchEdgeColor : [ none | flat | interp ] -or- {a ColorSpec}.
patchFaceColor : [ none | flat | interp ] -or- {a ColorSpec}.
axesZColor :
axesYColor :
axesXColor :
textColor :
figurePosition :
EDU»
```

After these defaults are specified, every figure you create will use these values for the properties. However, you can still change the properties of individual figures using the **SET** command.

7.10.3 Modifying Fonts with **UISETFONT**

One problem with changing an object's font is that a lot of fonts may be installed on a system, and some font names are archaic. The **UISETFONT** command opens a window that allows you to specify an object's font. The format is **uisetfont(h, 'Message Box Title')**, where *h* is the handle of the object whose font you want to change. Let's change the font of the *y*-axis label in the previous example:

```
EDU» uisetfont(hy,'Specify the font for y-axis.')
```

When we issue the command, the following dialog box appears:

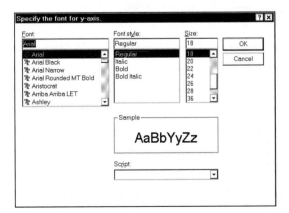

Select the properties you wish to change:

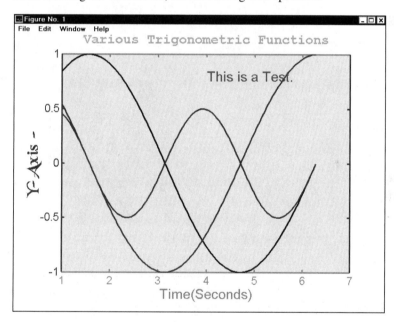

After clicking the **OK** button, the font changes as specified:

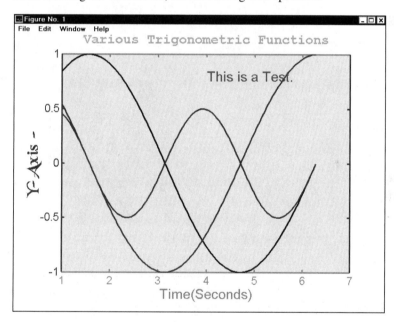

7.10.4 Modifying Colors with **UISETCOLOR**

The MATLAB command **UISETCOLOR** is similar to the **UISETFONT** command. The format is **UISETCOLOR**(h,'**Message Box Title**'). Where *h* is the handle of the object whose color you want to change. We'll change the color of the *y*-axis label in the previous example:

EDU» **uisetcolor(hy,'Specify the font color for y-axis.')**

When we issue the command, the following dialog box appears:

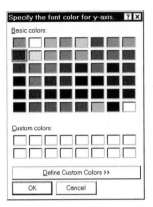

Select a color and click the **OK** button:

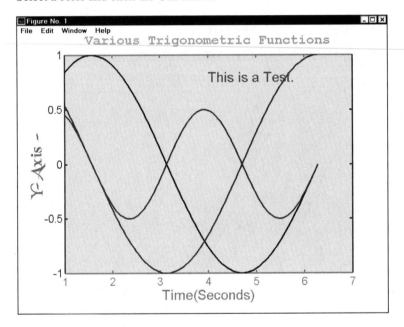

7.11 Problems

Problem 7-1 Write a MATLAB function that implements a command menu to facilitate plotting functions. The command menu is as follows:

1. Specify the name and path to a MATLAB m-file that implements an arbitrary function.
2. Specify the domain of the plot (the minimum and maximum values of the *x*-variable for the plot).
3. Plot the function.
4. Specify the **XLABEL**, **YLABEL**, and **TITLE** for the plot.
5. Toggle the grid on or off.

6. Zoom in on an area of the plot specified by the mouse.
7. Label the numerical values of a point on the plot. The point is specified by the mouse.
8. Close the current plot.
9. Print the current plot.
10. Exit.

These requirements must be placed on individual commands.

- Command 1 uses the function **UIGETFILE** to request a file with the extension .m.
- Command 3 uses the **PLOT** command.
- Commands 6 and 7 use message boxes to instruct the user how to select the points.
- Command 8 uses the **CLOSE** function.
- Command 9 uses the **PRINTDLG** function.
- Command 10 (exit) closes all open plots before exiting.

Problem 7-2 Add the following capabilities to the plotting command shell developed in Problem 7-1.

a. Find the minimum point of the function in the specified domain (specified in command 2 of Problem 7-1).
b. Find the maximum point of the function in the specified domain (specified in command 2 of Problem 7-1).
c. Find the integral of the function from $x = a$ to $x = b$ and display the result on the plot.

Here are some additional requirements for the above commands:

- For parts a and b, draw a line to the max or min point and display the message "Maximum at x=?, y=??" or "Minimum at x=?, y=??" where ? and ?? are the coordinates of the maximum or minimum point. The location of the text string should be chosen by the user with the mouse.
- For part c, use the MATLAB **QUAD8** function to find the integral. Place stars at points a and b on the plot and display the text message, "The integral from x=? to x=?? is ???" where ? and ?? are the two points, and ??? is the value of the integral. The location of the text string should be chosen by the user with the mouse.

Problem 7-3 Add the following capabilities to the plotting command shell developed in Problem 7-2.

d. For the given function, find x such that $f(x)$ is zero, where $f(x)$ is the function specified in command 1 of the plotting shell.
e. Allow the user to change the color and line width of the plot's trace.
f. Allow the user to specify the plot color.
g. Allow the user to change the title's color, font, and font size.
h. Allow the user to change the color, font, and font size of the y-axis label.
i. Allow the user to change the color, font, and font size of the x-axis label.
j. Exit—note that E is no longer exit.

Here are some additional requirements for the above commands:

- For item d, use the MATLAB function **FZERO** to find the solution. The user must select command 1 to specify the function before selecting command d. However, the user need not specify command 2 to select the domain of the function. If the user selects command d, the program must do the following:

(1) Ask the user for an initial guess for x.

(2) If the program cannot find a solution with the initial guess, it creates a message box that states that the zero cannot be found.

(3) If a solution is found, the program plots the function and draws a line indicating the location of the solution. When plotting the function, if command 2 of the plotting shell was chosen previously and the domain includes the solution $f(x) = 0$, the function is plotted for the domain specified in command 2. If the domain doesn't include the solution, then the function is plotted for a reasonable domain that includes the solution.

■ For items g, h, and i, use the **UISETFONT** and **UIGETFONT** functions.

Problem 7-4   Locate the directory on your system where wav files are stored (Windows comes with some built-in sounds that you can use). Write a script file to perform the following operations:

a. Read the wave file using the **WAVREAD** function.

b. Using the subplot function, create a window with two plots.

c. On the top plot, plot the sound signal versus time. You will have to create a time array for the sound. Note that the time can be created if you know the sampling frequency at which the sound was recorded.

d. On the bottom plot, create a histogram of the sound file with 20 bins.

Problem 7-5   The impedance of a parallel **RCL** circuit is $Z(\omega) = \dfrac{1}{\dfrac{1}{R} + j\omega C + \dfrac{1}{j\omega L}}$ where ω is the frequency in radians per second, R is the value of the resistor in ohms, C is the value of the capacitor in farads, and L is the value of the inductor in henrys. Create the following plots for $R = 100 \ \Omega$, $C = 1 \ \mu F$, and $L = 0.1 \ mH$:

a. Calculate $Z(\omega)$ for frequencies 1 kHz ≤ F ≤ 100 kHz. Note that $\omega = 2\pi F$, where F is the frequency in Hz.

b. Create a subplot with three plots in the same window.

c. In the top plot, plot the real part of $Z(\omega)$ versus ω as a dashed red line and the imaginary part of $Z(\omega)$ versus ω as a solid green line.

d. In the middle plot, plot the magnitude of $Z(\omega)$ versus ω as a dotted blue line.

e. In the bottom plot, plot the angle of $Z(\omega)$ in degrees versus ω as a yellow solid line.

f. Add a legend where needed. Give titles for all plots and label all axes appropriately.

MATLAB Applications

8

OBJECTIVES

- ☐ Use polynomials in MATLAB.
- ☐ Use polynomial and cubic Spline curve-fitting functions.
- ☐ Solve single-variable equations.
- ☐ Integrate single-variable functions.
- ☐ Solve first- and nth-order initial-value differential equations.
- ☐ Solve linear and nonlinear boundary-value differential equations.

In this chapter we solve several problems of interest to engineering and natural science students. Although the solution to these problems is covered the programming techniques are not, but we do use programming skills learned in earlier chapters. One strength of MATLAB is its numerous built-in functions for solving problems. Some of these functions require no programming, while others require new functions or short programs. For example, to solve a differential equation that changes, you need to write a function to implement the differential equation; the function must contain some control flow statements to implement the changes.

8.1 Polynomials in MATLAB

Polynomials in MATLAB are represented by row arrays. The elements of the row array represent the coefficients of the polynomial. The examples in Table 8-1 will make this clear.

TABLE 8-1 Examples of polynomials in MATLAB

Polynomial	MATLAB Representation
$x + 1$	[1 1]
$x - 1$	[1 −1]
$-x + 1$	[−1 1]
$x^2 + 3x + 7$	[1 3 7]
$x^2 + 3x$	[1 3 0]
$5x^2 + \sqrt{2}x - 1$	[5 1.414 −1]
$5x^4 + 3x^2 + x + 1$	[5 0 3 1 1]

Many functions in MATLAB require a polynomial as input and many functions return polynomials as output. You must know the form in which the polynomials are represented.

8.1.1   Evaluating Polynomials

Suppose we have a function $y = x^2 + 2x + 1$ and we want to evaluate the function for several values of x. In MATLAB, the polynomial is represented as [1 2 1]. How do we evaluate it for a specific value of x, say $x = 5$? We could use the following few lines of code:

```
EDU» poly=[1 2 1];
EDU» x=5;
EDU» y=poly(1)*x*x + poly(2)*x+poly(3)
y =
    36
```

To evaluate a polynomial, we must properly interpret the coefficients. Suppose the polynomial is of an arbitrary order:

$$a_1 x^n + a_2 x^{n-1} + a_3 x^{n-2} + \cdots + a_n x + a_{n+1}$$

In MATLAB, this polynomial is represented as $[a_1\ a_2\ a_3 \ldots a_n\ a_{n+1}]$. The length of the row array is $n + 1$. How do we evaluate this polynomial for an arbitrary value of x, say $x = 5$? We use this code:

```
poly=[1 2 -1 4 -5];
n=length(poly)-1;
x=5;
y=0;
for i = n:-1:1
  y=y+poly(n-i+1)*(x^i);
end
y = y+poly(n+1);
y
y =
   865
```

Duplicating this code whenever you need to evaluate a polynomial is tedious, so MATLAB provides the **POLYVAL** function for evaluating polynomials. The syntax of the **POLYVAL** function is y = polyval(*poly*, x) where *poly* is a row array that represents the polynomial and x is an array that contains the values for which you want to evaluate the function. Note that x can be a single value or an array of values.

```
EDU» poly=[1 2 -1 4 -5];
EDU» polyval(poly,5)
ans =
   865
EDU» polyval(poly,1)
ans =
    1
EDU» polyval(poly,-1)
ans =
  -11
EDU» polyval(poly,2.17)
ans =
   41.5815
```

We can evaluate the polynomial at all the individual points at the same time by creating a row array for *x*. **POLYVAL** returns an array of results:

```
EDU» x=[5, 1, -1, 2.17];
x =
    5.0000    1.0000   -1.0000    2.1700
EDU» polyval(poly,x)
ans =
  865.0000    1.0000  -11.0000   41.5815
```

8.1.2   Plotting Polynomials

Now that you know how to represent polynomials, you can easily plot them. Let's plot the function $y = (x - 1)(x + 1)$ for values of *x* from –2 to 2. The first thing we do is re-write the function as the polynomial $y = x^2 + 0x - 1$. MATLAB represents this polynomial as the array [1 0 –1]. We can now plot the function:

```
EDU» poly=[1 0 -1];
EDU» x=linspace(-2,2,100);
EDU» y=polyval(poly,x);
EDU» plot(x,y);
EDU» grid
```

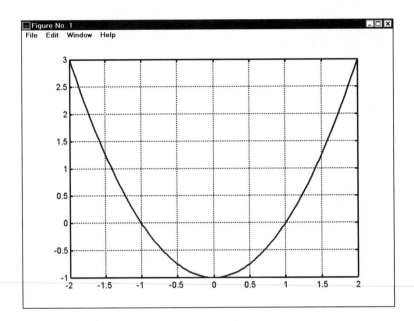

For our next example, let's plot the function $y = (x - 1)(x + 1)(x - 2)$. We first convert the function to a polynomial. The question is, how can we get this polynomial easily? Note that this third-order polynomial is expressed as the product of three first-order polynomials, $y = p_1 * p_2 * p_3$ where $p_1 = x - 1$, $p_2 = x + 1$, and $p_3 = x - 2$. In MATLAB, $p_1 = [1 -1]$, $p_2 = [1\ 1]$, and $p_3 = [1 -2]$. The MATLAB function **CONV** multiplies two polynomials and returns the resultant polynomial as a row array. For example, $p_1 * p_2 = (x - 1)(x + 1) = x^2 + 0x - 1$. In MATLAB, the result is returned as $[1\ 0 -1]$. Let's multiply the polynomials using MATLAB:

```
EDU» p1=[1 -1];
EDU» p2=[1 1];
EDU» conv(p1,p2)
ans =
     1     0    -1
```

We can now find the polynomial for $y = (x - 1)(x + 1)(x - 2)$.

```
EDU» p1=[1 -1];
EDU» p2=[1 1];
EDU» poly_x=conv(p1,p2);
EDU» p3=[1 -2];
EDU» poly=conv(p3,poly_x);
EDU» poly
poly =
     1    -2    -1     2
```

We obtain the same result with this code:

```
EDU» p1=[1 -1];
EDU» p2=[1 1];
EDU» p3=[1 -2];
EDU» poly=conv(p3,conv(p1, p2));
EDU» poly
poly =
     1    -2    -1     2
```

This result tells us that $y = (x - 1)(x + 1)(x - 2) = x^3 - 2x^2 - x + 2$. Now let's evaluate and plot the polynomial.

```
EDU» x=linspace(-1.5,2.5,100);
EDU» p1=[1 -1];
EDU» p2=[1 1];
EDU» p3=[1 -2];
EDU» poly=conv(p3,conv(p1, p2));
EDU» y=polyval(poly,x);
EDU» plot(x,y);
EDU» grid;
```

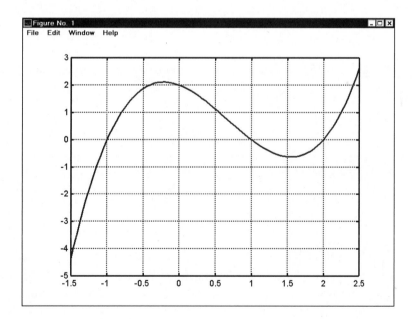

8.1.3 Roots of Polynomials

MATLAB provides a function called **ROOTS**, which finds the solution x to the equation $y = f(x) = 0$ when $f(x)$ is a polynomial. In the last example, we had

$y = (x - 1)(x + 1)(x - 2) = x^3 - 2x^2 - x + 2$. For this polynomial, function **ROOTS** finds the solution to the equation $x^3 - 2x^2 - x + 2 = 0$. In this example, we know there are three solutions $x = 2$, $x = 1$, and $x = -1$. Let's use the **ROOTS** function to find the solution:

```
EDU» poly=[1 -2 -1 2];
EDU» roots(poly)
ans =
    2.0000
    1.0000
   -1.0000
```

The **ROOTS** function returns a column array that contains the solution to the equation, also known as the roots of the equations. In this case, the roots are $x = 2$, $x = 1$, and $x = -1$. We know the answer to the previous polynomial, how about finding the roots of the equation $y = 4x^4 - 22x^3 + 28x^2 + 18x - 36$.

```
EDU» p=[4 -22 28 18 -36];
EDU» roots(p)
ans =
    3.0000
    2.0000
    1.5000
   -1.0000
```

The **ROOTS** function also finds the complex roots of a polynomial. Let's solve $y = x^2 + x + 1$.

```
EDU» roots([1 1 1])
ans =
  -0.5000 + 0.8660i
  -0.5000 - 0.8660i
```

This result says that the solution to $x^2 + x + 1 = 0$ is $x = -0.5 + 0.8660i$ and $x = -0.5 - 0.8660i$.

8.2 Curve Fitting

In many scientific applications, we will often measure data (input versus output for example) for which we will need to find a polynomial to describe the data. We may use the polynomial either to predict results or to determine a mathematical equation that describes the physics of our experiment. Suppose we have the data shown in the following plot:

```
EDU» FID=fopen('ran_data.bin','r');
EDU» [input, count] = fread(FID, [1 100], 'double');
```

```
EDU» [output, count] = fread(FID, [1 100], 'double');
EDU» status=fclose(FID);
EDU» plot(input,output, '+k')
EDU» grid
```

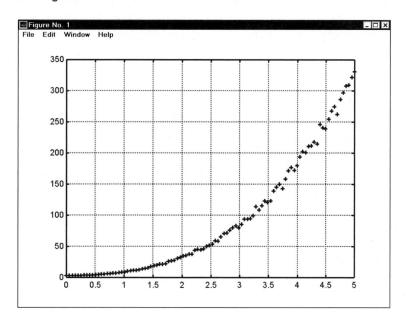

Let's use this data to curve fit a polynomial and a Spline.

8.2.1 Polynomial Curve Fitting

The MATLAB function **POLYFIT** fits an *n*th-order polynomial to data using a least squares minimization technique. For the data shown previously, we'll find a polynomial equation that best describes the data. Function **POLYFIT** has the syntax *poly* = polyfit(*x, y, N*) where *x* and *y* are the data, and *N* is the order of the polynomial we'll fit to the data. Let's try fitting first-, second-, third-, and fourth-order polynomials to data.

Let's start with the first-order polynomial. We'll use the polyfit function to find the best first-order fit for the given data, and then we'll plot the data and the polynomial on the same plot. In the following code segment, the line set(gca, 'PlotBoxAspectRatio',[1,0.7,1]) reduces the size of the plot in the *y*-dimension to 70% of the original size. This allows more room for the plot's title. Note that if you try to duplicate the next four plots on your computer, the plots shown here are screen captures of maximized windows on a monitor with 1152 × 864 resolution. You may need to change your aspect ratio slightly to get the title text to display properly on your computer.

```
EDU» poly1=polyfit(input, output, 1);
EDU» x=linspace(0,5,100);
```

```
EDU» y=polyval(poly1,x);
EDU» plot(input, output, '+k', x, y);
EDU» str=sprintf('Straight Line Fit\ny=%gx %g',poly1(1),poly1(2));
EDU» set(gca,'PlotBoxAspectRatio',[1,0.7,1])
EDU» title(str)
EDU» poly1
poly1 =
    60.7081   -58.8280
```

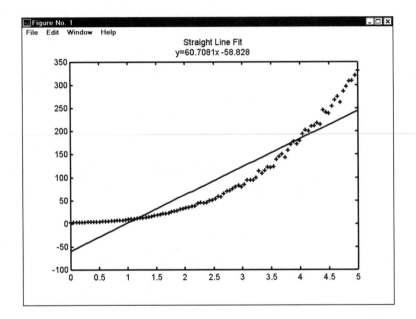

A straight-line fit, which is a first-order fit, doesn't appear to be a good fit for this data. The coefficients of the first-order polynomial are [60.7081 –58.8280]. Therefore, the equation of the line is $y = 60.7081x - 58.8280$.

Let's try a second-order fit. In the following **SPRINTF** statement, the character following the ^ displays as a superscript (the ^ character is not displayed).

```
EDU» poly2=polyfit(input, output, 2);
EDU» x=linspace(0,5,100);
EDU» y=polyval(poly2,x);
EDU» plot(input, output, '+k', x, y);
EDU» str=sprintf('Second Order Curve Fit\ny=%gx^2 %gx +
%g',poly2(1),poly2(2),poly2(3));
EDU» set(gca,'PlotBoxAspectRatio',[1,0.7,1])
EDU» title(str)
```

```
EDU» poly2
poly2 =
  17.8950   -28.7669   14.9813
```

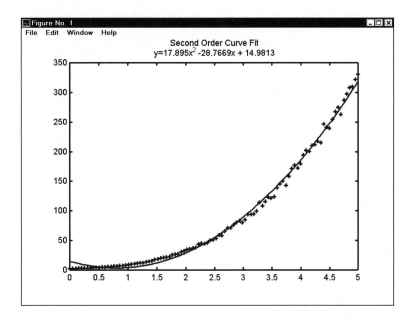

A second-order polynomial is a much better fit to our data. The coefficients of the second-order polynomial are [17.8950 –28.7669 14.9813]. Therefore, the equation of the second-order polynomial line is $y = 17.8950x^2 - 28.7669x + 14.9813$.

Next, we try a third-order fit:

```
EDU» poly3=polyfit(input, output, 3);
EDU» x=linspace(0,5,100);
EDU» y=polyval(poly3,x);
EDU» plot(input, output, '+k', x, y);
EDU» str=sprintf('Third Order Curve Fit\ny=%gx^3 + %gx^2 + ➥
%gx +  %g',poly3(1),poly3(2),poly3(3),poly3(4));
EDU» set(gca,'PlotBoxAspectRatio',[1,0.7,1])
EDU» title(str)
EDU» poly3
poly3 =
    1.9350    3.3822    0.1132    3.2513
```

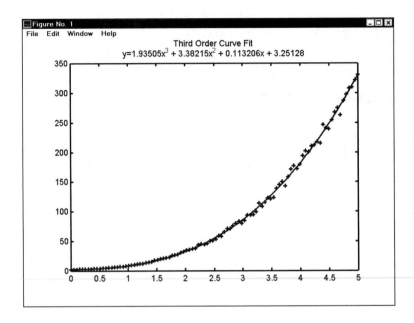

The coefficients of the third-order polynomial are [1.9350 3.3822 0.1132 3.2513]. Therefore, the equation of the third-order polynomial is $y = 1.9350x^3 + 3.3822x^2 + 0.1132x + 3.2513$.

And finally the fourth-order fit:

```
EDU» poly4=polyfit(input, output, 4);
EDU» x=linspace(0,5,100);
EDU» y=polyval(poly4,x);
EDU» plot(input, output, '+k', x, y);
EDU» str1='Fourth Order Curve Fit';
EDU» str2=sprintf('\ny=%gx^4 + %gx^3 %gx^2 + %gx + %g', ➡
poly4(1),poly4(2),poly4(3),poly4(4),poly4(4));
EDU» set(gca,'PlotBoxAspectRatio',[1,0.7,1])
EDU» title([str1,str2])
EDU» grid
EDU» poly4
poly4 =
    -0.1589    3.5245    -1.7100    5.7054    1.9165
```

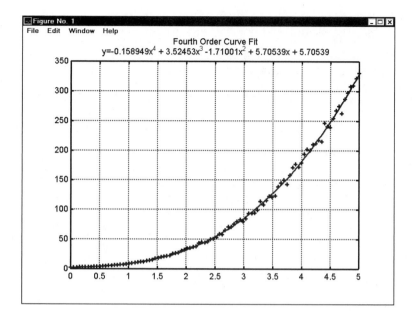

The coefficients of the fourth-order polynomial are [−0.1589 3.5245 −1.7100 5.7054 1.9165]. Therefore, the equation for the fourth-order fit is $y = −0.1589x^4 + 3.5245x^3 − 1.7100x^2 + 5.7054x + 1.9165$.

Typically you will use a polynomial curve fit to accommodate several trials of the same experiment—that is, when you've measured the output for the same input several times. Typically, your output will be slightly different each trial. Some experimental data is shown in Table 8-2.

TABLE 8-2 Experimental data

Input	Output	Input	Output	Input	Output
0	0.05	2	2.2	4	4.05
0	0.1	2	2.05	5	5.6
0	0.08	3	3.1	5	4.9
1	1.15	3	3.05	5	5.1
1	1.13	3	2.95	6	6.1
1	.95	4	4.2	6	6.03
2	2.1	4	4.2	6	6.12

Note that this data gives different output values for the same input. That is, for an input of 0, the output was measured three times, at 0.05, 0.01, and 0.08. This data is typical of what you might measure in an experiment. To increase your confidence in a measurement or experimental procedure, you should always perform the same experiment several times. You'll find that each measurement has some experimental error and yields slightly different results.

Let's plot the data in Table 8-2 with MATLAB:

```
EDU» x=[0 0 0 1 1 1 2 2 2 3 3 3 4 4 4 5 5 5 6 6 6];
EDU» y=[ .05 .1 0.08 1.15 1.13 0.95 2.1 2.2 2.05 3.1 3.05 2.95 ↦
4.2 4.2 4.05 5.6 4.9 5.1 6.1 6.03 6.12];
EDU» plot(x,y,'k*');
EDU» grid;
```

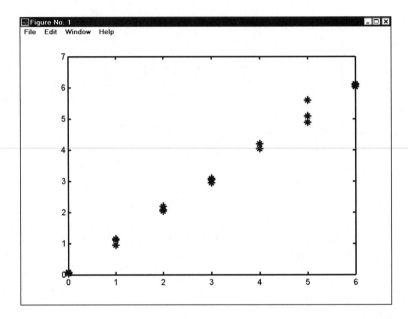

As you can see, for the same value of x, y has several different values.

We can now use a first-order polynomial curve fit to find the straight line that best describes the data. We'll plot the original data and the best-fit straight line on the same plot:

```
EDU» x=[0 0 0 1 1 1 2 2 2 3 3 3 4 4 4 5 5 5 6 6 6];
EDU» y=[ .05 .1 0.08 1.15 1.13 0.95 2.1 2.2 2.05 3.1 3.05 2.95 ↦
4.2 4.2 4.05 5.6 4.9 5.1 6.1 6.03 6.12];
EDU» line=polyfit(x, y, 1);
EDU» x_hat=linspace(0,6,100);
EDU» y_hat=polyval(line, x_hat);
EDU» plot(x, y, 'k*', x_hat, y_hat);
```

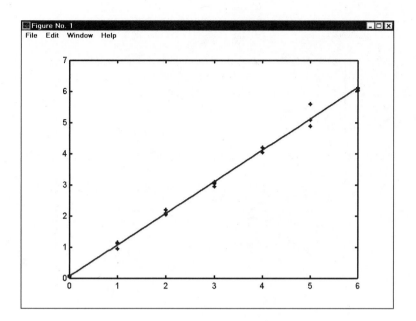

Note that the line doesn't go through every point of the data. This line minimizes the distance between all points and the line. A straight line can't go through all of the points, but a line can be found that minimizes the distance between all points and the line.

We can now estimate an output value for an input that wasn't measured. For example, what is reasonable output for an input of 2.5, which we didn't measure? We can predict a value using the best-fit line we generated from the polynomial curve fit. The predicted output value for input of 2.5 is

```
EDU» polyval(line,2.5)
ans =
    2.5999
```

Variable *line* is the polynomial coefficients generated from the curve fit of the previous example. We can also how this point on the plot:

```
EDU» xp=2.5;
EDU» yp=polyval(line,xp);
EDU» plot(x, y, 'kX', x_hat, y_hat, xp,yp,'ro')
```

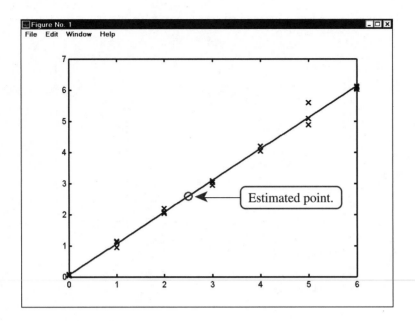

The fitted point at input 2.5 is marked by a red circle.

8.2.2 Cubic Spline

A cubic Spline is another method of fitting a curve to data. A Spline creates a line that goes through all the data points. Thus, a Spline can't be used if you have several output measurements for each input. Thus, a Spline doesn't work for the data of Table 8-2.

The syntax of the **SPLINE** function in MATLAB is

$$y = \text{spline}(in\_data, \ out\_data, \ x)$$

where *in_data* and *out_data* are the measured data, *x* is an input for which you wish to predict output, and *y* is the predicted output. Here's how a Spline works:

```
EDU» in=[1 2 3 4 5];
EDU» out=[1 3.3 1.9 4 6.7];
EDU» x=linspace(1,5,100);
EDU» y=spline(in, out, x);
EDU» plot(in, out, 'kX', x, y)
```

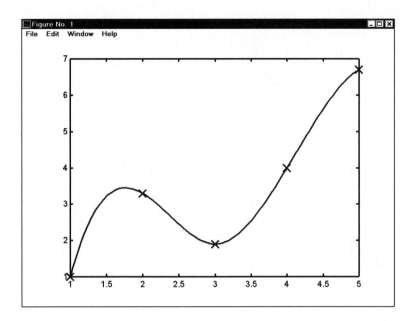

The original data is marked on the plot with black ×'s and the Spline is plotted as a continuous line. The Spline goes through all points. We can now ask the question, what can we expect for output if our input is 2.5? The **SPLINE** function gives this answer:

```
EDU» y=spline(in, out, 2.5)
y =
    2.4547
```

Let's create a plot to show the original data, the Spline fit, and the data point we just calculated. The estimated point is marked with a red circle.

```
EDU» in=[1 2 3 4 5];
EDU» out=[1 3.3 1.9 4 6.7];
EDU» x=linspace(1,5,100);
EDU» y=spline(in, out, x);
EDU» yp=spline(in, out, 2.5);
EDU» plot(in, out, 'kX', x, y, 2.5, yp, 'ro')
```

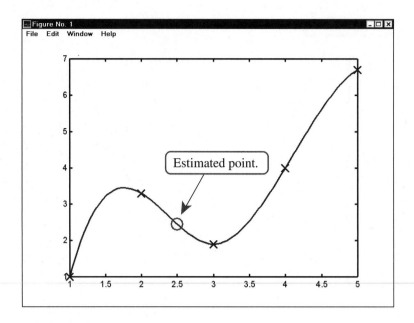

We can also use a Spline to fit a curve to the data from the polynomial curve fits in Section 8.2.1 on pages 435–439. Although the data has many points, there is only one output for each input, so a Spline will work. Since the output measurements have inputs from 0 to 5, the **SPLINE** function can only predict outputs for inputs in this range. It can't be used for inputs outside this range. We will now plot the input data used in Section 8.2.1 and the fitted outputs from the **SPLINE** function. The input data is plotted with black + signs. The output data is plotted as a line.

```
EDU» FID=fopen('d:\EXAMPLES\ran_data.bin','r');
EDU» [input, count] = fread(FID, [1 100], 'double');
EDU» [output, count] = fread(FID, [1 100], 'double');
EDU» status=fclose(FID);
EDU» x=linspace(0,5,100);
EDU» spline_out=spline(input, output, x);
EDU» plot(input, output, '+k', x, spline_out);
EDU» grid
```

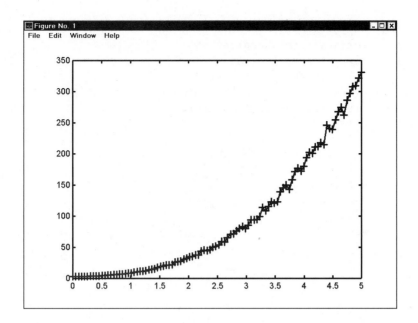

Note that the Spline goes through every point. Typically a Spline is used for much sparser data than the data in this example. **For large amounts of data, or where several output values exist for the same input, a polynomial curve fit should be used.**

Let's compare curve fits between a Spline and third-order polynomial for the two data sets we used in this section.

```
EDU» subplot(2,1,1)
EDU» FID=fopen('d:\EXAMPLES\ran_data.bin','r');
EDU» [input, count] = fread(FID, [1 100], 'double');
EDU» [output, count] = fread(FID, [1 100], 'double');
EDU» status=fclose(FID);
EDU» x=linspace(0,5,100);
EDU» spline_out=spline(input, output, x);
EDU» yp=spline(input, output,2.5);
EDU» h1=plot(input, output, '+k', x, spline_out, 2.5, yp, 'ro');
EDU» set(h1(3),'markersize',15);
EDU» text_str=sprintf('At x = 2.5, y = %g',yp);
EDU» h=text(2,200, text_str);
EDU» set(h, 'FontSize', 15, 'FontName', 'Arial');
EDU» title('Spline Fit');
EDU» grid
EDU» subplot(2,1,2)
EDU» poly=polyfit(input,output,3);
EDU» poly_out=polyval(poly,x);
EDU» yp=polyval(poly,2.5);
```

```
EDU» h2=plot(input, output, '+k', x, poly_out, 2.5, yp, 'ro');
EDU» set(h2(3),'markersize',15);
EDU» text_str=sprintf('At x = 2.5, y = %g',yp);
EDU» h=text(2,200, text_str);
EDU» set(h, 'FontSize', 15, 'FontName', 'Arial');
EDU» title('3rd Order Polynomial Fit');
EDU» grid
```

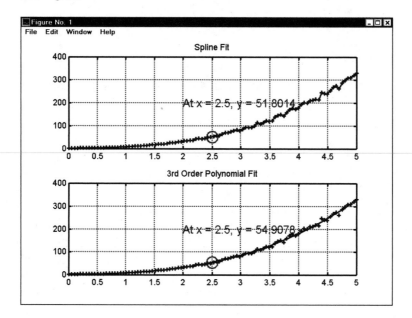

As you can see, the two curve fits are different, and they predict different output values for an input of $x = 2.5$. For a large amount of data, the polynomial curve fit appears to be more appropriate.

Now let's compare the two methods for sparse data.

```
EDU» subplot(2,1,1)
EDU» input=[1 2 3 4 5];
EDU» output=[1 3.3 1.9 4 6.7];
EDU» x=linspace(1,5,100);
EDU» spline_out=spline(input, output, x);
EDU» yp=spline(input, output,2.5);
EDU» h1=plot(input, output, '+r', x, spline_out, 2.5, yp, 'ro');
EDU» set(h1,'markersize',15);
EDU» text_str=sprintf('At x = 2.5, y = %g',yp);
EDU» h=text(2,5, text_str);
EDU» set(h, 'FontSize', 15, 'FontName', 'Arial');
EDU» title('Spline Fit');
EDU» grid
```

```
EDU» subplot(2,1,2)
EDU» poly=polyfit(input,output,3);
EDU» poly_out=polyval(poly,x);
EDU» yp=polyval(poly,2.5);
EDU» h2=plot(input, output, '+r', x, poly_out, 2.5, yp, 'ro');
EDU» set(h2,'markersize',15);
EDU» text_str=sprintf('At x = 2.5, y = %g',yp);
EDU» h=text(2,5, text_str);
EDU» set(h, 'FontSize', 15, 'FontName', 'Arial');
EDU» title('3rd Order Polynomial Fit');
EDU» grid
```

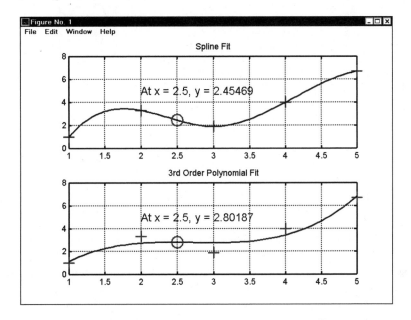

The third-order polynomial curve doesn't go through any points, while the Spline fit goes through all of them. For this data, it appears that the Spline fit is a better choice. However, a higher-order polynomial fit might be better. Let's try a polynomial curve fit for a fifth-order polynomial.

```
EDU» input=[1 2 3 4 5];
EDU» output=[1 3.3 1.9 4 6.7];
EDU» x=linspace(1,5,100);
EDU» poly=polyfit(input,output,5);
EDU» poly_out=polyval(poly,x);
EDU» yp=polyval(poly,2.5);
EDU» h1=plot(input, output, '+r', x, poly_out, 2.5, yp, 'ro');
EDU» set(h1,'markersize',15)
EDU» text_str=sprintf('At x = 2.5, y = %g',yp);
```

```
EDU» h=text(2,4.5, text_str);
EDU» set(h, 'FontSize', 15, 'FontName', 'Arial');
EDU» title('5th Order Polynomial Fit');
EDU» grid
```

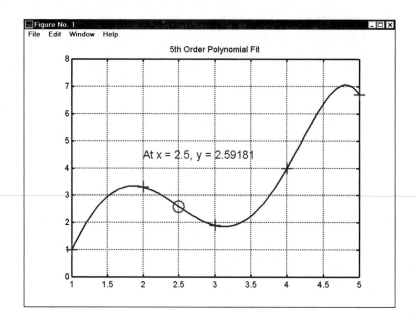

We see that with five data points, the fifth-order polynomial fit goes through all the data points, which demonstrates a general characteristic of curve-fitting polynomials: For n data points, an nth-order polynomial fit will go through all of the data points. However, for large amounts of data, it isn't usually a good idea to use a higher-order polynomial that goes through all the points because it will have large oscillations. Lower-order polynomials are best for curve fitting.

Now let's check the Spline fit; it produces a different curve than the fifth-order polynomial fit:

```
EDU» input=[1 2 3 4 5];
EDU» output=[1 3.3 1.9 4 6.7];
EDU» x=linspace(1,5,100);
EDU» poly=polyfit(input,output,5);
EDU» poly_out=polyval(poly,x);
EDU» spline_out=spline(input, output, x);
EDU» yp=spline(input, output,2.5);
EDU» h=plot(input, output, '+r', x, poly_out, x, spline_out);
EDU» set(h,'markersize',15)
EDU» legend('Data', '5th Order Poly Fit', 'Spline Fit',2)
EDU» grid
```

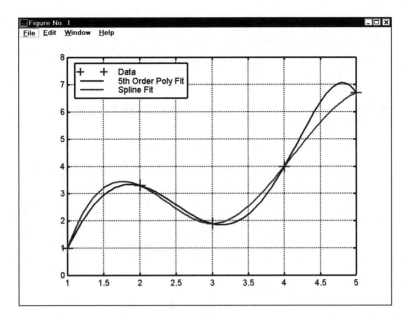

From these examples, you can see that you should base your choice of curve-fitting method (Spline or polynomial) and the polynomial degree for your curve fit on the data and the particular requirements of your application. Always compare different fits; this will help you decide which one bests suits your needs.

8.3 Solving Equations

The MATLAB function **FZERO** finds the roots of an equation. Its syntax is

$$\textbf{xact = fzero(fname, x)}$$

Variable *fname* is a text string that contains the name of a MATLAB function that implements the equation to be solved. Variable *x* is our initial guess of the *x*-value where the function may go to zero. Variable *xact* is the value of *x* where the function is zero within a specified tolerance.

Suppose we want to solve the problem $x^3 + 3x = \ln(x)$. We need to find the value of *x* that makes this equation true. This equation can be restated as $y = x^3 + 3x - \ln(x)$ where we find *x* such that $y = 0$, or $x^3 + 3x - \ln(x) = 0$. An equation of this form can be solved with the **FZERO** function. Thus, we can solve any single-variable equation if we rewrite the equation so that we're solving for zero.

Let's solve the equation $(x - 3)(x + 5) = 5$. First, we rewrite the function as $y = (x - 3)(x + 5) - 5$, then we write Function 8-1, which implements this function.

FUNCTION 8-1 First **FZERO** function

```
function y=fx2(x);
% This is file fx2.m
y=(x-3).*(x+5)-5;
```

We can now use the **FZERO** function to find the roots. We make an initial guess as to where we think a root is. Let's try $x = 4$:

```
EDU» format long
EDU» fzero('fx2', 4)
ans =
    3.58257569495584
```

We check this answer by placing the result into our function:

```
EDU» fx2(3.58257569495584)
ans =
    2.664535259100376e-015
```

Our result is something times 10^{-15}, which is a number very close to zero.

Next, we try the root closest to $x = -5$:

```
EDU» fzero('fx2', -5)
ans =
   -5.58257569495584
```

If we evaluate **fx2** at this point, we will see that the function is very close to zero.

Now that we know how to use the **FZERO** function, you can use it to solve more complicated problems. Let's solve the equation $x^3 + 3x + 5 = \sin(x)$. We first rewrite the equation as $y = x^3 + 3x + 5 - \sin(x)$, then define Function 8-2, which implements the equation.

FUNCTION 8-2 Second **FZERO** function

```
function y=fx1(x);
% This is file fx1.m
y=(x.^3) + 3*x+5 - sin(x);
```

Because we have no idea what this function looks like, to come up with a guess for x where the function is close to zero, we plot the function:

```
EDU» x=linspace(-5,5,100);
EDU» y=fx1(x);
EDU» plot(x, y);
EDU» grid;
```

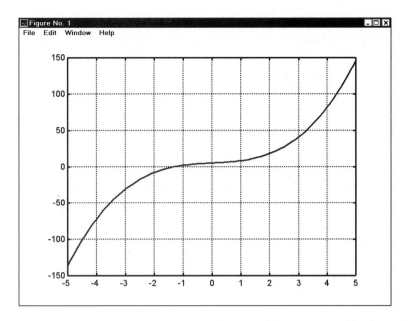

There appears to be a zero close to $x = 0$, so we tell the **FZERO** function to find the zero closest to $x = 0$.

```
EDU» format long
EDU» fzero('fx1', 0)
ans =
  -1.28273076924844
```

FZERO tells us that for $x = -1.28273076924844$, function $fx1(x)$ is such a small number that we can call it zero.

```
EDU» fx1(-1.28273076924844)
ans =
  -7.105427357601001e-015
```

8.4 Numerical Integration

MATLAB functions **QUAD** and **QUAD8** find numerical values of $\int_A^B f(x)\,dx$. Function **QUAD** uses an adaptive-recursive Simpson's rule, and **QUAD8** uses an adaptive-recursive Newton-Cotes eight-panel rule. Both methods are recursive in that $\int_A^B f(x)\,dx$ can be written as

$$\int_A^B f(x)\,dx = \int_A^{\frac{A+B}{2}} f(x)\,dx + \int_{\frac{A+B}{2}}^B f(x)\,dx$$

Thus, if you call function **QUAD** or **QUAD8** for $\int_A^B f(x)\,dx$, it calls itself twice to calculate

$$\int_A^{\frac{A+B}{2}} f(x)\,dx \qquad \text{and} \qquad \int_{\frac{A+B}{2}}^B f(x)\,dx$$

Then, when the function calls itself to calculate $\int_A^{\frac{A+B}{2}} f(x)\,dx$, it will split this integral into two portions and calculate

$$\int_A^{\frac{A+B}{2}} f(x)\,dx \quad \text{as} \quad \int_A^{A+\frac{A+B}{2}\over 2} f(x)\,dx + \int_{A+\frac{A+B}{2}\over 2}^{\frac{A+B}{2}} f(x)\,dx$$

These functions keep splitting the interval in half and calling themselves to calculate the integral of smaller and smaller intervals, until the interval is small enough to calculate the integral of the subinterval within a specified tolerance.

The syntax of the **QUAD** and **QUAD8** functions is

$$\textbf{area = quad8(\textit{fname, A, B, Tol})}$$

Variable *fname* is a text string that contains the name of the MATLAB function you want to integrate. As with the **FZERO** function, to integrate a function $f(x)$, you must create a MATLAB function that implements $f(x)$. Variables A and B are scalars that contain the interval of integration. Variable *Tol* specifies the accuracy of your result. If you don't specify a value for *Tol*, the default tolerance sets to 0.001. Note that if you make *Tol* too small or the interval from A to B too large, functions **QUAD** and **QUAD8** will reach their recursion limit and announce that there may be a discontinuity in the function you are integrating.

Let's calculate the integral from 0 to 1 of $f(x) = x^2 + 2x + 1$. First we create Function 8-3, which implements $f(x)$.

FUNCTION 8-3 Quadratic function for integration

```
function y=quadfunc(x);
% This is file name quadfunc.m
y=x.*(x+2)+1
```

Note that we wrote the function as $x*(x + 2) +1$ rather than $x*x + 2*x + 1$ because the latter method requires two multiplications while Function 8-3 requires only one.

The theoretical value of the integration is

$$area = \int_0^1 \left(x^2 + 2x + 1\right) dx = \left[\frac{x^3}{3} + x^2 + x\right]_0^1 = 2.3\overline{33}$$

Now let's calculate the integral with MATLAB:

```
EDU» format long
EDU» area=quad('quadfunc', 0, 1)
area =
    2.33333333333333
```

This use of the **QUAD** function omits the *Tol* variable, so the tolerance is set to the default value 0.001.

Numerical integration is most useful for finding integrals of functions that can't be integrated mathematically. An example would be $\int_{-\infty}^{\infty} e^{-x^2/2}\, dx$. Note that integration tables don't have an integral listed for this function, so we'll solve it numerically. First we define Function 8-4 to implement $e^{-x^2/2}$.

FUNCTION 8-4   Exponential function

```
function y=expsq(x);
% This is file expsq.m
y=exp(-0.5*x.*x);
```

Before we integrate any function, we should have an idea of what it looks like, so let's generate a plot of this function from –4 to 4.

```
EDU» x=linspace(-4,4,100);
EDU» y=expsq(x);
EDU» plot(x,y, 'LineWidth',2);
EDU» set(gca,'LineWidth',2,'FontName','Arial', 'FontSize',15)
```

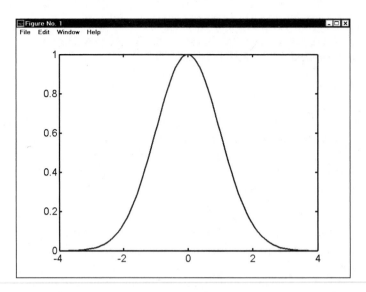

As you can see, most of the area is contained in the interval –3 to 3. Now that we know what it looks like, we can choose the interval of integration better.

Obviously, we can't calculate the integral from –∞ to +∞. However, we can calculate the integral from –A to +A, where A is a positive number. We can make A larger and larger and see how the result changes. If the integral is finite, as we make A larger, the result of the integral will converge on a value.

```
EDU» format long
EDU» area1=quad('expsq', -1, 1)
area1 =
    1.71125209288731
EDU» area1=quad('expsq', -2, 2)
area1 =
    2.39257528001565
EDU» area1=quad('expsq', -3, 3)
area1 =
    2.49993506967274
EDU» area1=quad('expsq', -4, 4)
area1 =
    2.50647235786976
EDU» area1=quad('expsq', -5, 5)
area1 =
    2.50662669677366
```

The result is converging on a number near 2.506. Note that if we make the interval of integration too large, the **QUAD** function will incorrectly determine that the integrand has a singularity:

```
EDU» area1=quad('expsq', -9, 9)
Warning: Recursion level limit reached in quad.  Singularity
likely.
```

```
> In d:\MATLAB\toolbox\MATLAB\funfun\quad.m (quadstp) at line 103
  In d:\MATLAB\toolbox\MATLAB\funfun\quad.m (quadstp) at line 141
  In d:\MATLAB\toolbox\MATLAB\funfun\quad.m (quadstp) at line 141
  In d:\MATLAB\toolbox\MATLAB\funfun\quad.m (quadstp) at line 141
  In d:\MATLAB\toolbox\MATLAB\funfun\quad.m (quadstp) at line 141
  In d:\MATLAB\toolbox\MATLAB\funfun\quad.m (quadstp) at line 141
  In d:\MATLAB\toolbox\MATLAB\funfun\quad.m (quadstp) at line 141
  In d:\MATLAB\toolbox\MATLAB\funfun\quad.m (quadstp) at line 141
  In d:\MATLAB\toolbox\MATLAB\funfun\quad.m (quadstp) at line 141
  In d:\MATLAB\toolbox\MATLAB\funfun\quad.m (quadstp) at line 141
  In d:\MATLAB\toolbox\MATLAB\funfun\quad.m (quadstp) at line 141
  In d:\MATLAB\toolbox\MATLAB\funfun\quad.m at line 73

Warning: Recursion level limit reached 100 times.
> In d:\MATLAB\toolbox\MATLAB\funfun\quad.m at line 81

area1 =
  2.50662712099876
EDU»
```

The function $e^{-x^2/2}$ does not contain a singularity, so the warning message is incorrect. The message is provided to handle integration of functions with a singularity, such as $\tan(x)$:

```
EDU» x=linspace(1,2,100);
EDU» plot(x,tan(x),'LineWidth',2);
EDU» set(gca,'LineWidth',2,'FontName','Arial', 'FontSize',15)
EDU» axis([1,2,-40,40]);
```

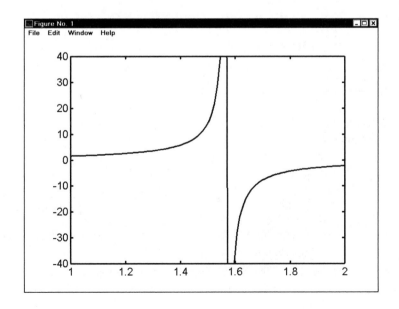

This function has a singularity at $\pi/2$. To integrate this function near the singularity, we would need to take very small slices to get an accurate answer. To reduce the slice size, functions **QUAD8** and **QUAD** call themselves recursively. For a rapidly changing function, such as $\tan(x)$, such small slices are required that a preset recursion limit is reached and the **QUAD** functions emit a message that there may be a singularity in the function.

Reaching the recursion limit indicates that the function has a singularity. However, the recursion limit can also be reached by specifying integration limits that are too large, such as we saw with the integration of $e^{-x^2/2}$.

Let's see what happens when we integrate the $\tan(x)$ function near the singularity. We'll integrate the function from $0.9\left(\dfrac{\pi}{2}\right)$ to $\dfrac{\pi}{2}$:

```
EDU» quad('tan',0.9*pi/2, pi/2)
Warning: Recursion level limit reached in quad.  Singularity
likely.
> In d:\MATLAB\toolbox\MATLAB\funfun\quad.m (quadstp) at line 103
  In d:\MATLAB\toolbox\MATLAB\funfun\quad.m (quadstp) at line 141
  In d:\MATLAB\toolbox\MATLAB\funfun\quad.m (quadstp) at line 141
  In d:\MATLAB\toolbox\MATLAB\funfun\quad.m (quadstp) at line 141
  In d:\MATLAB\toolbox\MATLAB\funfun\quad.m (quadstp) at line 141
  In d:\MATLAB\toolbox\MATLAB\funfun\quad.m (quadstp) at line 142
  In d:\MATLAB\toolbox\MATLAB\funfun\quad.m (quadstp) at line 142
  In d:\MATLAB\toolbox\MATLAB\funfun\quad.m (quadstp) at line 142
  In d:\MATLAB\toolbox\MATLAB\funfun\quad.m (quadstp) at line 142
  In d:\MATLAB\toolbox\MATLAB\funfun\quad.m (quadstp) at line 142
  In d:\MATLAB\toolbox\MATLAB\funfun\quad.m (quadstp) at line 142
  In d:\MATLAB\toolbox\MATLAB\funfun\quad.m at line 73
Warning: Recursion level limit reached 16 times.
> In d:\MATLAB\toolbox\MATLAB\funfun\quad.m at line 81
ans =
    4.175439133640823e+011
EDU»
```

Here the recursion limit is reached, and the answer for the area is a very large number. The recursion limit is useful in integrating functions with singularities. However, the warning message can also be triggered when integrating well-behaved functions if too large of an interval is used.

Recall from Section 3.6 that a recursive function is much slower than a non-recursive function. Functions **QUAD8** and **QUAD** use an adaptive-recursive algorithm that adjusts its slice size to calculate the integral according to the function it is integrating. For rapidly changing functions, a small slice size is used, and for slowly varying functions, larger slices are used. The slices are as large as possible to reduce the number of calculations, while still maintaining a specified tolerance. The down side of the recursive algorithm is that the resulting overhead needed for MATLAB to implement recursion slows down the calculation much more than the larger slices save.

Function 8-5 implements Simpson's rule; it splits the integration interval into n slices, where n is specified by the user.

FUNCTION 8-5   Integration function using Simpson's rule

```
function area = simpson(f_name,a,b,n)
% This is file simpson.m
% function area = simpson(f_name,a,b,n)
% This function uses Simpson's composite integration method
% to find the integral of the function specified by f_name from
% a to b with n sub intervals.
%
% Written by Marc E. Herniter, Electrical Engineering Department,
% College of Engineering and Technology, Northern Arizona University.

if (n/2) ~= floor(n/2)
    n = n + 1;   % make sure n is even.
end
h=(b-a)/n;
h2 = 2*h;

% add f(a) and f(b) to the area sum.
area = feval(f_name,a) + feval(f_name,b);

% Add up the odd terms.
x=a-h;
even_sum = 0;
odd_sum = 0;
for i = 1:2:n-1
    x=x+h2;
    odd_sum = odd_sum + feval(f_name,x);
end

% Add up the even terms.
x=a;
for i = 2:2:(n-2)
    x = x+h2;
    even_sum = even_sum + feval(f_name,x);
end
area = (h/3)*(area + 4*odd_sum + 2*even_sum);
```

We can now calculate the integral of $e^{-x^2/2}$ for different values of n (the number of slices) and see how long it takes to calculate the integral. First let's see how many slices we need to achieve an accurate result:

```
EDU» simpson('expsq', -8, 8, 10)    ◄——————— 10 slices
ans =
    2.75199846036978
```

```
EDU» simpson('expsq', -8, 8, 50)     ←——————— 50 slices
ans =
    2.50662827463100
EDU» simpson('expsq', -8, 8, 100)    ←——————— 100 slices
ans =
    2.50662827463100
EDU» simpson('expsq', -8, 8, 1000)   ←——————— 1000 slices
ans =
    2.50662827463100
```

As you can see, the results for $n = 50$, $n = 100$, and $n = 1000$ are exactly the same. Now let's compare how long it takes the **QUAD** and Simpson functions to run.

```
EDU» format long
EDU» tic
EDU» quad('expsq', -8, 8)
EDU» toc
ans =
    2.50662821804725
elapsed_time =
    1.59000000000000
```

The recursive function took 1.59 seconds on the author's system to calculate the integral to an accuracy of 0.001 (the default tolerance). Next, let's time the Simpson function with 100 slices.

```
EDU» tic
EDU» simpson('expsq', -8, 8, 100)
EDU» toc
ans =
    2.50662827463100
elapsed_time =
    0.11000000000000
```

The Simpson function took 0.11 second to run while the recursive **QUAD** function took nearly 1.59 seconds. Also, the answer produced by the **QUAD** function is not as accurate as that of the Simpson function. This is because we didn't specify a tolerance to the **QUAD** function, so it used the default tolerance of 0.001. We can specify a tighter tolerance, but we'll probably hit the **QUAD** function's recursion limit. Once again, the recursive function takes much longer to run than the nonrecursive function.

The problem with our Simpson function is that it doesn't find an answer to a specified tolerance. How can we create a function that uses the Simpson function but finds the answer to a specified tolerance? One method is calculate the integral twice, once for n slices and once for $2n$ slices. The solution for $2n$ slices will be more accurate than that for n slices. The difference between the two answers is an estimation of the error for n

slices. That is, an approximation of the error for the integral calculated with n slices is the absolute value of Simpson(n) – Simpson($2n$). Here's a sample solution:

```
Tol = 1e-10;
n=10;
error = 1;
area1 = simpson('expsq', -8, 8, n);
while error > Tol
   n=n*2;
   area2=simpson('expsq', -8, 8, n);
   error=abs(area1-area2);
   area1=area2;
end;
area2
area2 =
   2.50662827463100
n
n =
    80
```

Note that the answer given is the calculation for *area2* (calculated with n slices). We don't really know the error associated with this calculation. The error we calculated was for *area1* (calculated with $n/2$ slices). However, since *area2* uses n slices and *area1* uses $n/2$ slices, *area2* is obviously more accurate than *area1*. The **WHILE** loop finds *area1* such that the error associated with *area1* is less than *Tol*. Since *area2* is more accurate, its error is also less than the specified tolerance, so *area2* is the result displayed for the answer.

8.5 Differential Equations

MATLAB has functions to solve both initial-value and boundary-value problems. Several functions solve an nth-order initial-value problem of the form

$$\frac{dy^n (t)}{dt} = f(t,\ y(t),\ \dot{y}(t),\ \ddot{y}(t),\ \dddot{y}(t), \ldots)$$

with the initial conditions $y(t_0) = y_0$, $\dot{y}(t_0) = \dot{y}_0$, $\ddot{y}(t_0) = \ddot{y}_0$, $\dddot{y}(t_0) = \dddot{y}_0$, and so on. Other functions also solve both linear and nonlinear second-order boundary-value problems. A linear boundary-value problem is of the form

$$\frac{d^2y}{dx^2} - p(x)\ \frac{dy}{dx} - q(x)y - r(x) = 0$$

with the boundary conditions $y(a) = \alpha$ and $y(b) = \beta$. We are only placing constraints on the value of y at the endpoints. Here's an example of a nonlinear boundary-value problem:

$$\frac{d^2y}{dx^2} + \left(\frac{dy}{dt}\right)^2 + y - \ln(x) = 0$$

with the boundary conditions $y(1) = 0$ and $y(2) = \ln(2)$.

8.5.1 Initial-Value Problems

We will split our study of initial-value problems into first-order and nth-order problems. The same MATLAB functions are used for both types of problems. If you can use MATLAB to solve a first-order problem, you'll also be able to use it for nth-order problems.

First-Order Problems

Let's start by solving a simple first-order equation $dy/dt = -y + t + 1$ with the initial condition $y(0) = 1$. Note that the differential equation is of the form $dy/dt = f(t, y)$, where $f(t, y) = -y + t + 1$.

Before we use MATLAB, note that there is an analytic solution to this equation:

$$y(t) = t + e^{-t} \qquad \text{for } 0 \leq t \leq 1$$

We'll compare the analytic solution to the numerical solution generated by MATLAB.

The first thing we do is create Function 8-6, which calculates the derivative. We'll call this function iv1.m.

FUNCTION 8-6 Function to calculate the derivatives for a first-order initial-value problem

```
function ydot=iv1(t,y)
% This is function iv1.m
ydot=-y+t+1;
```

MATLAB provides several functions for solving initial-value problems. We'll look at two of them, **ODE23** and **ODE45**. Function **ODE23** uses second- and third-order Runge-Kutta methods. Recall that higher-order methods are more accurate than lower-order methods. Function **ODE23** calculates the solution using both second- and third-order methods. Since the third-order method's result is more accurate, the difference between the two results is an estimate of the error for the second-order method. Using this estimate, **ODE23** determines how accurate its calculations are, and adjusts the integration step size if the error is not within a specified tolerance. Function **ODE45** works the same as **ODE23**, except that it uses fourth- and fifth-order Runge-Kutta methods.

Both functions use a variable step size. Thus, when finding the solution from time t_1 to time t_2, we must calculate the solution at intermediate points. The time between two simulation points is called the step size. The smaller the step size, the more accurate the result. However, for a smaller step size, more intermediate results are calculated, and the simulation takes longer to complete. Functions **ODE23** and **ODE45** use as large a step size as possible while maintaining the accuracy below a specified tolerance.

As the simulation proceeds, the step size may have to be decreased to maintain the solution below the tolerance. If the step size is reduced a predefined number of times, the functions will indicate that there may be a singularity and end the simulation.

The syntax of these functions are

$$[t, \ y] \ = \ \text{ODE23}(F, \ [t_0, \ t_{\text{final}}], \ Y_0)$$
$$[t, \ y] \ = \ \text{ODE45}(F, \ [t_0, \ t_{\text{final}}], \ Y_0)$$

Variable F is a text string that contains the name of the MATLAB function that calculates the derivatives. For our example, $F = \texttt{'iv1'}$. Variable t_0 and t_{final} contain the starting and ending times for the simulation. Variable y_0 is the initial condition, $y(t_0) = y_0$.

The results are returned in variables t and y. Variable t contains the time coordinate of the points, and variable y the y-coordinates. We then plot the values (y versus t).

We can now simulate our function from $t = 0$ to $t = 1$:

```
EDU» yo=1;
EDU» [t,y]=ode45('iv1', [0,1],yo);
EDU» plot(t,y)
EDU» xlabel('Time(s)')
EDU» ylabel('Y(t)')
EDU» title('First Order Initial Value Problem')
```

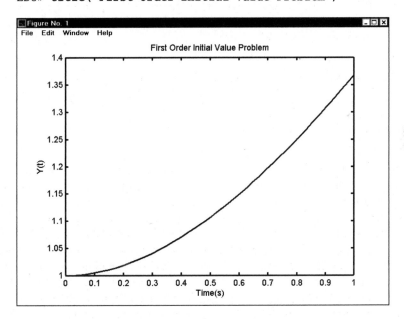

Let's compare the results of the numerical and theoretical solutions.

```
EDU» Yact=t+exp(-t); %theoretical solution
EDU» Error=abs(y-Yact);
EDU» Error'
```

```
ans =
  1.0e-008 *
  Columns 1 through 7
     0        0.3946   0.5704   0.2345   0.0297   0.3280   0.4878
  Columns 8 through 14
     0.1846   0.0538   0.2706   0.4158   0.1421   0.0730   0.2211
  Columns 15 through 21
     0.3531   0.1060   0.0881   0.1786   0.2985   0.0754   0.0997
...(not a complete listing)
```

The error is on the order of 10^{-9}.

As a second example, let's simulate a more difficult problem. In this example, the differential equation changes with time, and at a certain value of the output, so it illustrates why you need to know how to program in MATLAB to solve some problems:

$$\frac{dV_o(t)}{dt} = \begin{cases} 0 & \text{if} \quad t < 10 \times 10^{-6} \text{ s} \\ \frac{K_L}{C}\left(V_{DD} - V_o(t) - V_{TL}\right)^2 - \frac{K_D}{C}\left(V_{in} - V_{TD}\right)^2 & \text{if} \quad V_o(t) > 3 \text{ and } t \ge 10 \times 10^{-6} \text{ s} \\ \frac{K_L}{C}\left(V_{DD} - V_o(t) - V_{TL}\right)^2 - \frac{K_D}{C}\left[2\left(V_{in} - V_{TD}\right)V_o(t) - \left(V_o(t)\right)^2\right] & \text{if} \quad V_o(t) \le 3 \text{ and } t \ge 10 \times 10^{-6} \text{ s} \end{cases}$$

where K_L, K_D, C, V_{DD}, V_{TL}, and V_{TD} are constants. Note that the differential equation changes once at $t = 10$ μs and again when $V_o \le 3$. The initial condition for this equation is $V_o(0) = 7$. Function 8-7 calculates the derivative.

FUNCTION 8-7 Function to calculate the derivatives in which the differential equation changes with time and voltage

```
function Vo_dot=iv2(t,Vo)
%This is file iv2.m
% Define the constants for the differential equation:

KL=20e-6; KD=100e-6; VTL=3; VTD=2; VDD=10; Vin=5; C=1e-9;

% Calculate the derivative:

if t < 10e-6
   Vo_dot=0;
else
   if Vo>3
        Vo_dot=(KL/C).*(VDD-Vo-VTL).^2 - (KD/C).*((Vin-VTD).^2);
   else
        Vo_dot=(KL/C).*(VDD-Vo-VTL).^2 - (KD/C).*(2.*(Vin-VTD).*Vo -Vo.*Vo);
   end
end
```

We see that we can easily use MATLAB to describe this differential equation. Let's simulate the function from 0 to 40 μs with an initial condition of $V_o(0) = 7$.

```
EDU» [t,Vo]=ode45('iv2', [0, 40e-6], 7);
EDU» plot(t,Vo);
EDU» axis([0 40e-6, 0, 8]);
EDU» xlabel('Time');
EDU» ylabel('Voltage');
EDU» grid on
```

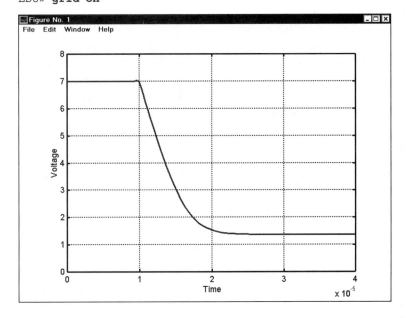

nth-Order Problems

Solving nth-order differential equations with MATLAB is the same as solving first-order equations if we write the higher-order equation as a system of n first-order differential equations. As an example, we'll convert a second-order differential equation into a system of two first-order equations:

$$\frac{d^2x}{dt^2} + 3\frac{dx}{dt} + x = 5$$

First, let $x_1 = x$; the differential equation becomes

$$\frac{d^2x_1}{dt^2} + 3\frac{dx_1}{dt} + x_1 = 5$$

We rewrite the equation as

$$\frac{d^2x_1}{dt^2} = 5 - 3\frac{dx_1}{dt} - x_1 \qquad \textbf{Eq. (8-1)}$$

Next, let

$$\frac{dx_1}{dt} = x_2 \qquad\qquad \textbf{Eq. (8-2)}$$

This is the first equation in our system. Next, since $dx_1/dt = x_2$, we can calculate dx_2/dt as

$$\frac{dx_2}{dt} = \frac{d^2x_1}{dt^2}$$

We substitute this equation into Eq. (8-1) and obtain

$$\frac{dx_2}{dt} = 5 - 3\frac{dx_1}{dt} - x_1$$

From Eq. (8-2), we know that $dx_1/dt = x_2$. Substituting this equation into the above equation gives the second equation:

$$\frac{dx_2}{dt} = 5 - 3x_2 - x_1 \qquad\qquad \textbf{Eq. (8-3)}$$

Thus, the second-order equation

$$\frac{d^2x}{dt^2} + 3\frac{dx}{dt} + x = 5$$

can be written as a system of two first-order equations:

$$\frac{dx_1}{dt} = x_2 \qquad \text{and} \qquad \frac{dx_2}{dt} = 5 - 3x_2 - x_1$$

where $x_1(t)$ in the system equals $x(t)$ in the original second-order differential equation.

As a second example, we'll rewrite a third-order differential equation as a system of three first-order equations. We start with this third-order equation:

$$\frac{d^3y}{dt^3} + 3\frac{d^2y}{dt^2} + 5\frac{dy}{dt} + 7y = 6$$

We let $x_1 = y$, and our equation becomes

$$\frac{d^3x_1}{dt^3} + 3\frac{d^2x_1}{dt^2} + 5\frac{dx_1}{dt} + 7x_1 = 6$$

Solving for the third derivative yields

$$\frac{d^3x_1}{dt^3} = -3\frac{d^2x_1}{dt^2} - 5\frac{dx_1}{dt} - 7x_1 + 6 \qquad\qquad \textbf{Eq. (8-4)}$$

Next, we let

$$\frac{dx_1}{dt} = x_2 \qquad\qquad \textbf{Eq. (8-5)}$$

This is the first equation in our system of equations. Taking the time derivatives of x_2, we get

$$\frac{dx_2}{dt} = \frac{d^2 x_1}{dt^2} \qquad \text{and} \qquad \frac{d^2 x_2}{dt^2} = \frac{d^3 x_1}{dt^3} \qquad \textbf{Eq. (8-6)}$$

Next, let

$$\frac{dx_2}{dt} = x_3$$

This is the second equation in our system of equations. From Eq. (8-6) we know that $dx_2/dt = d^2x_1/dt^2$. Thus

$$x_3 = \frac{d^2 x_1}{dt^2} \qquad\qquad \textbf{Eq. (8-7)}$$

Lastly, if we take the time derivative of x_3, we get

$$\frac{dx_3}{dt} = \frac{d^3 x_1}{dt^3} \qquad\qquad \textbf{Eq. (8-8)}$$

We can now substitute Eqs. (8-8), (8-7), and (8-5) into Eq. (8-4):

$$\frac{d^3 x_1}{dt^3} = -3\frac{d^2 x_1}{dt^2} - 5\frac{dx_1}{dt} - 7x_1 + 6 \qquad \textbf{Eq. (8-4)}$$

Making the substitutions $dx_3/dt = d^3x_1/dt^3$, $x_3 = d^2x_1/dt^2$, and $x_2 = dx_1/dt$, we get

$$\frac{dx_3}{dt} = -3x_3 - 5x_2 - 7x_1 + 6$$

This is one of our first-order equations. The remaining equations have already been specified:

$$\frac{dx_2}{dt} = x_3 \qquad \text{and} \qquad \frac{dx_1}{dt} = x_2$$

Thus, the third-order differential equation

$$\frac{d^3 y}{dt^3} + 3\frac{d^2 y}{dt^2} + 5\frac{dy}{dt} + 7y = 6$$

can be rewritten as a third-order system of equations:

$$\frac{dx_1}{dt} \quad = \quad x_2$$

$$\frac{dx_2}{dt} \quad = \quad x_3$$

$$\frac{dx_3}{dt} \quad = \quad -3x_3 - 5x_2 - 7x_1 + 6$$

where $y(t) = x_1(t)$. The original third-order equation has three initial conditions: $y(t_0)$, $dy/dt(t_0)$, and $d^2y/dt^2(t_0)$. The third-order system has the initial conditions, $x_1(t_0)$, $x_2(t_0)$, and $x_3(t_0)$, where

$$y(t_0) \quad = \quad x_1(t_0)$$

$$\frac{dy}{dt}(t_0) \quad = \quad x_2(t_0)$$

$$\frac{d^2y}{dt^2}(t_0) \quad = \quad x_3(t_0)$$

The method presented lets us take an nth-order differential equation of the form

$$\frac{d^n y}{dt^n} + k_1 \frac{d^{n-1}y}{dt^{n-1}} + k_2 \frac{d^{n-2}y}{dt^{n-2}} + \cdots + k_{n-2}\frac{dy}{dt} + k_{n-1}y = f(t,y)$$

and rewrite it as a system of n first-order equations:

$$\frac{x_1}{dt} \quad = \quad x_2$$

$$\frac{x_2}{dt} \quad = \quad x_3$$

$$\vdots \quad = \quad \vdots \qquad\qquad \textbf{Eq. (8-9)}$$

$$\frac{x_{n-1}}{dt} \quad = \quad x_n$$

$$\frac{x_n}{dt} \quad = \quad f_1(t, x_1, x_2, \ldots, x_{n-2}, x_{n-1}, x_n)$$

Because of its array capabilities, MATLAB solves a single first-order differential equation as easily as a system of n first-order equations. Recall that to solve the differential equation $y(t) = t + e^{-t}$, we first write Function 8-8, which returns the value of the derivative.

FUNCTION 8-8   Function to calculate the derivatives for a first-order initial-value problem (repeated)

```
function ydot=iv1(t,y)
% This is function iv1.m
ydot=-y+t+1;
```

To solve the differential equation, we pass the name of this function and the initial condition, $y(0)$, to the differential equation solver, **ODE45**. To solve an nth-order system, we first create a MATLAB function that returns the values of n derivatives as a column array, and then a second column array that contains the initial conditions for each first-order equation in the system.

An example clarifies the process. Let's solve the following fourth-order differential equation:

$$\frac{d^4y}{dt^4} + 5\frac{d^3y}{dt^3} + 7\frac{d^2y}{dt^2} + 6\frac{dy}{dt} + 7y = 8$$

This equation can be rewritten as the fourth-order system:

$$\frac{dx_1}{dt} = x_2$$

$$\frac{dx_2}{dt} = x_3$$

$$\frac{dx_3}{dt} = x_4$$

$$\frac{dx_4}{dt} = -5x_4 - 7x_3 - 6x_2 - 7x_1 + 8$$

To use MATLAB to solve this system, we write Function 8-9, which returns the value of the derivative for each equation. The values will be returned in a column array.

FUNCTION 8-9 Derivatives for a fourth-order initial-value problem

```
function xdot=iv3(t,x)
% This is function iv3.m
x1dot=x(2);
x2dot=x(3);
x3dot=x(4);
x4dot=-5*x(4) - 7*x(3) - 6*x(2) - 7*x(1) + 8;
xdot=[x1dot;x2dot;x3dot;x4dot];
```

We'll simulate this differential equation for 5 seconds with the initial conditions $x_1(0) = 1$, $x_2(0) = 0$, $x_3(0) = 0$, and $x_4(0) = 0$. Note that function **ODE45** returns results for $x_1(t)$, $x_2(t)$, $x_3(t)$, and $x_4(t)$. Remember that $x_1(t) = y(t)$ in the original fourth-order differential equation, so we'll only plot $x_1(t)$.

```
EDU» xo=[1;0;0;0];
EDU» [t,x]=ode45('iv3', [0, 5], xo);
```

We will first look at the data returned by **ODE45**:

```
EDU» x
```

```
x =
     1.0000          0          0          0
     1.0000     0.0000     0.0000     0.0001
     1.0000     0.0000     0.0000     0.0001
     1.0000     0.0000     0.0000     0.0002
     1.0000     0.0000     0.0000     0.0002

                    .
                    .
                    .

     1.1813    -0.1209    -0.0469     0.1454
     1.1679    -0.1251    -0.0307     0.1506
     1.1541    -0.1276    -0.0141     0.1536
```

As you can see, the function returns four columns of data. The first column is $x_1(t)$, the second column is $x_2(t)$, and so on.

We can now plot the first column versus time.

```
EDU» xo=[1;0;0;0];
EDU» [t,x]=ode45('iv3', [0, 5], xo);
EDU» y=x(:,1);
EDU» plot(t,y);
EDU» xlabel('Time');
EDU» ylabel('y(t)');
```

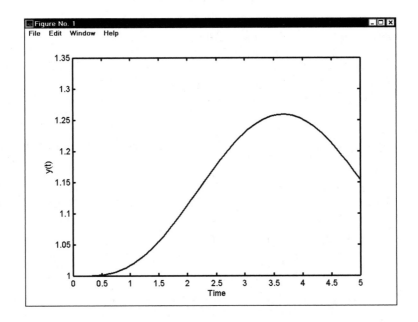

8.5.2 Boundary-Value Problems

In second-order boundary-value problem (BVPs), the relationship between the *y*- and *x*-coordinates is specified by a differential equation. One difference between a BVP and an initial-value problem is that for a second-order BVP, the values of *x* and *y* are fixed at two points in space. For a second-order initial-value problem, the values of *y(t)* and *dy/dt(t)* are fixed at the start of the simulation. We'll limit our discussion to second-order, or two-dimensional, boundary-value problems.

Boundary-value problems can be split into two groups: linear and nonlinear. A linear boundary-value problem is of the form

$$\frac{d^2y}{dx^2} - p(x)\frac{dy}{dx} - q(x)y - r(x) = 0$$

with the boundary conditions $y(a) = \alpha$ and $y(b) = \beta$. This equation is called linear because *y*, *dy/dx*, and d^2y/dx^2 appear by themselves, multiplied by a constant, or multiplied by a function of *x*. Note that the functions $p(x)$, $q(x)$, and $r(x)$ can be nonlinear functions of *x*.

Here are some examples of nonlinear problems:

$$\frac{d^2y}{dx^2} + \left(\frac{dy}{dx}\right)^2 + y - \ln(x) = 0 \qquad \text{with } y(1) = 0 \text{ and } y(2) = \ln(2)$$

and

$$\frac{d^2y}{dx^2} = \frac{1}{8}\left(32 + 2x^3 - y \cdot \frac{dy}{dx}\right) \qquad \text{with } y(1) = 17 \text{ and } y(3) = \frac{43}{3}.$$

The first equation is nonlinear because of the term $(dy/dt)^2$ and the second equation is nonlinear because of the term $y(dy/dt)$.

In MATLAB, the technique for solving the two types of boundary-value problems is the same, but different MATLAB functions are required. The nonlinear BVP solver solves both linear and nonlinear problems. However, it takes longer to solve linear problems with the nonlinear solver, so you should always use the linear solver for linear problems.

Linear Boundary-Value Problems

Let's solve the linear boundary-value problem

$$\frac{d^2y}{dx^2} + \frac{2}{x}\frac{dy}{dx} - \frac{2}{x^2}y - \frac{\sin(\ln(x))}{x^2} = 0 \qquad \text{with } y(1) = 1 \text{ and } y(2) = 2$$

We first rewrite the equation as an initial-value problem:

$$\frac{d^2y}{dt^2} + \frac{2}{t}\frac{dy}{dt} - \frac{2}{t^2}y - \frac{\sin(\ln(t))}{t^2} = 0$$

We must now break the second-order equation into a system of two first-order equations. See page 463 for a procedure to break an nth-order differential equation into a system of n first-order equations. **Always let $x_1 = y$ for all boundary-value problems.** You must choose $x_1 = y$ so that the BVP solver uses the boundary conditions correctly. For this particular problem, we get the system of equations

$$\frac{dx_1}{dt} = x_2$$

$$\frac{dx_2}{dt} = -\frac{2}{t}x_2 + \frac{2}{t^2}x_1 + \frac{\sin(\ln(t))}{t^2}$$

We now define Function 8-10, which calculates these derivatives.

FUNCTION 8-10 Function to calculate the derivatives for a linear boundary-value problem

```
function xdot=lbvp1(t,x)
% This is function Lbvp1.m

x1dot=x(2);
x2dot=-(2/t)*x(2) + (2/(t*t))*x(1) + sin(log(t))/(t*t);
xdot=[x1dot;x2dot];
```

We can now use the linear BVP solver, lin_bvp2. This function is available at the author's website at http://www.cet.nau.edu/meh/Books/Matlab.htm and has the syntax,

```
[x, y] = lin_bvp2(fprime, x₁, y₁, x₂, y₂, h, print_step).
```

Variable *fprime* is a text string that contains the name of the MATLAB function that calculates the derivatives: x_1 and y_1 are the coordinates of the first boundary point; x_2 and y_2 are the coordinates of the second boundary point. Variable h is the integration step size. The smaller the value of h, the more accurate the result and the longer the solution takes. Variable *print_step* is the frequency at which you want to save results. Suppose we are solving a problem from $x = 0$ to $x = 1$ and we set h to 10^{-6}. This causes the solver to calculate a solution 1,000,000 times for the specified interval. If every result is saved in arrays x and y, these variables would each be a column array with 1,000,000 elements. However, if we set h to 10^{-6} and *print_step* to 0.1, a solution is calculated 1,000,000 times, but over the range of 0 to 1, only 10 results will be saved. This allows us to use a very small integration step size to get accurate results, and still save the data in reasonable-sized arrays.

Let's solve our differential equation:

```
EDU» [x,y] = lin_bvp2('lbvp1',1,1,2,2, 0.001, 0.01);
EDU» plot(x,y);
EDU» xlabel('x')
EDU» ylabel('y(x)')
EDU» hold on
EDU» plot(1,1,'r*',2,2,'rp','MarkerSize',15,'LineWidth',2)
EDU» hold off
```

The two boundary points marked below with an asterisk and a star show that the solution does indeed match up with the two conditions.

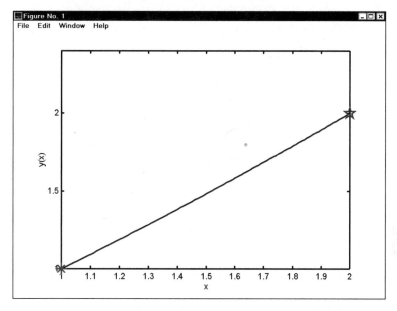

Our second example of a linear boundary-value problem is

$$\frac{d^2U}{dr^2} + \frac{2}{r}\frac{dU}{dr} = 0 \qquad \text{with } U(2) = 110 \text{ and } U(4) = 0$$

First, we rewrite the equation as a second-order initial-value problem with $x_1(t) = U(t)$:

$$\frac{d^2x_1}{dt^2} + \frac{2}{t}\frac{dx_1}{dt} = 0$$

Next, we write the second-order equation as a system of two first-order equations:

$$\frac{dx_1}{dt} = x_2$$

$$\frac{dx_2}{dt} = -\frac{2}{t}x_2$$

Then we write Function 8-11 to calculate these derivatives.

FUNCTION 8-11 Function to calculate the derivatives for the second linear boundary-value problem

```
function xdot=lbvp2(t,x)
% This is function Lbvp2.m
x1dot=x(2);
x2dot=-(2/t)*x(2);
xdot=[x1dot;x2dot];
```

Let's solve our differential equation:

```
EDU» [r,u] = lin_bvp2('lbvp2',2,110,4,0, 0.01, 0.01);
EDU» plot(r,u);
EDU» xlabel('r')
EDU» ylabel('U(r)')
EDU» hold on
EDU» plot(2,110,'r*',4,0,'rp','MarkerSize',15,'LineWidth',2)
EDU» hold off
```

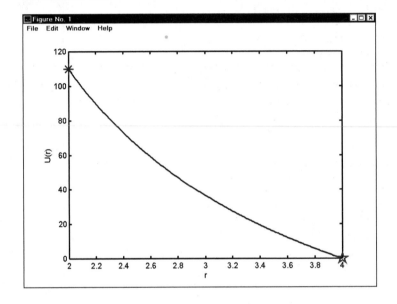

Nonlinear Boundary-Value Problems

Nonlinear boundary-value problems are solved like linear problems except that we use function n_linbvp.m instead of lin_bvp2.m. These functions are available at the author's website at www.cet.nau.edu/meh/Books/Matlab.htm.

Let's solve the nonlinear boundary-value problem:

$$\frac{d^2y}{dx^2} + \left(\frac{dy}{dx}\right)^2 + y - \ln(x) = 0 \qquad \text{with } y(1) = 0 \text{ and } y(2) = \ln(2)$$

First we rewrite the equation as an initial-value problem:

$$\frac{d^2y}{dt^2} + \left(\frac{dy}{dt}\right)^2 + y - \ln(t) = 0$$

We must now break the second-order equation into a system of two first-order equations. See page 463 for a procedure to break an nth-order differential equation into a system of n first-order equations. Always let $x_1 = y$ for all boundary-value problems. You must choose $x_1 = y$ so that the BVP solver uses the boundary conditions correctly. For this particular problem, we get the system of equations:

$$\frac{dx_1}{dt} = x_2$$
$$\frac{dx_2}{dt} = -(x_2)^2 - x_1 + \ln(t)$$

We next define Function 8-12, which calculates these derivatives.

FUNCTION 8-12 Function to calcultate the derivatives for a nonlinear boundary-value problem

```
function xdot=nlbvp1(t,x)
% This is function nlbvp1.m
x1dot=x(2);
x2dot=-x(2).*x(2) - x(1) + log(t);
xdot=[x1dot;x2dot];
```

We can now use the nonlinear boundary-value problems solver, nlin_bvp. This function was developed at Northern Arizona University and has the syntax,

$$[x, y] = n\_linbvp \ (fprime, \ x_1, \ y_1, \ x_2, \ y_2, \ h, \ print\_step).$$

Function n_linbvp is used the same as lin_bvp2. See page 470 for a description of the input and output variables.

Let's solve our differential equation:

```
EDU» [x,y] = n_linbvp('nlbvp1',1,0,2,log(2), 0.001, 0.01);
EDU» plot(x,y);
EDU» xlabel('x')
EDU» ylabel('y(x)')
EDU» hold on
EDU» plot(1,0,'r*',2,log(2),'rp','MarkerSize',15,'LineWidth',2)
EDU» hold off
```

The two boundary points with an asterisk and a star show that the solution does match up with the two conditions.

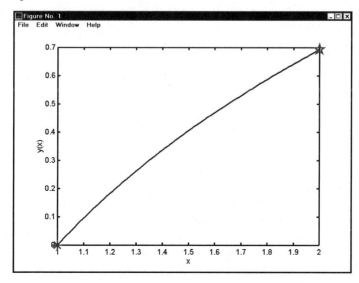

EXERCISE 8-1 Simulate the nonlinear boundary-value problem

$$\frac{d^2y}{dx^2} = \frac{1}{8}\left(32 + 2x^3 - y\frac{dy}{dx}\right) \qquad \text{with } y(1) = 17 \text{ and } y(3) = \frac{43}{3}.$$

Here is the solution:

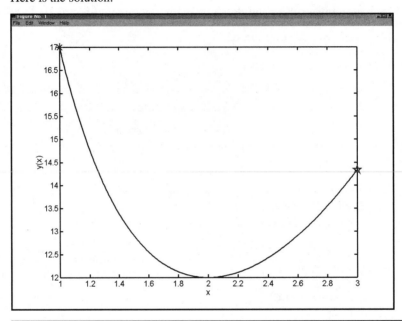

8.6 Problems

Problem 8-1 Solve the following differential equations with MATLAB:

$$\frac{dy}{dt} = \begin{cases} 0 & \text{for} & t < 1 \\ 1 & \text{for} & 1 \le t < 3 \\ y & \text{for} & 3 \le t < 3.5 \\ -1 & \text{for} & 3.5 \le t < 5 \\ \dfrac{y^2}{10} & \text{for} & 5 \le t < 6 \\ 0 & \text{for} & 6 \le t < 8 \\ -\dfrac{y^2}{10} & \text{for} & 8 \le t \end{cases} \qquad \text{with } y(0) = 0 \qquad \textbf{Eq. (1)}$$

$$\ddot{y} + 5\dot{y} - 0.06(\dot{y})^2 + 7y = \sin(t) \quad \text{with } y(0) = 0.1, \ \dot{y}(0) = 0, \text{ and } \ddot{y}(0) = 0 \quad \textbf{Eq. (2)}$$

Write two script files, one for each differential equation, that do the following:

a. Simulates the differential equation for 10 seconds.
b. Plots $y(t)$ from 0 to 10 seconds. Plots must include

- Title, Xlabel, Ylabel, and axes numbers set to 15-point Arial font. Font color is optional.
- The trace line width and axes line width set to 2 points.

c. Curve-fit to find the value for y at $x = 5.37543$.

- Determine which fit is best to use: a Spline fit or a polynomial fit.
- Mark the point on the same graph with a red plus sign.
- Set marker size to 15 points, and the line width of the marker to 2 points.

d. From the values of x and y from part c, create the text line At x = 5.37543, y = ???, where ??? is the value for y found in part c.

- Locate the text on the plot so that it doesn't obscure the trace.
- Set the text to red 15-point Arial font.

e. In another convenient location, display the text, Figure closing in 20 seconds. . . in green 15-point Arial font.
f. Below the text line of part e, display a number that states how many seconds before the plot closes. Start the number at 20 and decrement by 1 to 0. The number should change every second.
g. When 0 is displayed in part f, the figure should remain open for 1 second, and then close automatically.

Problem 8-2 Solve the following differential equations with MATLAB:

$$\frac{dy}{dt} = \begin{cases} 0 & \text{for} \quad t < 1 \\[2mm] 2 & \text{for} \quad 1 \le t < 3 \\[2mm] y & \text{for} \quad 3 \le t < 3.5 \\[2mm] -1 & \text{for} \quad 3.5 \le t < 5 \\[2mm] \dfrac{y^3}{100} & \text{for} \quad 5 \le t < 6 \\[2mm] 0 & \text{for} \quad 6 \le t < 8 \\[2mm] -\dfrac{y^2}{10} & \text{for} \quad 8 \le t \end{cases} \qquad \text{with } y(0) = 0. \qquad \textbf{Eq. (1)}$$

$$\ddot{y} + 4\dot{y} - 0.07(\dot{y})^2 + 6y = \cos(t) \quad \text{with } y(0) = 0.2, \ \dot{y}(0) = 0, \text{ and } \ddot{y}(0) = 0 \quad \textbf{Eq. (2)}$$

Write two script files, one for each differential equation, that

- simulates the differential equation for 10 seconds
- plots $y(t)$ from 0 to 20 seconds, plots labeled appropriately
- prints the value of $y(20)$

As a check, you should get the following results: At $t = 20$, $y(t) = 0.726248$ for the first differential equation, and $y(t) = -3.6572$ for the second differential equation.

Problem 8-3   The resistance of a resistor as a function of temperature is given by the equation

$$R(T) = R_{nom}\left[1 + \frac{\alpha}{10^6}(T - T_{nom}) + \frac{\beta}{10^6}(T - T_{nom})^2\right]$$

where T_{nom} is the nominal temperature ($T_{nom} = 25°C$) and R_{nom} is the nominal resistance measured at 25°C. The data in Table 8-3 is for a 1000-Ω resistor ($R_{nom} = 1000$):

a. Use a polynomial curve fit to find numerical values for α and β.
b. Plot the measured data and a line showing the curve fit. Show the measured data with red ×'s.
c. Use an appropriate title, and x- and y-labels.
d. On the plot, display the text line Alpha = ???, Beta = ??? and fill in the appropriate numerical values.
e. Add a legend to the plot.

TABLE 8-3 Resistance values for Problem 8-3

Temperature (°C)	Resistance (Ω)
25	1000
35	1003
45	1010
55	1021
65	1036
75	1055
85	1078
95	1105
105	1136
115	1171
125	1210

Problem 8-4   Write a function called simp_???.m, where ??? are your initials. The syntax of the function is **area = simp_???(fname, A, B, tol)**; *fname* is a text string containing the name of a MATLAB function that implements a mathematical expression that you would like to integrate. Function simp_??? calculates the integral from A to B of function *fname*. The returned result must be accurate to the specified tolerance *tol*. You may use the function called simpson.m located on the author's website at http://www.cet.nau.edu/meh/Books/Matlab.htm. Test your function with several known integrals.

Problem 8-5 The following voltage and current data have been measured for an unknown device. Note that at the same current, the voltage was measured several times.

$I = 0$, $V = 1$ mV, 3mV, –2 mV, 0

$I = 1$mA, $V = 1$ V, 1.03 V, 1.06 V, 1.1 V, 0.97 V, 0.93 V

$I = 2$mA, $V = 2.2$ V, 1.97 V, 2.05 V, 2.1 V, 1.95 V, 1.89 V

$I = 3$mA, $V = 3$ V, 3.1 V, 3.05 V, 2.98 V, 2.85 V, 3.001 V

$I = 4$mA, $V = 4$ V, 3.9 V, 4.2 V, 4.07 V, 4.13 V

$I = 5$mA, $V = 5.05$ V, 5.1 V, 4.97 V, 4.93 V, 5.06 V

Use a polynomial curve fit and determine the resistance of the device. For a two-terminal device, $V = IR$, where V is the voltage, R is the device resistance, and I is the current.

Problem 8-6 Using the data and polynomial curve fit from Problem 8-5, write a script file that does the following:

a. Plots the data with ×'s for 15 seconds. The plot must have a title, and x- and y-labels.
b. Closes the plot after 15 seconds.
c. Calculates the resistance using a polynomial curve fit and tells the user what the resistance is. This message should be displayed for 15 seconds with a message box.
d. Plots the measured data and the curve fit on the same plot. Show the measured data with ×'s, and display the curve fit as a line. This plot should be displayed until the user hits any key. When the user hits a key, the plot should close.

Problem 8-7 Solve the following differential equations with MATLAB:

$$\frac{dy}{dt} = \begin{cases} 0 & \text{for} & t < 1 \\ 1 & \text{for} & 1 \le t < 3 \\ 0.001y^t & \text{for} & 3 \le t < 7 \\ te^{-t} & \text{for} & 7 \le t < 9 \quad \text{with } y(0) = 0 \\ -1 & \text{for} & 9 \le t < 15 \\ 0 & \text{for} & 15 \le t < 17 \\ 2e^{-t/20} & \text{for} & 17 \le t \end{cases} \qquad \textbf{Eq. (1)}$$

$$\dddot{y} - 0.001\cos(t)\ddot{y} + 0.005(\ddot{y})^2 - 0.006\dot{y} + 0.007y = 0.1\sin(t) \qquad \textbf{Eq. (2)}$$
$$\text{with } y(0) = 0.1, \ \dot{y}(0) = 0, \ \ddot{y}(0) = 0, \text{ and } \dddot{y}(0) = 0$$

Write two script files, one for each differential equation. Name the script files f1_???.m and f2_???.m, where ??? are your initials. You'll also need to create two functions that return the values of the derivatives for each differential equation. Name these files iv1_???.m and iv2_???.m, where ??? are your initials. Each script file should also do the following:

a. Simulate the differential equation for 20 seconds
b. Plot $y(t)$ from 0 to 20 seconds. Plots should have

  ■ **TITLE, XLABEL, YLABEL,** and axes numbers set to 15-point Arial font, font color is optional
  ■ trace line width and axes line width set to 2 points

c. Start a **WHILE** loop that does the following:

  1. Asks the users if they want to locate and display the coordinates of a point on the graph. If the user types **y**, go to step 2. If the user types n, go to step **d**. If the user types anything else, emit an error message and repeat the question.
  2. Display a message box that asks the user to select a point on the trace with the mouse. Do not continue until the user clicks the **OK** button.
  3. Display the mouse cursors and then retrieve a single point.
  4. Using the x-coordinate of the point returned by the mouse, find the point in the differential equation solution nearest this x-coordinate. Use these coordinates as the selected point.
  5. Place a circle at the selected point (use the **PLOT** command and use the "o" marker—note that you will only plot a single point). Use a size 7-point red marker.
  6. Use a message box and ask users where they want to display the coordinates of the point.
  7. Draw a line from the coordinates of the marker to the selected point, and then display the text (???,???) where ??? and ??? are the x- and y-coordinates of the point. Display the line in red with a 1-point line width. Display the text in 11-point Arial, also in red.
  8. Go to step 1.

d. Close the figure and display a nice good-bye message.

Problem 8-8 Find three solutions to the equation

$$\frac{\sin(x)}{x} = 0 \qquad \text{for } x \geq 0$$

Problem 8-9 Find the first maximum and first minimum of the equation

$$f(x) = \frac{\sin(x)}{x} \qquad \text{for } x > 0$$

Problem 8-10 Write a program to find the real roots of the polynomial $y = ax^3 + bx^2 + cx + d$. Your program should do the following:

a. Request values of a, b, c, and d from the user.
b. Find all real roots for the function.

    c. Plot the function using a domain for *x* that clearly shows all of the real roots.

    d. Title and label the plot appropriately.

    e. Mark all roots with a red circle on the plot.

    f. Bring up a message box that either gives the numerical values of the roots (*x*- and *y*-coordinates), or displays the message, `There are no real roots`.

Problem 8-11   Use MATLAB to find the roots of the equation $(5x^4 + 10x^2 + 2)(x^3 - 3x^2 - 6x + 9)$.

Index

'Y, 20c', 1

Comma = fscanf(fid, x, 'Y, 1c', 1